DYNAMICAL CHAOS — MODELS AND EXPERIMENTS
Appearance Routes and Structure of Chaos in Simple Dynamical Systems

WORLD SCIENTIFIC SERIES ON NONLINEAR SCIENCE — SERIES A

Editor: Leon O. Chua
       University of California, Berkeley

Published Titles

Volume 1:   From Order to Chaos
              L. P. Kadanoff

Volume 6:   Stability, Structures and Chaos in Nonlinear Synchronization Networks
              V. S. Afraimovich, V. I. Nekorkin, G. V. Osipov, and V. D. Shalfeev
              Edited by A. V. Gaponov-Grekhov and M. I. Rabinovich

Volume 7:   Smooth Invariant Manifolds and Normal Forms
              I. U. Bronstein and A. Ya. Kopanskii

Volume 12: Attractors of Quasiperiodically Forced Systems
              T. Kapitaniak and J. Wojewoda

Forthcoming Titles

Volume 11: Nonlinear Dynamics of Interacting Populations
              A. D. Bazykin

Volume 13: Chaos in Nonlinear Oscillations: Controlling and Synchronization
              M. Lakshmanan and K. Murali

Volume 14: Impulsive Differential Equations
              A. M. Samoilenko and N. A. Perestyuk

Volume 15: One-Dimensional Cellular Automata
              B. Voorhees

Volume 16: Turbulence, Strange Attractors and Chaos
              D. Ruelle

Volume 17: The Analysis of Complex Nonlinear Mechanical Systems: A Computer Algebra Assisted Approach
              M. Lesser

Volume 18: Wave Propagation in Hydrodynamic Flows
              A. L. Fabrikant and Yu. A. Stepanyants

WORLD SCIENTIFIC SERIES ON NONLINEAR SCIENCE

Series A   Vol. 8

Series Editor: Leon O. Chua

DYNAMICAL CHAOS — MODELS AND EXPERIMENTS
Appearance Routes and Structure of Chaos in Simple Dynamical Systems

V. S. Anishchenko
Saratov State University
Russia

World Scientific
Singapore • New Jersey • London • Hong Kong

Published by

World Scientific Publishing Co. Pte. Ltd.
P O Box 128, Farrer Road, Singapore 9128
USA office: Suite 1B, 1060 Main Street, River Edge, NJ 07661
UK office: 57 Shelton Street, Covent Garden, London WC2H 9HE

Library of Congress Cataloging-in-Publication Data

Anishchenko, V. S. (Vadim Semenovich), 1943–
    Dynamical chaos : models and experiments : appearance routes and structure of chaos in simple dynamical systems / Vadim S. Anishchenko.
       p.   cm. -- (World Scientific series on nonlinear science.
Series A. Monographs and treatises ; v. 8)
    Includes bibliographical references and index.
    ISBN 9810221428
    1. Chaotic behavior in systems. I. Title. II. Series.
Q172.5.C45A537   1995
003'.85--dc20   94-48586
                                                                                                       CIP

Copyright © 1995 by World Scientific Publishing Co. Pte. Ltd.

All rights reserved. This book, or parts thereof, may not be reproduced in any form or by any means, electronic or mechanical, including photocopying, recording or any information storage and retrieval system now known or to be invented, without written permission from the Publisher.

For photocopying of material in this volume, please pay a copying fee through the Copyright Clearance Center, Inc., 222 Rosewood Drive, Danvers, Massachusetts 01923, USA.

Printed in Singapore.

To my dear

wife Tatiana

and daughter Aliona

thanking for their

patience and love.

V.S. Anishchenko

PREFACE

Dear reader! You are faced with a further book on the problem of dynamical chaos in dissipative systems. This subject has already been covered in a number of notable books. An impression might come to mind that the problem has been exhausted. However, let us not hasten this conclusion.

In studies of most of the books dedicated to dynamical chaos in dissipative systems, a group of questions can be separated which, in my opinion, deserve a more detailed analysis. The following issues are the most important:

1. The role of characteristic homoclinic trajectories and Poincaré structures which define general bifurcational mechanisms of the appearance and the main properties of chaotic attractors in the appropriate class of nonlinear systems.

2. Regularities and interplay of the bifurcational mechanisms of transition to chaos, as well as the statistical properties of chaotic attractors which are revealed under the multi-parameter bifurcational study of dynamical systems.

3. The influence of external and internal fluctuations on typical bifurcations, structure transformation, the scenarios of development of chaotic attractors and their statistical properties.

The use of the theory of robust hyperbolic attracting sets (the theory of "truly" strange attractors) is found to be insufficient for investigation of the above problems. The application of the concept of quasiattractors which include regular (periodic and quasiperiodic) subsets of trajectories, along with the "strange" ones appears to be more fruitful. The analysis of dynamical systems from the viewpoint of this quasiattractor conception is more constructive for the examination of experimental results.

In low-dimensional dynamical systems, the occurrence of quasiattractors is mainly due to three types of homoclinic trajectories, and namely: 1) a saddle-focus separatrix loop of the equilibrium state; 2) a homoclinic structure in the form of "Smale's horseshoe" that is realized under intersection between stable and unstable manifolds of a saddle limit cycle; and 3) homoclinic trajectories appearing from the breakdown of a resonant two-dimensional torus.

The study of the typical properties of quasiattractors engendered by the above types of homoclinic trajectories is the basic content of this book.

In Chapters 1 and 2, the elements of the stability and bifurcation theories are briefly outlined, as well as the methods for experimental research of dynamical chaos. The incorporation of this material pursues the goal of assisting young scientists in understanding the main part of the book and to spare them the necessity of referring to additional literature on the first stage of examination.

Chapter 3 attends to an original system with chaotic dynamics (a modified oscillator with inertial nonlinearity) along with its mathematical model. A simple electronic circuit with one-and-a-half degree of freedom due to a number of its attributes has proved to be extremely convenient for numerical and experimental investigation of a low-dimensional chaos.

The objective of Chapters 4 and 5 is to analyze in more detail, both numerically and experimentally, the dynamics of an autonomous oscillator under variation of the control parameters of the system.

In Chapters 6 and 7, the typical hierarchies of instabilities are discussed which accompany nonlinear phenomena as the regimes of quasiperiodic oscillations with two or three independent frequencies are destroyed. For a basic model the inertial nonlinearity oscillator is used. The successive increase in dimension is achieved for the models under study by introducing an external periodic force or using a system of coupled oscillators.

Chapter 8 is devoted to the problem of synchronization of chaotic oscillations. A more simple class of chaotic attractors is considered for which a pronounced basic frequency in the power spectrum is typical (Shilnikov's attractors). The feasibility of generalization for the concepts of the classical oscillation theory on the external and mutual synchronization of periodic oscillations is shown for the case when these attractors are synchronized.

Recent findings on chaotic oscillations in the well-known Chua's circuit are presented in Chapter 9. Unlike the inertial nonlinearity oscillator, Chua's circuit is described by the equations with distinct symmetry properties and is characterized by three equilibrium states. These properties are responsible for the appearance of attractors with more complicated structure permitting a number of novel effects (e.g. the phenomenon of stochastic resonance) to be observed in this system. Chapter 9 has been written in collaboration with A. B. Neiman and M. A. Safonova.

Chapters 10 and 11 deal with the results of research into the influence of fluctuation on the bifurcations of regular and chaotic attractors. The foremost conclusion is here a justification of a strong sensitivity of particular chaos regimes in quasihyperbolic systems to small external perturbations. These chapters has been written together with A. B. Neiman.

Finally, Chapter 12 is devoted to the problem of reconstruction of dynamical systems with account of homoclinic trajectories and noise. This chapter is written together with M. A. Safonova.

The exposition and discussion of most of the results obtained is performed by starting from a detailed comparison between the data of theory, numerical simulation and full-scale experiment. This allows many typical regularities to be obviously interpreted from the viewpoint of physics and the role of fluctuations, involved in a physical experiment, to be evaluated.

Finally, a distinctive feature of this book is that it has been written based on the results of original studies performed in the Laboratory of Nonlinear Dynamics of the Saratov State University for the past decade under my supervision.

The book's peculiarities briefly listed above, which make it different from books in print, lead me to hope that you, dear reader, will read this book with particular interest and to your benefit.

I express my profound thanks to V. V. Astakhov, T. E. Vadivasova-Letchford, M. A. Safonova, D. E. Postnov and A. B. Neiman, my students and colleagues, and collaborators in most of the works which have provided the basis for this book.

I am sincerely grateful to Professor L. O. Chua from Berkeley University, Professor J. Kurths from Potsdam University and their colleagues in USA and Germany for an inestimable help and support rendered to our laboratory's team. Their assistance has highly fostered the completion of work on this book.

I would like to thank Prof. Yu. L. Klimontovich, Prof. L. P. Shilnikov, Prof. V. N. Belykh, Prof. Yu. I. Neimark, Prof. M. I. Rabinovich, Prof. V. Ebeling, Prof. F. Moss, Prof. S. P. Kuznetsov, and Prof. L. Schimansky-Geier for numerous discussions and their noteworthy scientific works which undoubtedly have made an impact on the formation of my notions in the field of nonlinear dynamics and statistical physics.

Lastly, the author wishes to express his gratitude to Mrs. S. M. Bormasova and Mrs. N. B. Smirnova-Yanson for great efforts in translation of the text of the book and to Miss O. V. Sosnovtseva for her great assistance in preparation the manuscript for the press.

The work on this book was partly financed by ISF Long-Term Research Grant NRO 000, the State Committee of Russia in Higher Education (Grant 93-8.2-10), the Physical Society of America (individual grant) and Linkage High Technology Grant NATO N HTECH. LG 93 07 49.

Vadim S. Anishchenko

CONTENTS

Preface   vii

Contents   xi

Chapter 1. Stability and Bifurcation of Dynamical Systems   1

- 1.1 Linear analysis of stability. Variational equations 1
- 1.2 Spectrum of Lyapunov characteristic exponents of phase trajectories of dynamical systems 2
- 1.3 Stability of equilibrium states 5
- 1.4 Stability of periodic solutions. Limit cycle multipliers 6
- 1.5 Stability of quasiperiodic and chaotic solutions 8
- 1.6 Discrete-time systems. Poincaré map 10
- 1.7 Stability of discrete system solutions 13
- 1.8 Structural stability and bifurcations 15
- 1.9 Bifurcation of equilibrium states 16
    - 1.9.1 Bifurcation of codimension one - a double equilibrium point 16
    - 1.9.2 Bifurcation of codimension two - a triple equilibrium point 17
    - 1.9.3 Limit cycle birth bifurcation 18
    - 1.9.4 Nonlocal codimension-one bifurcations. The separatrix loop of saddle equilibrium state 20
- 1.10 Bifurcations of periodic solutions 21
    - 1.10.1 Saddle-node bifurcation of limit cycle 22
    - 1.10.2 Period doubling bifurcation of cycle 23
    - 1.10.3 Two-dimensional torus birth bifurcation 25
    - 1.10.4 Symmetry breaking bifurcation 27
    - 1.10.5 Nonlocal periodic motion bifurcations accompanied by a period becoming infinite 29
- 1.11 Nonlocal bifurcations in the vicinity of double-asymptotic trajectories . . . 31

Chapter 2. Numerical Methods of Chaos Investigations — 33

2.1 Experimental approach to investigation of nonlinear system dynamics 33
2.2 Calculation of the Poincaré map 35
2.3 Numerical analysis of periodic solutions and their bifurcations 42
2.4 Numerical analysis of statistical properties of attractors 50
2.5 Algorithms for calculating the spectrum of Lyapunov characteristic exponents (LCE) 55
2.6 Method of numerically calculating the singular solutions 60
2.7 Dimension calculating algorithms 64

Chapter 3. Inertial Nonlinearity Oscillator. Regular Attractor Bifurcations — 69

3.1 General equations of one-and-a-half-freedom-degree oscillators 69
3.2 Statement of equations for a modified oscillator with inertial nonlinearity 73
3.3 Periodic oscillation regimes in the oscillator and their bifurcations under variation of the parameters 79
    3.3.1 Andronov-Hopf bifurcation 80
    3.3.2 Limit cycle bifurcations 83

Chapter 4. Autonomous Oscillation Regimes in Oscillator — 90

4.1 Two-parametric analysis of transition to chaos via the cascade of period doubling bifurcations 90
4.2 The Poincaré map 96
4.3 System dynamics in the supercritical range of parameter values. Hysteresis and transition to chaos via intermittency induced by fluctuations 104
4.4 Interaction of chaotic attractors. Intermittency of "chaos-chaos" type 112
4.5 Dissipative nonlinearity influence on attractor bifurcations 116

Chapter 5. Quasiattractor Structure and Properties and Homoclinic Trajectories of Autonomous Oscillator — 122

5.1 Oscillator dynamics in the vicinity of a homoclinic trajectory of saddle-focus separatrix loop type 122
5.2 Role of homoclinic saddle cycle trajectories in the chaotic attractor bifurcations 131
5.3 Physical interpretation of exciting the nonperiodic oscillations in oscillator with inertial nonlinearity 138

5.4	On the dimension of an attractor	141

Chapter 6. Two-Frequency Oscillation Breakdown — 146

6.1	General problem statement	146
6.2	Bifurcation diagram of nonautonomous oscillator in the vicinity of basic resonance. Computer simulation	148
6.3	The bifurcation diagram of system (6.1). Full-scale experiment	151
6.4	Two-dimensional torus doubling bifurcation. Soft transition to chaos	155
6.5	Bifurcation mechanism of torus-chaos birth under two-frequency oscillation breakdown	159
6.6	Universal quantitative regularities of soft transition to chaos via two-dimensional torus breakdown	164

Chapter 7. Breakdown of Two- and Three-Frequency Quasiperiodic Oscillations — 174

7.1	Transitions to torus-chaos in the system of two coupled oscillators	174
7.2	Qualitative description of bifurcations in the system of coupled oscillators by using a model map	180
7.3	Transitions to chaos via three-frequency quasiperiodic oscillations	188

Chapter 8. Synchronization of Chaos — 199

8.1	Introduction and definition of the problem	199
8.2	Experimental system and its mathematical model	200
8.3	Methods of investigation	203
8.4	Forced synchronization of chaos	205
8.5	Bifurcational mechanisms of synchronization in the region of chaos	210
8.6	Mutual synchronization of symmetrically coupled oscillators	211
8.7	Evolution of distribution density of phase spectra difference in the process of synchronization	216

Chapter 9. Nonlinear Phenomena and Chaos in Chua's Circuit — 219

9.1	Definition of the problem	219
9.2	Chua's circuit	220
9.3	Chaos-chaos intermittency and 1/f noise in Chua's circuit	231
9.4	Dynamics of non-autonomous Chua's circuit	241
9.5	Stochastic resonance in Chua's circuit	248
9.6	Confirmation of the Afraimovich-Shilnikov torus-breakdown theorem via Chua's torus circuit	255

Chapter 10. Bifurcations of Dynamical System in the Presence of Noise — 268

10.1 Some methods of stochastic calculus 268
10.2 Influence of external noise on the bifurcations of equilibrium state 278
10.3 Period doubling bifurcations in the presence of noise 291

Chapter 11. Chaos Structure and Properties in the Presence of Noise — 302

11.1 Introduction 302
11.2 Regimes of dynamical chaos under the influence of noise with finite intensity 304
11.3 Hyperbolic and quasihyperbolic attractors under the influence of colored noise 308
11.4 Bifurcations of chaotic attractors in the presence of noise 312
11.5 Transitions in chaotic systems induced by noise 319
11.6 Statistical properties of intermittency in quasihyperbolic systems in the presence of noise 325

Chapter 12. Reconstruction of Dynamical Systems from Experimental Data — 337

12.1 Introduction 337
12.2 Methods and algorithms 338
12.3 Results of map (12.1) reconstruction 342
12.4 Reconstruction of system (12.2) 345
12.5 Comparison of qualitative characteristics of reconstructed attractors with original ones 350
12.6 Reconstruction of differential system 354

Bibliography — 361

Index — 381

CHAPTER 1

STABILITY AND BIFURCATION OF DYNAMICAL SYSTEMS

1.1. LINEAR ANALYSIS OF STABILITY. VARIATIONAL EQUATIONS

Let a dynamical system be described by the equations:

$$\dot{x}_i = f_i(x_1, x_2, \ldots x_N; \mu_1, \mu_2, \ldots, \mu_m), \quad i = 1, 2, \ldots, N \tag{1.1}$$

or in vector form

$$\dot{\mathbf{x}} = F(\mathbf{x}, \mu), \tag{1.2}$$

where the right hand parts f_i are nonlinear differentiable functions depending on parameters μ_k. Consider the system (1.1) to have no special symmetry features as being *a general system*.

Let $\mathbf{x}^0(t)$ be a particular solution of the system with the stability to be investigated. Introduce into consideration variables $y_i(t)$ defining a small deviation from the particular solution

$$y_i(t) = x_i(t) - x_i^0(t), \quad i = 1, 2, \ldots, N. \tag{1.3}$$

Substitute (1.3) into (1.1) to obtain

$$\dot{y}_i = f_i(x^0 + y) - f_i(x^0)$$

or

$$\dot{y}_i = \sum_{j=i}^{N} \frac{\partial f_i}{\partial x_j} y_j + \mathbb{O}(y_i), \tag{1.4}$$

where the derivatives f_i are taken at the points of the particular solution $x_i = x_i^0$. The nonlinear terms $\mathbb{O}(y_i)$ tend to zero faster than does the sum of linear components when perturbations y_i decrease.

The stability of the particular solution of the nonlinear system $x^0(t)$ is defined by the stability of the linearized system (1.4):

$$\dot{y}_i = \sum_{j=1}^{N} \frac{\partial f_i}{\partial x_j} y_j. \tag{1.5}$$

Equations (1.5) are called *the variational equations* [1.1,1.2-1.5]. They can be written in matrix form:

$$\dot{\mathbf{y}} = A(t)\mathbf{y}, \tag{1.6}$$

where $A(t)$ is a square matrix with coefficients to be determined by the derivatives:

$$a_{i,j}(t) = \frac{\partial f_i}{\partial x_j}\bigg|_{x_i = x_i^0}; \quad i,j = 1,2,...,N. \tag{1.7}$$

There exists *a fundamental system* of solutions for the linear matrix equation (1.6). *A fundamental matrix* of solutions $Y(t)$ formed by N linearly independent solutions of system (1.6) satisfies the matrix equation:

$$\dot{Y}(t) = A(t)Y(t). \tag{1.8}$$

An arbitrary solution of system (1.6) can be presented in the form:

$$\mathbf{y}(t) = Y(t) \cdot \mathbf{y}(t_0), \tag{1.9}$$

if $Y(t)$ is normalized as $t = t_0$, i.e., $Y(t_0) = E$.

An autonomous linear system (1.6) is *stable according to Lyapunov*, if and only if any solution of (1.9) is bounded. Hence, the coefficients of the fundamental matrix $Y(t)$ should be bounded for a stable system.

1.2. SPECTRUM OF LYAPUNOV CHARACTERISTIC EXPONENTS OF PHASE TRAJECTORIES OF DYNAMICAL SYSTEMS

A Lyapunov characteristic exponent or simply the characteristic exponent of a function $\Phi(t)$ is a real number to be defined by the relation [1.1,1.2,1.6]:

$$L[\Phi(t)] = \overline{\lim_{t \to \infty}} \left[\frac{1}{t} \ln|\Phi(t)|\right], \tag{1.10}$$

where an upper limit is denoted by a bar and $|\Phi(t)|$ is the function modulus. For the function $\Phi(t) = exp[\alpha(t)]$, we have $L = \alpha$. The concept of characteristic exponents provides a technique for evaluating a function growth degree as compared with the exponent function.

For the linear system (1.6) having the arbitrary matrix $A(t)$ the characteristic exponents of nontrivial solution can be introduced similarly [1.2-1.6]:

$$\lambda_i = \overline{\lim_{t \to \infty}} \left[\frac{1}{t} \ln \|y^i(t)\| \right], \quad i = 1,2,\ldots,N, \tag{1.11}$$

where $y^i(t)$ and $\|0\|$ are the i-th fundamental solution of system (1.6) and a norm, respectively. According to the definition, the characteristic exponents are real and, since the matrix $A(t)$ is bounded, finite. The numbers λ_i are called the *generalized characteristic exponents* of an arbitrary linear system of type (1.6). For the variational system describing the evolution of perturbations $y(t)$ near the particular solution $x^0(t)$ of the nonlinear system (1.2), the set of λ_i is called the Lyapunov characteristic exponents of particular solution (or phase trajectory) $x^0(t)$ of nonlinear system (1.2). The set of numbers put into order with respect to decrease forms the so-called Lyapunov characteristic exponents' spectrum (LCE spectrum) of the phase trajectory $x^0(t)$ being one of the most important characteristics for the nonlinear system solution [1.6-1.8].

We call the first and the highest exponent λ_1 the *largest LCE spectrum exponent* of solution. In case of equality $\lambda_1 = \lambda_2 = \ldots = \lambda_k (k < N, \lambda_{k+j} < \lambda_k, j = 1,\ldots,N - k)$, there would be k largest exponents available.

The LCE spectrum signature[1] will be different depending on the form of matrix $A(t)$. The stability of the particular solution will be treated differently, too.

For a linear autonomous dynamical system, the sum of characteristic exponents of the spectrum of its solutions is not smaller than the upper limit of the average matrix trace value [1.2]:

$$\sum_{i=1}^{N} \lambda_i \geq \overline{\lim_{t \to \infty}} \frac{1}{t} \int_{t_0}^{t} Sp A(\tau) d\tau. \tag{1.12}$$

The equality sign is valid for the systems being correct according to Lyapunov.

A matrix determinant, as the determinant of a linear operator, is independent of

---

[1] The signature is to be interpreted as an ordered sequence of three symbols +, 0, - corresponding to positive, zero or negative real numbers of the LCE spectrum of solution $x^0(t)$ [1.3,1.5].

basis choice and presents the volume of N-dimensional parallelepiped. The parallelepiped is built in the phase space of the system on the vectors with coordinates given by the columns of the matrix.

The evolution of phase volume in time obeys Ostrogradsky-Liouville's formula:

$$V(t) = V(t_0) \exp\left[\int_{t_0}^{t} SpA(\tau)d\tau\right], \qquad (1.13)$$

where $V(t) = \det Y(t)$ is the phase volume.

Consider the divergence of the phase velocity vector for the nonlinear system of equations (1.1):

$$divF = \sum_{i=1}^{N} \frac{\partial \dot{x}_i}{\partial x_i} = \sum_{i=1}^{N} \frac{\partial f_i}{\partial x_i}. \qquad (1.14)$$

For the variational equations, expression (1.14) takes the form:

$$divF = \sum_{i=1}^{N} a_{ii}(t) = SpA(t). \qquad (1.15)$$

Thus, the divergence in a linear approximation coincides with the trace of matrix $A(t)$. The system phase volume nearby the particular solution will evolve in time according to the expression:

$$V(t) = V(t_0) \exp[t \ \overline{(divF)}] = V(t_0) \exp[t \cdot \sum_{i=1}^{N} \lambda_i], \qquad (1.16)$$

where time-averaging is denoted by the bar.

Consider the average relative evolution rate of a small phase volume to obtain:

$$\frac{1}{V(t)} \frac{d}{dt}[V(t)] = \sum_{i=1}^{N} \lambda_i. \qquad (1.17)$$

Thus, the rate of the phase volume evolution in the vicinity of trajectory $\mathbf{x}^0(t)$ is characterized by the sum of its LCE spectrum exponents. The sum of the LCE's is negative for the systems with a negative divergence, i.e., for dissipative ones. The limit volume of an attractor in the phase space is zero. Let, for example, all $\lambda_i < 0$. The damping of perturbations along all the eigenvectors takes place in the system. The limit state of the system is a stable equilibrium in this case. A limit set, the attractor, is a fixed point in the phase space.

The largest LCE is zero and all the others are negative for a limit cycle[2]. The attractor is a closed curve in the phase space of the system.

A strange attractor regime is realized in dissipative systems only and is featured by the presence of positive exponents in the LCE spectrum. The attractor is localized in the finite region of the phase space and involves a Cantor set of hypersurfaces in this complicated case. The limit set of trajectories, corresponding to the attractor, is not a manifold.

The phase volume of the system does not change in time if the sum of the LCE spectrum exponents is zero. The system is conservative and has no attractors involved in. The phase volume in time would increase in case of the positive divergence of a vector field. Such a regime being stationary is non-real from the physical point of view. The phase volume growth can be observed, however, within a finite time interval showing the relaxation process of transition to a new limit regime (or escape to infinity).

1.3. STABILITY OF EQUILIBRIUM STATES

If the particular solution is an equilibrium state, i.e., if it does not depend on time, the right hand parts of equations (1.1) become zero:

$$f_i(x_1^0, x_2^0, \ldots, x_N^0, \mu) = 0. \tag{1.18}$$

The coordinates of possible equilibrium states, corresponding to fixed points in the system's phase space are determined by the roots of algebraic equations (1.18). The matrix A of the variational system is independent of time at the fixed point and its general solution has the form:

$$y(t) = exp(A\ t)\ y(t_0). \tag{1.19}$$

The solution is *stable according to Lyapunov* if the eigenvalues of linearization matrix defined by the roots of the equation:

$$det[A - sE] = 0 \tag{1.20}$$

---

[2] That the largest LCE spectrum exponent is zero for a stationary periodic solution was proved first by A.A.Andronov [1.9-1.10]. At least one of the LCE spectrum exponents is always zero for the arbitrary bounded solution $x_0(t)$ of the autonomous system (1.1) which does not tend in time toward a singular point.

are characterized by the real parts $Re\ s_i \leq 0$ [1.1-1.5]. Solution $\mathbf{y}(t)$ is *asymptotically stable* if all the s_i satisfy the strict inequality $Re\ s_i < 0$. This means that the arbitrarily small perturbations of the equilibrium point $\mathbf{x}^0$ are damped and tend to zero as $t \to \infty$.

As seen from definition (1.11), the LCE spectrum of a stable stationary solution consists of the negative numbers $\lambda_i = Re\ s_i(\mu)$, $i = 1,2,..,N$ put into order with respect to decrease. The negativity of the largest LCE spectrum exponent would be a condition of the asymptotic stability of solution. The equilibrium is unstable if at least one of the eigenvalues is positive in its real part. The condition $Re\ s_i(\mu) \neq 0$ highlights a case of robust equilibrium states being stable or unstable in a certain finite region of the variation of parameters μ. The situation, when the largest LCE spectrum exponent of the steady-state solution $\mathbf{x}^0$ becomes zero, corresponds to bifurcation and requires special analysis. Note, that the LCE spectrum signature of the system attractor, being a stable equilibrium, has the form:

$$"-", "-", "-",...,"-".$$

The loss of stability by the solution (the state) would mean transition from a robust stable attractor to an attractor of some other type through the non-robust state at the bifurcation point. With this, either one of the eigenvalues or a complex-conjugate pair becomes zero in its real part.

1.4. STABILITY OF PERIODIC SOLUTIONS. LIMIT CYCLE MULTIPLIERS

Any *periodic* particular solution of system (1.1) is distinguished by the condition

$$\mathbf{x}^0(t) \equiv \mathbf{x}^0(t + T), \qquad (1.21)$$

where T is the period of solution.

The stability of periodic solution is defined by examining an appropriate variational system which will be periodic, too [1.2]

$$\dot{\mathbf{y}} = A(t)\ \mathbf{y},\ A(t) \equiv A(t + T). \qquad (1.22)$$

It is not difficult to make sure that, if $Y(t)$ is a normalized fundamental matrix of the solutions of system (1.22), the matrix $Y(t + T)$ is fundamental, too, and the following relation is valid [1.2]:

$$Y(t + T) = Y(t) \cdot Y(T). \qquad (1.23)$$

The matrix $Y(t)$ is called the *matrix or the operator of monodromy*. The solution of the variational equations defines, due to (1.23), a linear map correlating the value of the perturbation $y(t + T)$ in a period to the arbitrary value of perturbation $y(t)$:

$$y(t + T) = y(t)Y(T). \qquad (1.24)$$

The monodromy matrix is independent of time. The eigenvalues ρ_i of monodromy matrix $Y(t)$, i.e., roots of the equation

$$det[Y(T) - \rho E] = 0 \qquad (1.25)$$

are called the *multipliers of periodic solution* $x^0(t)$ and define its stability. Really, the action of the monodromy operator (1.24) is as follows: the initial perturbation of a periodic solution, considered in projections onto the eigenvectors, is multiplied by an appropriate multiplier ρ_i in period T. It means that the damping of perturbations is to be answered by the requirement $|\rho_i| < 1$.

Any multiplier ρ_i is answered by the nontrivial solution $\xi(t + T) = \rho_i \xi(t)$ of system (1.22) and, on the contrary, the fulfillment of the equality serves as the multiplier definition. Hence, an important conclusion follows: the periodic solution $x^0(t)$ has at least one of multipliers equal to $+1$ [1.2].

Multipliers, as the eigenvalues of monodromy matrix, satisfy the following equations:

$$\sum_{i=1}^{N} \rho_i = Sp Y(t), \qquad \prod_{i=1}^{N} \rho_i = det\, Y(t) > 0, \qquad (1.26)$$

which are highly useful for the analysis of numerical results.

The LCE spectrum of periodic solution is defined, according to (1.11), by the multipliers:

$$\lambda_i = ln\, |\rho_i|/T. \qquad (1.27)$$

One of the spectrum exponents is always zero and corresponds to a unit multiplier. The periodic solution is stable if all the other multipliers belong, in the complex plane of multiplier values, to the interior of the unit circle, i.e., $|\rho_i| < 1$, $i = 1,2,...,(N - 1)$. The signature of the LCE spectrum of the stable limit cycle is:

"0", "−", "−", ..., "−".

If perturbations are growing along one or several eigenvectors, then the corresponding multipliers $|\rho| > 1$ and positive exponents would appear in the LCE spectrum of periodic solution. The periodic solution with the multipliers, lying partially inside the unit circle and partially outside it, is unstable and is called the *saddle solution*.

The robust stable and saddle periodic solutions $x^0(t)$ of the system (1.2) exist in a certain region of parameter values and are defined by the presence of only one zero exponent in the LCE spectrum. The emerging to the unit circle of a multiplier or the complex-conjugate pair of multipliers of the limit cycle additionally is a criterion of structural instability. With this, additional zero exponents indicative of bifurcation appear in the LCE spectrum of periodic solution.

1.5. STABILITY OF QUASIPERIODIC AND CHAOTIC SOLUTIONS

The solutions $x^0(t)$ in the form of *quasiperiodic and aperiodic (chaotic)* oscillations become possible with the dimension of the phase space of system (1.2) increasing up to $N \geq 3$. The appropriate system of the variational equations is described by quasiperiodic or aperiodic matrix $A(t)$ in these cases. The study of the stability of such particular solutions becomes, unlike the steady-state and periodic ones, more complicated and is performed, as a rule, by using numerical computation.

Let a particular solution $x^0(t)$ be the quasiperiodic function:

$$x^0(t) = x^0[\phi_1(t), \phi_2(t),...,\phi_p(t)], \qquad (1.28)$$

where $\phi_n(t) = \omega_n t$, $n = 1,2,...,p$. The function $x^0(\phi)$ has a period of 2π with respect to each of arguments ϕ_n:

$$x^0(\phi_n + 2\pi) \equiv x^0(\phi_n), \qquad n = 1,2,...,p. \qquad (1.29)$$

For the quasiperiodic function, the equality of type (1.21) is not fulfilled. Quasiperiodic oscillations are not periodic generally. The solution $x^0(t)$ is called the *ergodic quasiperiodic oscillation* if there exist no rational relations between the frequencies ω_n.

The stability of quasiperiodic solutions is described by the LCE spectrum. The linearization matrix $A(t)$ of the variational equations is quasiperiodic and the Lyapunov characteristic exponents λ_i are defined only in the limit as $t \to \infty$ [1.7-1.8]. In practice, one can restrict himself to the finite time, depending on the rate of convergence of functions $\lambda_i(t)$ to the limits of λ_i in every specific case and obtain the LCE spectrum exponent values with a certain given accuracy. The periodicity of solution $x^0(t)$ with respect to each of arguments ϕ_n results in the LCE spectrum of quasiperiodic oscillation comprising p zero exponents. If the solution is asymptotically stable according to Lyapunov, then p zero exponents of the LCE spectrum would be the largest and all the others strictly negative ones. Here, the invariant manifold of dimension p, called the p-dimensional torus, is a system attractor in the phase space. If only one of the exponents of the LCE spectrum of solution on a multidimensional torus becomes positive (in the presence of p zero exponents), then the quasiperiodic solution is unstable according to Lyapunov.

The simplest example of a two-frequency quasiperiodic solution is the regime of periodic amplitude modulation of a harmonic signal. Similar oscillation regimes arise under the interaction of nonlinear oscillators, by periodically affecting a selfoscillating system and in other cases which are realized in systems with at least 1,5 degrees of freedom. The solution for the regime of two-frequency oscillations can be presented in the ideal case of harmonic signals as follows:

$$x(t) = B_0(1 + m\ sin\Omega t) \cdot sin(\omega_0 t + \psi), \qquad (1.30)$$

where ω_0 is a main oscillation frequency and Ω is the frequency of a modulation signal not related with ω_0 rationally. An attractor in the form of *two-dimensional ergodic torus* corresponds to the stationary regime of oscillations with two independent frequencies. The LCE spectrum signature of the two-dimensional torus will be:

"0", "0", "-", "-", ..., "-".

If a particular solution $x^0(t)$ is aperiodic, but bounded for $pt \to \infty$, then it corresponds to the regime of chaotic oscillation. At least one positive exponent appears and there exists at least one zero exponent in the LCE spectrum of such a solution [1.7,1.8]. The situation differs from all the cases discussed above in principle. The presence of positive exponents in the LCE spectrum is indicative, in definition, of the instability of solution according to Lyapunov. In what sense can we speak about the stability of chaotic solution? The question is nontrivial but it can be answered quite definitely.

The aperiodic solution $x^0(t)$ corresponding to the regime of chaotic oscillations conforms to the bounded attracting limit set of trajectories in the phase space which is called a strange attractor. The set of trajectories is featured by instability according to Lyapunov, on the one hand, but due to the boundedness of the solution it is to be *stable according to Poisson*. The exponential divergence of nearby phase trajectories and bounded attractor size lead inevitably to the trajectory returning sooner or later into the vicinity of any initial state in the attractor that is as small as you like, but finite. There is available a logically justified chain of phenomena becoming complicated: equilibrium - phase point is fixed in time; limit cycle - phase trajectory returns to any point of the attractor strictly in a period; quasiperiodic oscillation - period is absent but there is a regular return of phase trajectory into the given vicinity of initial state; and finally, strange attractor - there is a return being, however, irregular in time and having the character of a random sequence in time. Thus, chaotic trajectories may be called stable, if there exists a limit set, the attractor, with a certain basin of attraction which has inside trajectories unstable according to Lyapunov, but stable according to Poisson.

The linearization matrix $A(t)$ of the variational equations system for the chaotic solution $x^0(t)$ would be aperiodic, but bounded. Therefore, the limit in (1.11) exists as $t \to \infty$ and defines the LCE spectrum of solution. The LCE spectrum signature for a strange attractor with the most simple structure will have the form:

$$"+", "0", "-", "-", .., "-".$$

The analysis of the variational equations realized relative to different solution types of the nonlinear system (1.2) enables classification of different attractor types to be introduced based on the stability concepts according to Lyapunov and according to Poisson. The limit set which corresponds to the particular type of stable solution and attracts phase trajectories from a certain region of initial conditions is an attractor. The attractor is regular or simple if the phase trajectories on the attractor are stable according both to Lyapunov and to Poisson. The attractor is strange if trajectories on the attractor, being stable according to Poisson, are unstable according to Lyapunov. There exists a finite number of regular attractors: equilibrium states, periodic motions and quasiperiodic motions. All other possible types of attractors are strange. There are no sufficient data at present to introduce the complete classification of strange attractors.

1.6. DISCRETE-TIME SYSTEMS. POINCARÉ MAP

The problem of the stability of a differential system solution can be put and solved for discrete-time systems in a similar way. Discrete-time systems can be considered to be independent when describing, for example, ecological processes but they can be obtained uniquely from differential systems by passing to the Poincaré point maps [1.3-1.5,1.11].

Consider a certain motion regime of a differential system which is described by a trajectory Γ in the phase space $\mathbb{R}^N$ of equations (1.2). Introduce into consideration a certain hypersurface S with the dimension $N-1$ in the phase space. The surface is assumed to be crossed by the phase trajectory Γ successively and transversely (at non-zero angle). Surface S is called the *Poincaré secant* for the phase trajectory Γ.

Trajectory Γ generates a certain *point map* on secant S which brings uniquely (but not in one-to-one way) the nearest point $\mathbf{x}(k + 1)$ following $\mathbf{x}(k)$ in correspondence with any point $\mathbf{x}(k)$ lying in the intersection of Γ with S. For illustration, an example of construction of point map is shown in Fig.1.1 for $N = 3$. The sequence of map points is given by intersections of Γ with S in one direction. The discrete set $\{x(k)\}$, $k = 0,1,2,...$ obtained on the cross-section is called the *Poincaré section* for trajectory Γ.

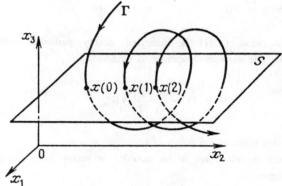

Fig.1.1. Method of the Poincaré section construction.

The law of correspondence between preceding and subsequent intersection points is called the *return, or Poincaré, map*. In general, the Poincaré map is specified by the nonlinear discrete equation of the dimension being equal to that of section $(N - 1)$. Thereby, the $(N - 1)$-dimensional phase space being a crossing hypersurface S is brought in a unique correspondence to the nonlinear dynamical system (1.1). The sets of points $x(k)$ on the section become phase trajectories. Every subsequent point $\mathbf{x}(k + 1)$ is obtained by applying the nonlinear transformation P to the preceding point $\mathbf{x}(k)$:

$$\mathbf{x}(k + 1) = P[\mathbf{x}(k),\mu], \qquad (1.31)$$

μ is the vector of parameters, which in coordinate form is written as follows:

$$x_i(k + 1) = P_i[x_i(k),\mu_1,...,\mu_m], \quad i = 1,2,...(N-1). \qquad (1.32)$$

Thus, the analysis of a differential dynamical system comes to that of the study of the corresponding Poincaré map. The structure of the dynamical system is defined uniquely (but not in one-to-one way) by the structure of the point map engendered by the system [1.11].

The nonlinear equation (1.31) is a discrete analog of differential system. It can be considered, however, as mentioned, irrespective of the initial differential system[3].

Particular solutions, namely, stationary, periodic, quasiperiodic and chaotic sequences $x^0(k)$ can exist in discrete systems as well.

The stability of the particular solution $x^0(k)$ is studied on the base of a suitable variational equation. If a small deviation $y(k) = x(k) - x^0(k)$ is introduced into consideration and written in coordinate form as

$$y_i(k) = x_i(k) - x_i^0(k), \quad i = 1,2,\ldots,(N-1), \tag{1.33}$$

and the input equation (1.31) is linearized nearby particular solution, then the following linear discrete variational equation is obtained:

$$y_i(k+1) = \sum_{j=1}^{N-1} \frac{\partial P_i}{\partial y_j} y_j(k), \tag{1.34}$$

where derivatives are taken at the points of particular solution.

The variational equations are to be recorded in vector form similarly to the case of differential systems:

$$y(k+1) = M(k,\mu) \, y(k), \tag{1.35}$$

where $M(k,\mu)$ is the square linearization matrix with coefficients m_{ij} prescribed by the corresponding derivatives (1.34). It is not difficult to make sure that it follows from (1.35) that:

$$y(k+1) = \prod_{i=1}^{k} M(i,\mu) y(1). \tag{1.36}$$

By analogy with differential systems, we can define the characteristic Lyapunov exponents of a particular solution of the discrete system (1.34) to be

$$\lambda_i = \overline{\lim_{k \to \infty}} \{\tfrac{1}{k} \ln \|y^i(k)\|\}, \tag{1.37}$$

---

[3]There exists a rigorous functional interconnection between continuous time t and discretization points t_k by the difference $t_{k-1} - t_k = \Delta(k)$ for the Poincaré maps (see Fig.1.1). For discrete model systems, the interconnection is lost and $\Delta = 1$ is proposed. This causes, in particular, map P on the section to be in a unique correspondence with specific phase trajectory but the inverse is incorrect.

where $y^i(k)$ is the i-th solution of fundamental system of solutions $Y(k)$ of equations (1.35).

1.7. STABILITY OF DISCRETE SYSTEM SOLUTIONS

The set of numbers $x^0 = (x_1^0, x_2^0, \ldots, x_{N-1}^0)$, which are independent of discrete time and satisfy the initial nonlinear equation (1.31), is called the fixed point of the map or its steady-state solution. The linearization matrix of the variational equations will not depend on k, either. The stability of steady-state solution is defined by the eigenvalues of matrix M:

$$det[M - \rho E] = 0, \qquad (1.38)$$

i.e., by the fixed point multipliers ρ_i. The steady-state solution is asymptotically stable if all the multipliers are strictly smaller than unity in modulus. The corresponding LCE spectrum of the system attractor in the form of a stable equilibrium state will be defined by totality of λ_i put into order with respsct to decrease:

$$\lambda_i = ln|\rho_i| < 0. \qquad (1.39)$$

The solution $x^0(k)$ is periodic if the following condition is fulfilled:

$$x^0(k) \equiv x^0(k + n),$$

with n as a period. In this case, $x^0(k)$ is called the period-n solution of the discrete system or n-cycle of the map. The linearization matrix is period-n, too, i.e., $M(k) \equiv M(k + n)$.

The monodromy matrix analog would be here the matrix M_n independent of discrete time:

$$M_n = M(n) \cdot M(n - 1) \cdot \ldots \cdot M(1) . \qquad (1.40)$$

For periodic solutions, the discrete time is to be measured by the integer number of periods $k = r \cdot n$ allowing one to write:

$$y(k) = M_n^{(r)} \cdot y(1); \quad M_n^{(r)} = \prod_{i=1}^{r} M_n^i . \qquad (1.41)$$

The multipliers ρ_{ni} of the linearization matrix of the map M_n n-cycle are calculated similarly to (1.38):

$$det[M_n - \rho_n E] = 0 \qquad (1.42)$$

and define the stability of n-periodic solution. It is easy to make sure that, by definition (1.37), the LCE spectrum of n-periodic solution would consist of:

$$\lambda_i = \frac{1}{n} \ln |\rho_{ni}|. \qquad (1.43)$$

An asymptotically stable n-cycle of the map is answered for by multipliers $|\rho_{ni}| < 1$ for any $i = 1,2,...,(N - 1)$. Thus, the LCE spectrum of the stable n-cycle would involve negative numbers only.

If the discrete equation (1.31) is the Poincaré map of some differential system, then the fixed point of the map corresponds to a simple limit cycle in this system. The presence of period-n point in the map corresponds to period n, the more complicated limit cycle of the differential system. In contrast to flow systems, in maps steady-state and periodic solutions are defined by n-cycle ($n = 1,2,...$) in general case. The LCE spectrum signature for these solutions is the same for the stable fixed point either of period 1 or of any other period $n = 2,...,$: "–", "–", "–",...,"–".

The linearized matrix of the Poincaré map M_n is, in general case, including $n = 1$, an analog of the monodromy matrix of an arbitrary periodic solution of original differential system. A remarkable property of the Poincaré map is that the eigenvalues of the linearization matrix ρ_{ni}, $i = 1,2,...,(N - 1)$, complemented by the unit multiplier $\rho_N = 1$, are strictly equal to the eigenvalues of the monodromy matrix $Y(T)$ of the differential system (1.2). On this account, the stability of periodic oscillation regimes is quantitatively described in differential systems by the multipliers of the n-cycle of the Poincaré map. The LCE spectrum of the Poincaré map, complemented by one zero exponent, provides, respectively, the LCE spectrum of periodic solution of the differential system generating an appropriate map.

The map multipliers describe the evolution of projections of the particular solution perturbation vector onto the eigenvectors of the linearization matrix in a period. The sense of the multiplier concept is lost if particular solution is not n-cycle, but the quasiperiodic or chaotic sequence $x^0(k)$. The task of stability analyzing becomes complicated and requires the direct calculation of the LCE spectrum exponents in the limit as $k \to \infty$ by using numerical methods, as in case of aperiodic solutions of differential systems. Calculation algorithms are discussed in Chapter 2 and here it should be noted that the signature of the LCE spectrum of quasiperiodic and chaotic attractors in maps is similar to that of the he corresponding attractors of flow systems deducting one zero.

1.8. STRUCTURAL STABILITY AND BIFURCATIONS

The investigation of conditions for the structure of attracting limit sets to be maintained under perturbation is a problem connected with introducing the concepts of *robustness, structural stability and topological equivalence* of dissipative dynamical systems. From the physical point of view, mathematical models seem to be real when, under small perturbations, they do not change qualitatively the structure of partition of the parameter space into the regions corresponding to different types of solution. The idea of system robustness, introduced by A.A.Andronov and L.S.Pontryagin for two-dimensional systems [1.12], turned out to be of extreme importance and use in physics. It is quite clear that it is practically impossible to write an exact system of equation corresponding to a specific physical system which lives in an arbitrary regime. Since, a model dynamical system must exhibit qualitative properties which are hold under small perturbations.

Differential dynamical systems are structurally stable if the small perturbations of evolution operator, differentiable at least once, lead to topologically equivalent solutions. The latter means, as a matter of fact, that a perturbed flow can be transformed into the unperturbed one using a certain continuous substitution of phase coordinates.

The sensitivity of dynamical system to small perturbation is defined by its state. Sometimes, the effect of perturbing factors is small and in other situations, a distinct difference of perturbed motion character was observed as compared with the initial one. The state of the system (or the type of motion) is stable in the former case and unstable in the latter one. As shown above, the task of stability theory is just to indicate the features and to formulate the criteria allowing one to justify with a certain confidence if the system behavior under study would be stable or unstable.

The mathematical description of the most interesting physical tasks provides differential equations depending on parameters. The parameter variation can result in the loss of stability by one motion regime and in the system's transition to another state. An example is periodic oscillation arising in Van-der-Pole oscillators when exceeding a generation threshold. This phenomenon is called bifurcation and a parameter value, at which it takes place, is called the bifurcation point.

The hierarchy of replacement of some stable system states by the others when varying the control parameters results in a sequence of phase transitions from certain robust (structurally stable) regimes to the other robust ones. It is performed through a structurally unstable state at the bifurcation point.

The stability theory, briefly described above, forms the mathematical basis of elementary bifurcation theory. The theory of stability and bifurcations allows one to consider the problem of the phase space partition of dynamical system into characteristic and topologically equivalent trajectories, to analyze partition structure, and to reveal the regions in the parameter space having the limit sets of specific types. Thereby, the possibility is practically provided to construct *bifurcation diagrams* illustrating the rearrangement mechanisms of motion regimes in the phase space of dynamical system by varying its parameters. The totality of the above questions forms an object of the modern qualitative theory of dynamical systems involving, naturally, the stability and bifurcation theories [1.13-1.23].

1.9. BIFURCATIONS OF EQUILIBRIUM STATES

1.9.1. Bifurcation of codimension one - a double equilibrium point.
Consider a dynamical system described by one first-order differential equation on the straight line x:

$$\dot{x} = F(x,\mu). \tag{1.44}$$

Let $x^0(\mu)$ be a structurally stable equilibrium state, i.e., $s(\mu) \neq 0$, where $s(\mu) = F'_x(x^0,\mu)$. The model equation describing dynamics near a singular point will be presented in this case by the linearized equation (1.44) in the form:

$$\dot{y} = sy. \tag{1.45}$$

It is seen from solution $y = y_0 exp(s \cdot t)$ that the stability of x^0 is defined by the sign of the eigenvalue s, i.e., by the sign of derivative $F'_x(\mu)$.

With certain parameter values, the eigenvalue s in the equilibrium state can become zero:

$$s(\mu) = F'_x(x^0,\mu) = 0. \tag{1.46}$$

Assume the second derivative to be other than zero:

$$a(\mu) = F''_{xx}(x^0,\mu)/2 \neq 0. \tag{1.47}$$

Then, x^0 is a two-fold root of the initial equation (1.44). The model system for a given bifurcation will have the form:

$$\dot{y} = b_1(\mu) + a(\mu)y^2, \tag{1.48}$$

where b_1 is a certain parameter.

Let $a > 0$ for clearness. Then, there are two equilibrium states (stable and unstable) in system (1.48) when $b_1 < 0$. They merge into one point when $b_1 = 0$ and disappear when $b_1 > 0$. If the manifold $F(x,\mu) = 0$ is represented in the combined parameter and phase coordinate space (Fig.1.2), then a feature of "fold" type would be visible with its projecting onto the parameter space [1.21-1.22]. The bifurcation "double equilibrium point" has codimension one since it is distinguished by the only one bifurcational condition (1.46). This bifurcation is quite often in application and is called the *saddle-node bifurcation*.

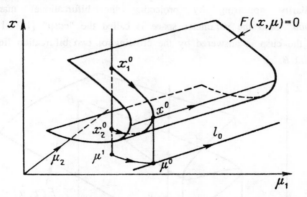

Fig.1.2. Saddle-node bifurcation manifold. The peculiarity of a fold type.

1.9.2. Bifurcation of codimension two - a triple equilibrium point. Change the parameter values of system (1.44) by moving in the parameter space along the line l_0 (see Fig.1.2) corresponding to the bifurcational condition (1.46). With certain parameter values, the quantity $a(\mu)$ can become zero:

$$a(\mu) = F''_{xx}(x^0,\mu)/2 = 0, \quad s(\mu) = F'_x(x^0,\mu) = 0. \quad (1.49)$$

With this, the third derivative $F'''_{xxx}(x^0,\mu) \neq 0$. Two bifurcational conditions (1.49) are fulfilled simultaneously here. The bifurcation of codimension two is realized and x^0 is the three-fold root of equation (1.44). The model system for bifurcation given is presented in the form:

$$\dot{y} = b_1 + b_2 \cdot y + b(\mu)y^3, \quad b(\mu) \neq 0. \quad (1.50)$$

There can exist either one or three robust steady-state solutions in system (1.50). Fig.1.3 shows the bifurcational manifold being a triple equilibrium point. For the parameter values lying in a dashed region inside the characteristic triangle (for example, point *A*), the system has three steady-state solutions. One of them, x_2^0, is always unstable and the two others, x_1^0 and x_2^0, are stable. There exists only one equilibrium state outside this region of parameter values. Three equilibrium states merge into one equilibrium point at point B.

A hysteresis is observed on the left of point B for the motion along parameters transversely to the bifurcation lines l_0: the equilibrium will be lost due to the merging and vanishing of equilibrium states x_1^0 and x_2^0 in the forward direction and due to that of another pair of equilibrium states x_2^0 and x_3^0 in the reverse direction.

A peculiarity appearing by projecting the bifurcational manifold *"triple equilibrium point"* onto the parameter space is called the "cusp" [1.21-1.22]. As seen from Fig.1.3, the cusp is answered by the crossing of two bifurcation lines of folds l_0 at the cusp point B.

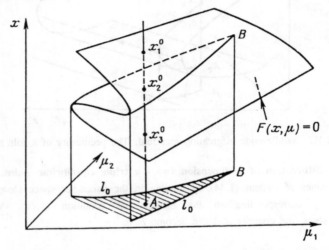

Fig.1.3. Manifold of "triple equilibrium" bifurcation. The peculiarity of the cusp type.

1.9.3. Limit cycle birth bifurcation. The bifurcation of codimension one can be realized in dynamical systems with the dimension $N \geq 2$ when the real parts of the complex-conjugate pair of the eigenvalues of the stationary solution's linearization matrix become zero. This bifurcation corresponds to the excitation of selfoscillation. It is called the *Andronov-Hopf bifurcation* [1.9,1.16,1.18-1.19].

Let a pair of the complex-conjugate eigenvalues $s_{1,2}$ of the dynamical system's equilibrium point on the plane be purely imaginary at certain parameter values, i.e.,

$$Re\ s_{1,2}(\mu_0) = 0, \quad Jm\ s_{1,2} \neq 0. \tag{1.51}$$

The model system, describing bifurcation locally, depends on one parameter and is to be written in complex form as [1.9,1.19]:

$$\dot{z} = (b_1 + j\omega)z + L_1 z |z|^2, \qquad (1.52)$$

where $\omega(\mu_0) \neq 0$, $L_1(\mu_0) \neq 0$.

The quantity $L_1(\mu)$ is called the *first Lyapunov quantity* of equilibrium state. It defines the stability of a periodic regime arising as a result of Andronov-Hopf bifurcation.

Consider the bifurcation diagrams on the plane. The stability of the equilibrium point $z = 0$ of system (1.52) is lost by b_1 passing through zero independently of the first Lyapunov quantity sign. The focus, being stable as $b_1 < 0$, is transformed into the unstable one as $b_1 > 0$.

Let $L_1(\mu_0) < 0$. In this case, the loss of stability by the equilibrium state is accompanied by the birth of a small stable limit cycle with the size growing with parameter evolution as a square root of supercriticality. The period of the cycle is defined by the relation:

$$T = 2\pi/\omega(\mu_0); \quad \omega(\mu_0) = |Jm\ s_{1,2}(\mu_0)|. \qquad (1.53)$$

They say about a *soft* limit cycle birth bifurcation in this case. Fig.1.4 shows the reconstruction of phase portrait for the soft Andronov-Hopf bifurcation.

Fig.1.4. Soft Andronov-Hopf bifurcation.

Consider the case as $L_1(\mu_0) > 0$. The equilibrium state before the bifurcation point is surrounded by the unstable closed trajectory bounding the basin of attraction of stable focus. By approaching the bifurcation point, the unstable cycle is contracted to the equilibrium state. The cycle vanishes at the bifurcation point by merging with the point of equilibrium which becomes unstable. Another regime is set in the system which is highly different from the regime having undergone bifurcation. They say about *an abrupt loss of stability* [1.9,1.19]. The rebuilding of the system's phase portrait is indicated in Fig.1.5 for the case of abrupt Andronov-Hopf bifurcation.

Note, that the condition of the first Lyapunov quantity other than zero provides the birth ($L_1 < 0$) or the death ($L_1 > 0$) of the only limit cycle in the system.

1.9.4. Nonlocal codimension-one bifurcations. The separatrix loop of saddle equilibrium state. Consider the case of the nonlocal bifurcation of a robust saddle equilibrium state being of importance for further discussion. This is the formation of a specific phase trajectory when one of the emerging separatrixes of saddle $x^0(\mu)$ returns into the saddle by forming the *"separatrix loop"* Γ_0. The fulfillment of this purely geometric condition in the system's parameter space is conformed by the bifurcational manifold of codimension one.

Fig.1.5. Abrupt Andronov-Hopf bifurcation.

The system's phase portrait rebuilding near the separatrix loop is described by the splitting function of separatrixes when the parameters are varied relative to the bifurcational manifold. Consider the case of a two-dimensional phase space where a one-dimensional secant S is introduced nearby the saddle. Define a coordinate η on the secant as it is shown in Fig.1.6.

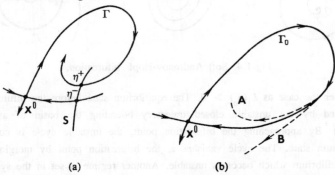

Fig.1.6. a). Determination of separatrix splitting function $H(\mu)$; b). Possible cases of separatrix loop Γ_0 destruction.

Let the difference between the coordinates of intersection of the entering and emerging separatrixes with secant S: $H(\mu) = \eta^+ - \eta^-$ be called the splitting function $H(\mu)$ of separatrixes.

If the splitting corresponds to case A in Fig.1.6, then $H(\mu) < 0$ and in case B $H(\mu) > 0$. The zero value of the function $H(\mu) = 0$ would be responsible for the realization of separatrix loop.

The condition of robustness for the given bifurcation is when the quantity $\sigma(\mu)$ being called the *saddle quantity* and given in the form

$$\sigma(\mu) = \text{Sp } A(\mu) = s_1(\mu) + s_2(\mu) \neq 0 \qquad (1.54)$$

differs from zero.

The value of function $H(\mu)$ can be considered as a bifurcation parameter. Let $\sigma < 0$. Then, the only stable limit cycle arises from loop Γ_0 as $H(\mu) < 0$ (the case of loop breaking toward A in Fig.1.6). If $H(\mu) > 0$ (loop breaking toward B), nothing is born from the loop. Loop Γ_0 is called unstable at the values of the saddle quantity $\sigma > 0$. The unstable limit cycle can only arise from it under the splitting of separatrixes [1.13, 1.20, 1.23].

We have confined ourselves by considering the typical bifurcations of equilibrium states which are of great importance in understanding the mechanisms of chaotic dynamics evolution. With this, the dimension of the system's phase space was chosen to be minimally possible for realizing the given bifurcation. The system was assumed to be of general status, i.e., to have no properties sensitive to small variations of control parameters. The bifurcations described take place in multidimensional systems, too, though new bifurcation types can appear along with these ones. Specific bifurcations can be generated due to several special properties of dynamical systems, such as, for example, the symmetry.

1.10. BIFURCATIONS OF PERIODIC SOLUTIONS

Problem statement. The question on the stability and bifurcations of periodic trajectories can be considered both directly with respect to the differential equations when the particular solution is answered by the *limit cycle*, and by analyzing the stability of the fixed points of the corresponding Poincaré map. The method of analyzing the map is more instructive and convenient to use for numerical investigation.

Consider the task of typical local bifurcations of periodic motions to be solved in terms of the Poincaré map. Let Γ be a phase trajectory in N-dimensional space. It corresponds to the periodic solution of the autonomous system of differential equations (1.1) depending on the set of parameters μ. Introduce into consideration the secant surface S. Let the point x_0 of intersection of Γ with S be the fixed point of the Poincaré map (1.31)[4].

The stability of the fixed point x^0 is completely described, as it was shown, by the eigenvalues of the map linearization matrix, i.e., by the multipliers of cycle $\rho_i(\mu)$, $i = 1,2,..,(N - 1)$ satisfying the equation (1.38). Cycle Γ is stable (asymptotically stable) if all the multipliers $|\rho_i| \leq 1$ ($|\rho_i| < 1$) at fixed parameter values. The multipliers evolve in magnitude when varying the system's parameters and one or several multipliers can become unity in modulus when achieving a certain critical value $\mu = \mu^*$. The emerging to the unit circle of at least one of multipliers corresponds to the bifurcation which leads to the topological rebuilding of the structure of phase trajectories in the vicinity of cycle Γ [1.2-1.5,1.9-1.10,1.16,1.19-1.24].

Consider the cases of stability lost by cycle when one multiplier or the complex-conjugate pair of multipliers is passing to the unit circle when varying parameters. We would remind you that one of the monodromy matrix multipliers is a unity, $\rho_N = 1$. The system of basis vectors in the section can be always chosen in such a manner that the eigenvector of multiplier ρ_N would be tangent to trajectory Γ at the fixed point of map x^0. Here, the cycle stability is defined by the map multipliers from which the largest one in modulus is to be analyzed (there are two of them for complex-conjugate pairs, naturally). For simplicity, one of the parameters of the system $\mu_i = \mu$ is assumed to be varied. Call it the control parameter[5].

1.10.1 Saddle-node bifurcation of limit cycle.
The largest multiplier of cycle $\rho(\mu^*)$ becomes $+1$ when the parameter achieves the critical value $\mu = \mu^*$. What comes

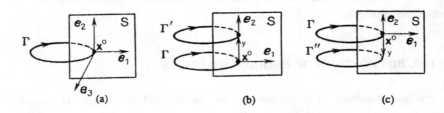

Fig.1.7. Saddle-node bifurcation of the periodic solution.

---

[4]To be convenient, the case of a simple limit cycle is considered with the only fixed point on the section. All results will remain, however, valid as well in the case of n-cycle in the map.

---

[5]In general, the number of control parameters is determined by bifurcation codimension. One can confine himself by one parameter when analyzing the bifurcation of codimension one.

about with this? Illustrate this situation using an example of a three-dimensional phase space. The cycle, crossing the secant at the fixed point x^0, is presented in Fig.1.7(a). Let e_1 and e_2 be the eigenvectors lying in the secant surface. The third eigenvector e_3 is tangent to trajectory Γ at the fixed point. The eigenvector e_1 is answered by multiplier $\rho_1(\mu)$ and the vector e_2 by $\rho_2(\mu)$. Consider $\rho_2(\mu) < 1$ at the parameter values near the critical one. This means that any trajectory, perturbed near cycle Γ toward e_2, approaches x^0 in the secant surface with a certain winding number. $\rho_1(\mu^*) = +1$ toward e_1 at the critical point. Two cases shown in Fig.1.7(b) and (c) are possible depending on the orientation of small initial perturbation y with respect to the eigenvector e_1. At the bifurcation point, either a pair of cycles Γ' and Γ'' is born or they merge and vanish. The bifurcation of the pair of cycles' birth (vanishing) is visualized by a model one-dimensional map on Lamerey's diagram. A certain one-dimensional model map $x(k+1) = P[x(k),\mu]$ and its evolution with varying control parameter are plotted in Fig.1.8. The structural instability is indicated at the critical point μ^* by the dashed line as $P'_x(x^0,\mu) = +1$. For one-dimensional map, the linearization matrix is composed of

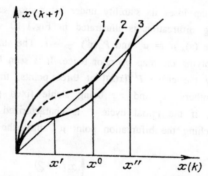

Fig.1.8. Saddle-node bifurcation in the standard one-dimensional map.

one term $P'_x(x^0)$ being the multiplier of the fixed point. The tangency of the map plot to bisectrix disappears (case 2 in Fig.1.8) or two intersection points, x' and x'', appear when varying the parameter μ. In the former case, two cycles perish by merging at the critical point, and in the latter one, a pair of fixed points is born, the stable x' and saddle x'', from the phase trajectories closeness.

The limit cycle bifurcation in the map considered is similar to the bifurcation "loss of equilibrium" for steady-state solutions. It is called also the saddle-node bifurcation of periodic solution.

1.10.2. Period doubling bifurcation of cycle. In the main case, multiplier $\rho(\mu^*)$ becomes -1 at the critical point $\mu = \mu^*$ provided that $\rho'_\mu \neq 0$. Cycle Γ continues to exist as the saddle one as $\mu > \mu^*$ and a limit cycle is born in its vicinity with a period close to the double one. Consider a suitable picture in the phase space presented in Fig.1.9. Cycle Γ, at the critical point, crosses S at point x^0. Specify a small increment y toward e_1 and follow the perturbed trajectory Γ′. The vector of increment will reverse the direction (multiplier is -1) in one revolution along trajectory Γ′ nearby Γ remaining unchanged in modulus within the linear approximation. The perturbed trajectory Γ′ will cross the surface S at point x_2 and would be closed, having made a revolution more, by returning to the initial point x_1.

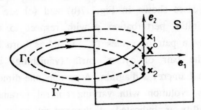

Fig.1.9. Period-doubling bifurcation for the limit cycle Γ.

The fixed point x^0 of a map loses its stability under bifurcation and forms a period-2 cycle. The period doubling bifurcation is illustrated in Fig.1.10 with one-dimensional map as an example. In case (a), $\mu = \mu^*$ and $P'_x(x^0) = -1$. The stable period-2 cycle is obtained for $\mu \geq \mu^*$ by applying the map operator twice. It is seen from Fig.1.10(b) that the bisectrix is crossed by the curve $P^{[2]}[x(k)]$ at three points, the first of them, x^0, being unstable and two others, x_1 and x_2, being stable fixed points. Mathematical results give evidence that, if the initial cycle Γ is characterized by the finite basin of attraction while approaching the bifurcation point μ^*, then the born limit period-2 cycle would be stable.

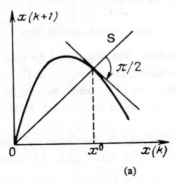
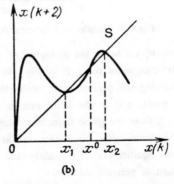

Fig.1.10. Period-doubling bifurcation in one-dimensional map for the parameter values $\mu = \mu^*$ (a) and $\mu \geq \mu^*$ (b).

The period doubling bifurcation can be illustrated by means of observing the time dependence of one of the system's phase coordinates. Let it be $x_1(t)$. In Fig.1.11, the realization $x_1(t)$ of the periodic motion Γ is shown at the moment of the period doubling bifurcation $\rho(\mu^*) = -1$ by dashed line. The realization $x_1(t)$ for $\mu \geq \mu^*$, corresponding to the born period-2 cycle is indicated by solid line.

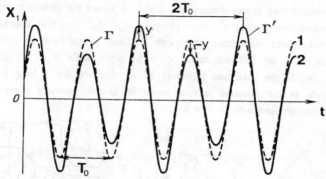

Fig.1.11. Period-doubling bifurcation of the cycle Γ displayed with the time series $x_i(t)$ as an example for the parameter values $\mu \leq \mu^*(1)$ and $\mu \geq \mu^*(2)$.

1.10.3. Two-dimensional torus birth bifurcation. This bifurcation is realized when a pair of complex-conjugate multipliers is emerging to the unit circle when varying the parameter. At the critical point μ^*, the following relation takes place:

$$\rho_{1,2}(\mu^*) = |\rho_{1,2}| exp(\pm j\Psi), \qquad (1.55)$$

where

$$|\rho_{1,2}(\mu^*)| = 1, \; \Psi(\mu^*) \neq 0, \, \pi, \, \pi/2, \, 2\pi/3.$$

The situation on the plane of complex multiplier values corresponds to Fig.1.12. The fixed point of the Poincaré map becomes unstable and an *invariant closed curve* in

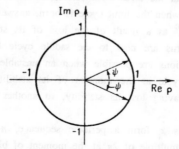

Fig.1.12. Multipliers of the limit cycle Γ at the moment of torus birth bifurcation.

map is also softly born or is contracted to a point near it. With this, an *invariant closed two-dimensional surface* is also softly born or is contracted to a limit cycle in the vicinity of cycle Γ in the system's phase space. Consider this bifurcation in detail with the three-dimensional phase space as an example. Let condition (1.55) take place for the Poincaré map fixed point x^0. The stability of x^0 is described by two multipliers since the Poincaré map is two-dimensional. Fig.1.13 shows the picture to be obtained by assuming perturbation y near x^0. New points x_i, $i = 1,2,...,\infty$ will appear on circle L with every subsequent intersection of Γ with S. These points form an infinite sequence on the circle with the constant radius $r = |y|$ since angle Ψ is, in general, not a multiple of 2π. If the perturbed trajectory Γ' is observed for a long time, then the invariant circle in the Poincaré section would be covered densely everywhere by the points of its intersections with surface S.

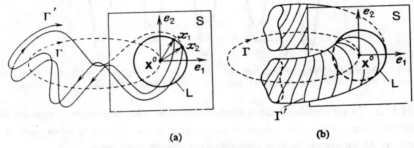

Fig.1.13. Two-dimensional torus birth from the cycle Γ.

The circle is invariant in the sense that any point on it passes to the point of the same circle. Fig.1.13(b) demonstrates the picture realized a lot of time later. The surface, whose Poincaré section presents circle L, will be covered densely by the trajectory Γ' everywhere. The surface is called the two-dimensional torus. All the trajectories in the vicinity of cycle Γ, having lost its stability, will be brought nearer and placed on the torus surface with time in case of the soft birth of two-dimensional torus. Note, that the torus birth bifurcation for maps can be interpreted as Andronov-Hopf bifurcation when the limit cycle is born, answered by the *invariant circle* L, from the fixed point x^0 as a result of the loss of its stability. Stable trajectories on the two-dimensional torus are near to the saddle cycle Γ at the soft torus birth bifurcation. Abrupt bifurcations are possible when an unstable two-dimensional torus is "sticking" into the limit cycle at the moment of its stability loss. Abrupt transition from the limit cycle Γ, having lost its stability, to another regime is experimentally observed here.

The intersection points x_i form a periodic sequence, *n*-cycle for L, if angle Ψ turns out to be an integer multiple of 2π at the moment of bifurcation. They say about *resonance phenomenon on torus* in this case. The trajectory, having done a finite number of revolutions, becomes closed and is a more complicated but periodic motion, the resonant limit cycle on torus. Strong and weak resonances are distinguished between in mathematics. The resonance order is defined by quantity q if multipliers are written in the form:

$$\rho_{1,2} = |\rho_{1,2}|exp(\pm j2\pi \frac{p}{q}), \quad \Psi = 2\pi \frac{p}{q}, \tag{1.56}$$

where $\frac{p}{q}$ is the Poincaré winding number. The picture of stability loss is influenced by the strong resonances $q = 1,2,3$ and 4 to a particularly great extent.

An example of the two-dimensional torus soft birth bifurcation is the appearance of the stable periodic modulation of initial periodic oscillation when modulation frequency Ω and the frequency of modulated oscillations ω_0 are either multiple (resonance) or incommensurate. The motion on two-dimensional torus is called *ergodic* in the latter case. Time dependencies for any of the phase variables $x_i(t)$ of the initial system of equations (1.2) have, in quasiperiodic oscillation regime, the form which is qualitatively shown in Fig.1.14.

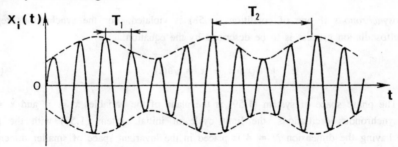

Fig.1.14. Time series $x_i(t)$ for the two-frequency oscillation regime.

It was assumed everywhere, when considering the qualitative interpretation of periodic motion bifurcation that $\mu = \mu^*$, and the evolution of the perturbed trajectory was analyzed. Topologically equivalent regimes were assumed to be realized for small deviations from the critical value. The regimes are observed, however, as $\mu \neq \mu^*$ only and depend on the level of supercriticality.

1.10.4. Symmetry breaking bifurcation. Consider the bifurcation of limit cycles being specific for dynamical systems with some symmetry properties. Let the largest of multipliers of the limit cycle Γ become $+1$ at the bifurcation point: $\rho(\mu^*) = +1$. All the other multipliers lie inside the unit circle. The cycle does not disappear with exceeding the critical value of μ^* but it becomes saddle. Two stable cycles are born at the critical point in the vicinity of cycle Γ (or a pair of saddle cycles merges with cycle Γ)[6]. This bifurcation differs from the saddle-node bifurcation in general system principally. The following simple example will make this clear. Consider a system of two identical symmetrically coupled nonlinear oscillators. The equations of the system can be written in the form:

$$\ddot{x}_1 + G(\mu,x_1)\dot{x}_1 + x_1 = \gamma\theta(x_1,x_2),$$
$$\ddot{x}_2 + G(\mu,x_2)\dot{x}_2 + x_2 = \gamma\theta(x_1,x_2). \tag{1.57}$$

Here, x_1, x_2 are the variables performing oscillation, $G(\mu,x)$ is a certain nonlinear function providing the limit cycle existence in every subsystem, μ is the nonlinearity parameter, γ is the coupling coefficient, θ is the function defining the character of symmetrical coupling.

Equations (1.57) are invariant with respect to the substitution of x_1 by x_2. At some parameter values, selfoscillations in system (1.57) can be *synchronous*, i.e., they can satisfy the requirement:

$$x_1 = x_2, \quad \dot{x}_1 = \dot{x}_2 \tag{1.58}$$

and *asynchronous* if one of conditions (1.58) is violated. For the synchronous regime, the selfoscillation process is to be described by the equation:

$$\ddot{x} + G(x,\mu)\dot{x} + x = \gamma\theta(x). \tag{1.59}$$

The phase space of system (1.59) is the plane of the variables $x = x_1$ and $\dot{x} = x_2$. For synchronous oscillations, the limit cycle of initial system (1.57) with the phase space having the dimension $N = 4$ is placed in the invariant space of smaller dimension. It is a two-dimensional surface in the four-dimensional space. If the synchronous cycle is asymptotically stable, then the Poincaré map multipliers $|\rho_i| < 1$, $i = 1,2,3$. The stability of the synchronous cycle Γ is lost differently with varying the control parameter μ or γ. The case is of interest when the instability of cycle is accompanied by growing perturbations in the direction transversal to the invariant subspace. One of the multipliers of cycle Γ with the eigenvector, lying not in the invariant plane, takes the value $+1$ as $\mu = \mu^*$. Cycle Γ becomes saddle in the four-dimensional space with increasing $\mu > \mu^*$ but it can remain stable in the invariant surface. Two mirror-symmetrical (with respect to the invariant plane) cycles, Γ' and Γ'', are branched off at the bifurcation point μ^* from the synchronous cycle lying in the invariant plane. This is illustrated in Fig.1.15.

Thus, the evidence of the symmetry breaking bifurcation is an experimentally

---

[6] Several limit cycles can be born in the general case of additional degenerations.

recorded limit cycle bifurcation with one of the multipliers becoming + 1 when the cycle in the system does not disappear but becomes saddle. Synchronous oscillations become unstable and asynchronous asymmetrical oscillation regimes appear in the system.

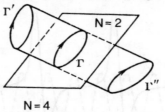

Fig.1.15. Bifurcation of the birth of cycles Γ' and Γ'' at the moment of stability loss for the symmetrical limit cycle Γ.

1.10.5. Nonlocal periodic motion bifurcations accompanied by a period becoming infinite. Consider two limit cycle bifurcations which one can encounter most frequently. Both of them are defined by the cycle period $T(\mu)$ which tends to infinity by approaching the bifurcation point μ^*. The first case is due to the periodic motion vanishing by its *"sticking"* into the separatrix loop of the saddle fixed point as it is shown in Fig.1.16. The separatrix contour Γ_0 is unstable since any trajectory, lying nearby, leaves the vicinity of the loop.

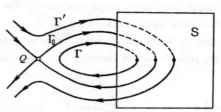

Fig.1.16. Bifurcation of the cycle Γ disappearance via its "sticking" into the separatrix loop Γ_0 of the saddle.

The stable limit cycle is increased in its size when approaching the critical point and, as a result, "captures" the saddle equilibrium state by sticking into the loop of separatrix. The following dependence is characteristic for the period of cycle $T(\varepsilon)$

$$T(\varepsilon) = c\, ln(\varepsilon^{-1}) + c_0. \tag{1.60}$$

If cycle $\Gamma(\mu)$ is stable as $\mu < \mu^*$, then multipliers tend to zero as $\varepsilon \to 0$, too. The bifurcation of the periodic motion disappearance is answered here by the typical time - evolution of any of the cycle phase coordinates which is illustrated qualitatively in Fig.1.17. The periodic motion nearby bifurcation looks like a certain sequence of pulses with the frequency which tends to zero while approaching the critical point. At the critical point the realization has the form of a single pulse, i.e., becomes aperiodic.

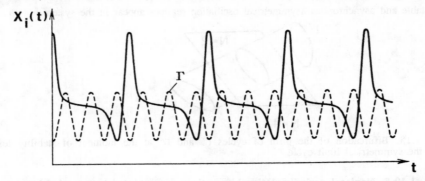

Fig.1.17. Typical dependence of the cycle Γ phase coordinate on time near the separatrix loop Γ_0 of the saddle.

The aperiodic trajectory can be obtained by using numerical experiment only. In full-scale experiments, only the increase of oscillation period according to (1.60) is recorded. Oscillations are abruptly lost nearby the critical point. The trajectory perturbed has, as a rule, no more points of intersection with the secant as the parameter exceeds the critical value in numerical experiments. The loss of cycle and the abrupt transition to a new motion regime are recorded.

The second bifurcation, which leads to the cycle period becoming infinity at the critical point, is due to the periodic motion vanishing at the moment of the birth of a

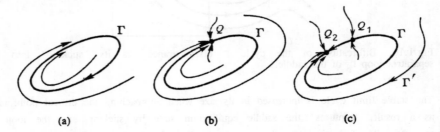

Fig.1.18. Cycle disappearance bifurcation due to the emergency of the saddle-node fixed point on the cycle.

steady singular point of *saddle-node type* on the cycle. The cycle period tends to infinity when approaching the critical value of parameter according to the rule:

$$T(\varepsilon) = c \cdot \varepsilon^{-1/2}. \tag{1.61}$$

The cycle multipliers tend to zero as in the previous case. The phase portrait rebuilding is shown qualitatively in Fig.1.18 for this bifurcation. The cycle is asymptotically stable for the values of $\mu < \mu^*$. A structurally unstable equilibrium state Q of saddle-node type is born on the cycle at the bifurcation point μ^*. The saddle-node is splitted into a saddle Q_1 and the stable Q_2 nodes when the parameter exceeds the critical value, $\mu \geq \mu^*$.

1.11. NONLOCAL BIFURCATIONS IN THE VICINITY OF DOUBLE-ASYMPTOTIC TRAJECTORIES

Consider the dynamical system of dimension $N = 3$ with the steady-state solution in the form of a saddle singular point depending on parameters roughly. A special solution of the system is assumed to be available at μ^* in the form of trajectory Γ_0 which is double-asymptotic to the singular point. The *saddle separatrix loop* Γ_0 is evidently not a robust solution and collapses under an arbitrary small parameter deviation from the bifurcational value. What happens with this?

As it was already discussed in 1.9.4, the birth of the only limit cycle (stable or unstable) is possible in the two-dimensional case. The coming out from the phase plane into the space of three and more dimensions results in qualitatively new phenomena. The rigorous analysis of multidimensional dynamical system bifurcations at the collapse of saddle and saddle-focus separatrix loop was carried out by L.P.Shilnikov in a series of remarkable papers [1.25]. These papers have played an important role in the understanding and explanation of the dynamical stochastization of the selfoscillations phenomenon.

Consider the main results, confining ourselves by a three-dimensional case for simplicity. Let the eigenvalues of the linearization matrix of the saddle equilibrium state s_i, $i = 1,2$ and 3 satisfy the condition

$$s_3 > 0, \quad \mathrm{Re}\, s_{1,2} < 0. \tag{1.62}$$

Introduce into consideration two saddle quantities:

$$\sigma_1(\mu) = \max_{i=1,2} \mathrm{Re}\, s_i(\mu) + s_3(\mu), \tag{1.63}$$

$$\sigma_2(\mu) = 2\mathrm{Re}\, s_1(\mu) + s_3(\mu)$$

and consider the rough case to be realized when σ_1 and σ_2 differ from zero at the bifurcation point and in its small vicinity. The loop Γ_0 collapses under the parameter value deviation from the bifurcational one, μ^*. Bifurcations, realized thereby, are defined by three cases of principle:

1. $\sigma_1(\mu^*) < 0$. The only stable limit cycle is born from loop Γ_0. The situation is similar to that considered earlier for the two-dimensional phase space. Stable periodic oscillations with finite amplitudes and relatively large periods are excited abruptly in the system.

2. $\sigma_1(\mu^*) > 0$. The eigenvalues $s_{1,2}$ are complex-conjugate. A *denumerable set* of saddle periodic motions is born with the collapse of loop in its vicinity. They do not settle, however, the whole set of possible trajectories in the vicinity of the loop collapsed. Apart from them, just one more of two limit cycle systems is realized depending on the sign of the saddle quantity σ_2. If $\sigma_2(\mu^*) < 0$, then *stable periodic motions* exist in the vicinity of loop Γ with varying μ in the denumerable set of intervals of the parameter $\Delta\mu_i$ values. These cycles can be observed experimentally depending on the sizes of intervals of the parameter values where they exist and on the basin of attraction in the phase space. If $\sigma_2(\mu^*) > 0$, then there exists a denumerable set of intervals $\Delta\mu_i$ where absolutely unstable periodic regimes (limit cycles with multipliers lying outside the unit circle) take place. The periodic regimes mentioned are impossible to be observed in real systems.

3. If $\sigma_1(\mu^*) > 0$, then all the eigenvalues s_i are real numbers. Moreover, the *only periodic solution* being of *saddle type only* can be born from loop Γ_0 with its collapse.

Thus, periodic solutions being both stable and saddle, and absolutely unstable can be born from the separatrix loop of saddle equilibrium state Γ_0 under a small parameter deviation from the bifurcational one. There may be either one solution or an infinite set of them. The case, as $\sigma_1(\mu^*) > 0$, leading to the birth of the countable set of saddle periodic motions, is of particular interest from the viewpoint of understanding the possibility of chaos realization. A *non-trivial hyperbolic subset* of trajectories arises in the vicinity of loop being a condition required for a strange attractor to exist.

CHAPTER 2

NUMERICAL METHODS OF CHAOS INVESTIGATION

2.1. EXPERIMENTAL APPROACH TO INVESTIGATION OF NONLINEAR SYSTEM DYNAMICS

In general case, the evolution problem as applied to the non-linear dynamical system

$$\dot{\mathbf{x}} = F(\mathbf{x}, \mu) \qquad (2.1)$$

is not solvable analytically. The solution of equation (2.1) can be found by using either computation or analogous simulation. Various problems appear at the numerical study of specific non-linear systems from the mathematical point of view that require special algorithms and calculation programs. The top priority problem is, however, a *numerical integration* of (2.1) to find the dependence $\mathbf{x}(t,\mu)$ under given initial conditions, i.e., the solution of *Cauchy problem*. The methods for numerically integrating a system of ordinary differential equations enable one, as a rule, to construct the solution with different values of initial conditions and parameters. The possibility to attack *regular unstable solutions* is of principle importance since the solutions are not realized in full-scale experiments. The evolution of such regimes with varying a parameter is often significant so long as it defines the ance mechanisms of structurally stable regimes observed experimentally.

The coordinates of the fixed points and their dependence on parameters are determined for the (2.1) by solving the algebraic equations $F(\mathbf{x},\mu) = 0$ using numerical methods. This problem, solved with varying the parameter many times, serves as a base for constructing the characteristic bifurcation lines in the parameter space of the system. The *Poincaré map* in the secant hypersurface is found by suitable methods. It demonstrates visually the character of solution and is informative when investigating the selfstochastic oscillation regimes in small-dimensional systems.

Calculation algorithms based on the linear algebra methods, that use the numerical integration results, allow one to solve the problem of the *stability* of solutions and their *bifurcations*. Thereby, they make it possible to explore the process of the structure rebuilding for the phase space partition into trajectories with variation of the parameters.

The techniques of statistically processing a set of selected (truncated) realizations $x_T(t,\mu)$ are applied to examine the "physical" properties of chaotic oscillations such as, for example, the integral power, its frequency distribution, the degree of the "fortuity" of the process (selfcorrelation) and others. They enable calculating the distribution function $p(x)$, correlation coefficient $R_x(\tau)$, the power spectrum of the process $S_x(f)$, the distribution function $p(x)$ moments of interest (for example, $<x>$, $<x^2>$, dispersion and others) and the dependence of the above characteristics on parameters.

The study of fluctuation effect on system dynamics is of great importance. A strange attractor as influenced by fluctuations may not change its structure and properties practically while the regular regime can be replaced by the chaotic one (chaotic oscillations induced by fluctuations). Reverse picture can be observed as well: the strange attractor can collapse being replaced by the regular regime and the limit cycle properties would not evolve as a whole if fluctuations are small enough. The mathematical simulation and numerical analysis of the above phenomenon require formulation and solution of either *stochastic differential equations*, when a random influence is introduced into the right hand parts of system (2.1) in the form of a random number producer with the given statistics, or a suitable *Fokker-Plank equation*.

The very unexpected situations arise during investigation of particular systems, therefore all the algorithms required are impossible to be enumerated. Here, the basic algorithms will be considered only which allow typical problems to be solved [2.1-2.4].

The methods of physical experiments are of great importance, too [2.4, 2.5]. A crucial feature of physical experiment is that it illustrates the *structurally stable oscillation regimes* in a system whose mathematical model is always approximate. The physical experiment enables one to solve the adequacy problem of the mathematical description of a real system and to set the application boundaries of approximate mathematical description. If the above is performed, then the behavior of solutions can be examined which are found numerically with variation of the parameters. A strange attractor can be observed, for example, and, if there would be some of them in the phase space, the more probable attractors can be ascertained.

Physical experiments are carried out with the help of devices assigning currents and voltages, i.e., electrical signals, to the phase coordinates of the system under study. The signals are investigated by using various instrumentation. If a physical model is not able to provide this condition immediately (for example, in the course of experiments with liquid flows), various techniques of the isomorphic transformation of the initial phase coordinates of the system into corresponding electric signals are used. So, for example, light-reflecting balls made of special material are added to the liquid during investigation of turbulent flows to determine the velocity of particle motion. The balls move together with the liquid whose velocity is measured by the Doppler shift of the reflected signal frequency of monochromatic laser.

Experimental investigations become simpler, more evident and can be more cost-effective than numerical ones by operating the radiophysical models of dynamical systems and, to be precise, the real electric selfoscillating systems with the dynamical system (2.1) as a model. The analysis of the system dynamics is experimentally simplified and fastened by varying its parameters. It is desirable, however, to control the results, obtained for some characteristic regimes, by using calculation. This is due to violation of the adequacy conditions of physical and mathematical models which occasionally takes place when varying the system parameters. Here, non-linear effects can be "linked up" which are not included in equations, but are observed by varying the real model parameters.

There is another condition to be kept in mind when studying the non-linear system dynamics and, namely, the fundamental dependence of one or another solution $\mathbf{x}(t,\mu)$ of (2.1) on specific initial conditions $\mathbf{x}(t_0,\mu)$. It is difficult and sometimes not feasible to control the initial conditions. Furthermore, when varying the parameters, one solution (one particular dynamic regime) can lose its stability and be replaced by another solution whose basin of attraction is near to that of the first regime. This phenomenon is difficult to observe and it leads to false results at the end [2.4].

It is seen from the above consideration that neither numerical nor physical methods for studying the non-linear system dynamics on the whole are to be preferred since every of them has both advantages and shortcomings. It is reasonable to apply both approaches whenever possible since they complement each other naturally.

In the present chapter, attention is concentrated on discussing the algorithms which form the basis for the most useful numerical techniques to analyze the evolution mechanisms and properties of strange attractors. As to the methods of physical experiment, they, unlike numerical ones, do not call for large modifications as applied to the investigation of chaotic oscillation regimes and do not need special attention.

2.2. CALCULATION OF THE POINCARÉ MAP

Consider a certain autonomous dynamical system:

$$\dot{x}_i = f_i(x_1, x_2, \ldots, x_N, \mu_1, \ldots, \mu_k). \tag{2.2}$$

In the phase space $\mathbb{R}^N$ of system (2.2), specify the $(N-1)$-dimensional surface S

which satisfies the conditions imposed on the *Poincaré secant* with the equation having the form:

$$S(x_1,...,x_N, \mu_1,...,\mu_k) = 0. \qquad (2.3)$$

The *Poincaré map* transforms the points which belong to the surface S into the points of the same surface, i.e., it maps secant S onto itself. For any point, $x_n \in S$, the transformed point $P(x_n) = x_{n+1} \in S$. To find the map P, one must solve the system of equations (2.2) at the noted values of parameters μ_i by setting the following initial conditions:

$$x_i(0) = x_i^0, \quad i = 1,2,...,N, \quad x^0 \in S \qquad (2.4)$$

and successively find the points of intersection of the solution obtained (trajectory Γ emanating from x^0) with surface S. Two independent problems appear. First, the calculation of the trajectory $x(t)$ of system (2.2) at the given parameter values and under initial conditions (2.4) is to be performed. The second problem is to determine the coordinates of points of intersection of the trajectory and the secant, i.e., to construct the Poincaré map (section)[1]. The trajectory can be calculated by using any of the known numerical integration techniques. For example, Runge-Kutt methods, in particular, that of the fourth order, are employed to investigate chaotic systems. To define the coordinates of points of intersection of Γ with S, the value of function $S(x,\mu)$ is to be calculated at every time step when integrating the system (2.2) until the sign of $S(x,\mu)$ changes. Let, for example, the sign of $S(x,\mu)$ change within the time interval from t_k to $t_k + \Delta t = t_{k+1}$ where Δt is the numerical integration step. Fig.2.1 makes the situation clear.

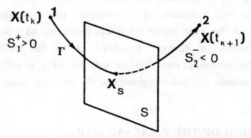

Fig.2.1. Precise determination of the trajectory intersection point with the secant surface.

---

[1]The difference in the terms Poincaré "section" and "map" arises in case of the loss of information on evolution operator, for example, at the simple geometric representation of a set of points on the section (or its projection).

The point x_s of intersection of the trajectory with the secant is to be specified. This problem can be solved by using interpolation methods. Successively dividing the integration step by two, one can apply the dichotomy technique and complete the calculation as the difference $|S_1^+ - S_1^{-1}|$ is reduced relative to the specified magnitude. There are no problems of principle, but higher requirements to x_s, which must be determined exactly, need additional computation and a more complicated algorithm. This results in the increase of machine time consumption, particularly, in case of a high system dimension N.

A simple and effective method is due to Henon [2.6]. Introduce into consideration the variable $x_{N+1} = S(x_1, x_2, \ldots, x_N)$. Then, taking into account (2.2) we get:

$$\frac{dx_{N+1}}{dt} = \frac{dS}{dt} = \sum_{i=1}^{N} \frac{\partial S}{\partial x_i} \cdot \frac{dx_i}{dt} \equiv \sum_{i=1}^{N} \frac{\partial S}{\partial x_i} \cdot f_i. \tag{2.5}$$

Now, the secant surface is defined by the following equation:

$$x_{N+1} = 0. \tag{2.6}$$

Consider a new system obtained by adding to (2.2) the equation:

$$\dot{x}_{N+1} = f_{N+1}(x_1, \ldots, x_N), \quad f_{N+1} = \sum_{i=1}^{N} f_i \frac{\partial S}{\partial x_i}. \tag{2.7}$$

Transform the obtained system of $(N + 1)$ equations so that a new variable x_{N+1} introduced into consideration becomes formally equivalent to variable t in (2.2). For this purpose, divide N first equations by the last one (2.7) and reverse the last equation. The obtained system of equations will be as follows

$$\frac{dx_1}{dx_{N+1}} = \frac{f_1(x)}{f_{N+1}(x)},$$

$$\ldots\ldots\ldots\ldots$$

$$\frac{dx_N}{dx_{N+1}} = \frac{f_N(x)}{f_{N+1}(x)}, \tag{2.8}$$

$$\frac{dt}{dx_{N+1}} = [f_{N+1}(x)]^{-1}.$$

As a result of integrating the system (2.2), function S changes its sign at the time moment $t = t_k$ as compared to the sign which it has had at the time moment $t = t_{k+1}$.

Now, the point $x_s \in S$ can be found by integrating the system (2.8) at one step only with respect to the new variable x_{N+1}:

$$\Delta x_{N+1} = -S_2, \qquad (2.9)$$

where $S_2 = S(x_1, x_2, \ldots, x_N)$ at the moment $t = t_{k+1}$ corresponding to point 2 (Fig.2.1) on trajectory Γ. Initial conditions to integrate (2.8) at step (2.9) are, respectively:

$$x_i(S_2) = x_i(t_{k+1}), \quad t(S_2) = t_{k+1}, \quad i = 1,2,\ldots,N. \qquad (2.10)$$

Having integrated the system (2.8) at one step (2.9), we hit the secant surface S immediately. The error of determination of intersection point is strictly equal to the given error of integrating the system (2.2) at one step and is minimal. Only the methods admitting an arbitrary time step are used to integrate the system (2.8) at a step $\Delta x_{N+1} = -S_2$ which can evolve with every new intersection of the surface S with trajectory Γ. It is better to use the same method both for trajectory calculation (when solving the system (2.2)) and to find the Poincaré map (when solving the system (2.8)). While making up the program which realizes the Henon algorithm described, there is no need to write equations (2.2) and (2.8) separately. They can be written in general form:

$$\begin{aligned}\frac{dx_i}{d\tau} &= q \cdot f_i(x_1, \ldots, x_N), \\ \frac{dt}{d\tau} &= q, \quad \tau = x_{N+1}, \quad i = 1,2,\ldots N.\end{aligned} \qquad (2.11)$$

It follows from equations (2.11) that systems (2.2) and (2.8) are obtained as $q = 1$ and $q = (f_{N+1})^{-1}$, respectively. Thus, the Henon algorithm, used together with one of numerical integration techniques, enables a set of intersection points of trajectory Γ with secant S to be easily obtained.

The Poincaré section method is the most visual as $N = 3$, where the set of points of intersection lies on a two-dimensional surface. For $N \geq 4$, the multi-dimensional Poincaré section, as represented graphically, loses its obviousness. In this case, it is more convenient to analyze the projections of a multi-dimensional section on the two-dimensional surfaces of interest. The multi-dimensional Poincaré section is given by the set of points $x \in S$. The totality of any two coordinates, x_l and x_m, $l \neq m$, $l,m = 1,2,\ldots,(N-1)$ is chosen from that set to represent the Poincaré section's two-dimensional projection onto the plane of coordinates selected. The Poincaré section (both multi-dimensional one and its projections) comprises a set of fixed points in case of the periodic solutions of initial system (2.2). A certain chaotic (pseudorandom) set of points appears on the secant in strange attractor regime with the number of points growing as the integration time increases. In some particular cases, the chaotic set can be placed along a thin band being close to one-dimensional curve in structure. Then, *one-dimensional* map (or one-dimensional return function) may be easily calculated, which is numerically built up for one of the chosen coordinates x^i of the Poincaré map. The return function $\Phi(x)$ is a mapping of a segment into itself:

$$x^i_{n+1} = \Phi(x^i_n, \mu), \quad n = 1, 2, \ldots \qquad (2.12)$$

and is graphically presented as one-dimensional curve composed of the set of points. The analysis of the one-dimensional return map allows a number of principle questions to be answered concerning the dynamics features of the initial equation system (2.2).

As an example, consider the Poincaré map calculation results for a three-dimensional two-parametric system introduced in Ref.[2.7]

$$\dot{x} = y(z - 1 + x^2) + \gamma x, \quad \dot{y} = x(3z + 1 - x^2) + \gamma y,$$
$$\dot{z} = -2z(v + xy), \qquad (2.13)$$

where x, y and z are the phase variables and γ and v are parameters being varied. The

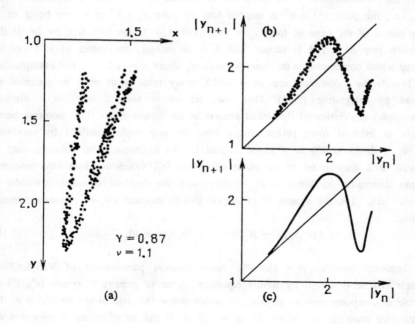

Fig.2.2. Construction of one-dimensional model map for the system (2.13).

system is dissipative as $\gamma < \nu$ and is defined by the constant divergence $2(\gamma - \nu) < 0$. Fig.2.2(a) shows the two-dimensional Poincaré section in the plane $z = const$ for $\gamma = 0.87$ and $\nu = 1.1$ which is indicative of predominant flow contraction along x and expansion along y. A model map, calculated for variable y as $|y_{n+1}| = \Phi(y_n)$, is shown in Fig.2.2(b). With increase of the contraction degree, the map can be approximately replaced by the one-dimensional map (Fig.2.2(c)) making the analysis much simpler.

The method of the Poincaré section is used as well to analyze the dynamics of *nonautonomous systems*. Introduce a periodic force into the first equation of system (2.2):

$$\dot{x}_1 = f_1(x_1, x_2, \ldots, x_N, \mu) + B\, sin(p\tau),$$
$$\dot{x}_i = f_i(x_1, x_2, \ldots, x_N, \mu), \quad i = 2, 3, \ldots, N \tag{2.14}$$

Here, B and p are the amplitude and frequency of harmonic force, respectively, τ is the time. The number of phase coordinates of system (2.14) remains equal to N. The dimension of the phase space (2.14) becomes $(N + 1)$, however, due to periodic force. How is the Poincaré map to be defined in this case? One of the methods is as follows. As seen from (2.14), the N-dimensional space of phase coordinates x_i, $i = 1, \ldots, N$ is mapped into itself, i.e., the point $x_i^{n-1} \in \mathbb{R}^N$ is mapped into the point $x_i^n \in \mathbb{R}^N$ in a time being equal to the period of the external force $t_0 = 2\pi/p$. For $p = 1$, the time is $t_0 = 2\pi$. If the integration step $\Delta t = t_0/k$ is chosen with k as an integer, the values of $\{x_i^n\}$ can be recorded which correspond to the time moments nt_0 where $n = 1, 2, \ldots$. For example, for $p = 1$ with the calculation step $\Delta t = 2\pi/10$, every tenth point must be recorded as obtained by integrating (2.14). Data file, set up in such a way, is a discrete (stroboscopic) description of dynamical process in the system at the time moments being multiple to external force period. Every time the *new surface* defined by condition $t_n = nt_0 = 2\pi n/p$ (2.15) is a secant surface in this technique. The Poincaré map is presented as a discrete set of the phase coordinate $\{x_i^n\}$ values at given time moments and has dimension N. What is to be done with the data set obtained? Consider a particular case. Let the system (2.2) be the two-dimensional ($N = 2$) nonautonomous equation:

$$\dot{x}_1 = f_1(x_1, x_2, \mu) + B\, sin\tau, \quad \dot{x}_2 = f_2(x_1, x_2, \mu), \tag{2.16}$$

(for simplicity consider $p = 1$). The synchronization phenomenon of Van-der-Pole oscillator can be described by similar equations. A set of n pairs of points $\{x_1^n, x_2^n\}$ is available in computer memory in this case which define the state of system (2.16) at the discrete time moments $nt_0 = n \cdot 2\pi$ ($n = 1, 2, \ldots$). If the set of points is presented on the plane of variables x_1, x_2, then we acquire a map of the two-dimensional phase space into itself in the external force period. The map is a projection of the set of intersection points of trajectory Γ with the secant planes onto the coordinate plane of our interest. Fig.2.3 shows synchronization at the external force frequency when the initial point (x_1^0, x_2^0) is mapped into itself in the time intervals $\Delta t = 2\pi$.

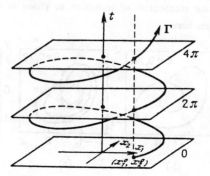

Fig.2.3. Poincaré map for the nonautonomous system (2.16).

The number of the Poincaré map points, being finite for an arbitrary large n, corresponds to regular (periodic) regimes as in the autonomous case. The set of points $\{x_1^n, x_2^n\}$ on the plane appears as the random one in the strange attractor regime. This chaotic set can look highly amusing and will evolve with variation of the system parameter μ. The obviousness of the Poincaré map is lost as $N > 2$ and its two-dimensional projections are to be built in those cases.

Consider one of the examples where the dynamics of a nonautonomous three-dimensional system is analyzed by the above technique [2.8]:

$$\dot{x} = \dot{y}, \quad \dot{y} = -0.125(x^2 + 3z^2) x - k_1 y + B \cos\tau ,$$
$$z = -k \cdot 0.125(3x^2 + z^2) z + B_0 , \quad (2.17)$$

where k_1, k_2 and B_0 are the system parameters. For $k_1 = k_2 = 0.5$ and $B_0 = 0.03$, the system was integrated for various amplitudes B and the limit sets were calculated in the three-dimensional map of the phase space into itself in the external force period $t_0 = 2\pi$. An interesting phenomenon of the quasiperiodic solution bifurcations was discovered by the authors with this. As it was already noted, an invariant closed curve in the Poincaré map on the plane corresponds to a two-dimensional torus, and its bifurcations are adequate to the torus bifurcations. Its evolution in the projection of the Poincaré three-dimensional map onto the plane of variables x, y in the external force period is shown in Fig.2.4. As seen from the figures, the invariant closed curve undergoes, with growing force intensity B, a series of period doubling bifurcations completed by a strange attractor birth. Further investigations have demonstrated the number of two-dimensional torus doublings, preceding the birth of chaos, to be finite at finite amplitudes B and the mechanism of transition to chaos to be associated with the torus breakdown regularities here.

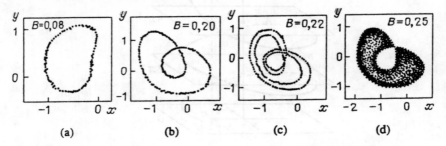

Fig.2.4. Limit sets of system (2.17) in the map on the plane.

The set of intersection points of trajectory Γ with the fixed secant S, selected in a certain way, can be considered as the Poincaré map of a nonautonomous system, i.e., the section is to be introduced into $\mathbb{R}^N$ with the help of equation (2.3), as in the autonomous case.

2.3. NUMERICAL ANALYSIS OF PERIODIC SOLUTIONS AND THEIR BIFURCATIONS

A usable class of the ordinary differential equation system solutions is formed by *time-periodic solutions* $x(t) \equiv x(t + T)$ called, for convenience, the *cycles*. These solutions and their evolution with variation of the parameters, besides being of interest in themselves, contribute to the explanation of many strange attractor birth mechanisms.

A number of independent calculation problems arises under the cycle investigation. First of all, the problem of finding the periodic solutions with different periods at the fixed system parameter values and that of defining their stability character are to be solved. Secondly, the evolution of a particular cycle family with variation of the parameter is to be examined and possible *bifurcations* responsible for the loss of the cycle stability are to be found. Finally, the problem of numerically constructing *bifurcation diagrams* in the system parameter space is to be settled which define the existence region and the character of stability loss by the cycle at the region boundaries [2.9].

In practice, there appear more questions to be answered, for example, with regard to the need of construction of *stable and unstable* saddle cycle *manifolds*, of revealing the *homoclinic trajectories*, and so on. The above problems are, however, the most general and we shall confine ourselves to their consideration here.

Apply to the autonomous system of equations (2.1) setting a flow in space $\mathbb{R}^N$ again.

1. How is the periodic solution of system (2.1) to be found? Approximate data on the cycle position in the phase space and on its stability character are necessary to do this. For example, if the cycle is born softly from the equilibrium point according to the *Andronov-Hopf bifurcation theorem*, then the limit cycle is to be found near the unstable singular point with a small shift in parameter from the bifurcation point. Other situations encounter that permit qualitative interpretation and can indicate approximately cycle shape and region of its localization in the phase space. If theoretical a priori considerations are unable to help, then, unfortunately, we have to inspect experimentally the entire phase space regions by varying the initial conditions and calculating trajectories every time and integrating the system (2.1).

To find the numerically cycle, we can proceed from the fact that the point of intersection of cycle Γ in $\mathbb{R}^N$ with the secant surface S is the fixed point x^* of the Poincaré map $P(x_n)$. Let $x(t)$ be the periodic solution of system (2.1) having a period T. Trajectory Γ in the phase space crosses the secant S at point x^* satisfying the condition

$$x^* = P(x^*). \qquad (2.18)$$

Suppose point x_0 on secant S, which is close to the fixed point x^*, to be known from any considerations. It can be found by solving the equation

$$x = P(x) \qquad (2.19)$$

with the help of the Newton iteration procedure

$$x_n = x_{n-1} - \frac{P(x_{n-1}) - x_{n-1}}{M(x_{n-1}) - E} \qquad (2.20)$$

where $P(x)$, $M(x)$, and E are the Poincaré map on S, the map linearization at point x, and the unit matrix, respectively. Point x^0 being close to x^* is specified as an initial approximation. The iteration procedure convergence *does not depend* on the character of the cycle stability! The method of simple iterations of map P can be used if the cycle is known to be stable in advance.

The fixed point x^* is calculated with the given accuracy ε if the following condition is fulfilled:

$$|x_n - x_{n-1}| \leq \varepsilon. \qquad (2.21)$$

When x^* is found, its coordinates $x(0) = x^*$ are chosen as initial conditions. Then, system (2.1) is integrated from the initial point to the moment $t = T$ when trajectory Γ crosses S again as $x = x^*$, i.e., the following condition is fulfilled

$$x^* = x(0) = x(T), \qquad (2.22)$$

which defines the period of cycle Γ to be found.

As a result of numerical integration, the whole cycle is constructed in the phase space or its projection is obtained onto the subspace of interest with the dimension, as a rule, $N = 1$ (realization of one of the phase coordinates) or $N = 2$ (cycle projection onto a chosen plane of two variables).

If the desired cycle Γ is known to cross the secant surface k times, then the fixed point x^* must have the corresponding periodicity. This fact is to be taken into account when writing equations (2.18)-(2.20).

After the cycle has been found, it is examined for its stability defined by the *multipliers of cycle* ρ_i or (the same) by the *multipliers of the fixed point* x^* of the Poincaré map in the linear approximation. Two ways of numerical solution of this problem are possible. We can define either the eigenvalues of *monodromy operator* $Y(T)$ or that of the Poincaré map linearization at the fixed point x^*. The cycle found is asymptotically stable according to Lyapunov and is the limit cycle of system (2.1) if all the multipliers ρ_i ($i = 1,2,...,N - 1$) of the Poincaré map are lying inside the unit circle.

2. After the cycle has been found and its stability character has been defined, we proceed with solving the second task. The cycle is known for the specific parameter value $\mu = \mu_0$ of system (2.1) and the fixed point $x^*(\mu_0)$ belonging to the cycle is given on secant S. The evolution of this solution should be numerically followed with varying the parameter μ from the value $\mu = \mu_0$ to $\mu_k = \mu_0 + k\Delta\mu$ (k and $\Delta\mu$ are integer and the step of parameter variation). To accomplish this, the fixed point $x^*(\mu_k)$ ($k = 0,1,2...$) on cycle $\Gamma(\mu_k)$ is revised, in general (for various k), from the initial point $x^0(\mu_k)$ by using the Newton technique. The point $x^*(\mu_k)$ is then employed as an initial approximation to define the initial point $x^0(\mu_{k+1})$ which is close to the cycle $\Gamma(\mu_{k+1})$

$$x^0(\mu_{k+1}) = x^*(\mu_k) + \frac{dx^*(\mu_k)}{d\mu}\Delta\mu, \qquad (2.23)$$

where $dx^*/d\mu$ is the *"derivative of the cycle with respect to a parameter"* determined from the equation

$$P(\mathbf{x}^*, \mu) = \mathbf{x}^*(\mu) \qquad (2.24)$$

as a derivative of the following implicit function

$$\frac{d\mathbf{x}^*}{d\mu}\bigg|_{\mu=\mu_k} = \frac{\partial P}{\partial \mu}(M-E)^{-1}. \qquad (2.25)$$

Thus, the periodic solution of system (2.1) is found for various parameter values $\mu = \mu_k$, $k = 0,1,2,...$, i.e., the evolution of a robust cycle in the system phase space is followed by variation of the parameter.

The multipliers of cycle $\rho_i(\mu)$ are calculated at every calculation step when varying the parameter. Their dependence on μ provides an exhaustive information on the stability character of the family of cycles $\Gamma(\mu)$. The emerging of one (or several) multiplier(s) onto the unit circle leads to bifurcation and defines the bifurcational value of parameter (*the bifurcation point*) μ^*. This value can be determined with the given accuracy ε, when required. The solution of the second problem, which is based on the continuous dependence of the system (2.1) solution on the parameter, gives a quite complete information about the way of birth (vanishing) of a particular cycle family, about the region of its existence and the evolution of stability character under the parameter μ variation.

3. Proceed to a more common case of the solutions of system (2.1) depending on two parameters

$$\dot{\mathbf{x}} = F(\mathbf{x}, p, q). \qquad (2.26)$$

State a problem to construct the bifurcation lines on the plane of parameters p and q corresponding to a certain type of stability loss by the family of cycles $\Gamma(p,q)$. Having fixed, for example, $p = p_0$ and varying the parameter q in the above way (cycle $\Gamma(p_0,q)$ is assumed to be found for the value $p = p_0$ and a certain q), determine the value $q = q_0$ when the largest of the cycle Γ multipliers $\rho(p_0,q_0)$ emerges onto the unit circle. The emerging is indicative in particular of the period doubling bifurcation $[\rho(p_0,q_0) = -1]$. Cycle $\Gamma(p_0,q_0)$ and the Poincaré map's fixed point on S, $\mathbf{x}^*(p_0,q)$, corresponding to it, are known at this point of the parameter plane. The plane point (p_0,q_0) belongs to the *bifurcation line l* (in this case to the period doubling bifurcation line)[2]. To refine the point, one needs to verify fulfillment of the corresponding bifurcation condition. In case of doubling, $\rho = -1$, the following

---

[2]Bifurcation lines in plane or corresponding hypersurfaces in R^k highlighted by one bifurcarion condition of equality type describe the bifurcations of *codimension one*.

equation is to be satisfied:

$$det[M(p_0,q_0) - E] = 0. \qquad (2.27)$$

To construct a bifurcation line, take a new value of $q_1 = q_0 + \Delta q$ and, by using the data at the fixed point $x^*(p_0,q_0)$ as an initial approximation, solve by means of the Newton technique the following equation:

$$x = P(x,p,q_1) \qquad (2.28)$$

along with the corresponding bifurcational equation (for doublings, this is (2.27) with respect to x and p). As a result, a new bifurcation point $(p_1,q_1) \in l$ is found. In this way, a bifurcation line is built in the plane of p, q which, being crossed by the family of cycles $\Gamma(p,q)$, causes its stability to be lost in a certain manner. The last depends on the form of the bifurcational equation, for example, (2.27).

The same family of cycles $\Gamma(p,q)$ can lose its stability in several ways with varying p and q. The construction of the corresponding bifurcation lines l_i provides an extensive information on the system's behavior when varying the control parameters. In cases, where equations (2.1) depend on $\mu \in \mathbb{R}^k$, $k = 3,4,...$, the algorithm admits generalization and allows one to build the appropriate hypersurfaces in $\mathbb{R}^k$ corresponding to the certain types of stability loss under codimension one bifurcations. There are no crucial problems, as well, in generalizing the algorithm to the bifurcational situations of codimension two (for example, the construction in $\mathbb{R}^3$ of a bifurcation line corresponding to the birth of a resonant two-dimensional torus with the given Poincaré winding number). It is, however, clear that such tasks will require to substantially increase the time of computation.

To illustrate the application of the above algorithms, consider the procedure of constructing the cycle period doubling bifurcation lines for one of the Rössler systems

$$\begin{aligned}\dot{x} &= -(y + z), \\ \dot{y} &= x + ey, \\ \dot{z} &= b + xz - \mu z,\end{aligned} \qquad (2.29)$$

where the parameter $b = 0.2$ is fixed. At first, find a stable limit cycle of the system by choosing the parameter values $e = 0.2$ and $\mu = 2.5$. The cycle exists, it has the form of one-rotational curve in the phase space (the cycle of period T_0) and is stable. Follow the evolution of this cycle by varying the parameter μ ($e = 0.2$). At the point $\mu \approx 2.83$, cycle T_0 suffers the bifurcation of period doubling since its largest in modulus multiplier becomes -1. Then, construct the bifurcation line l satisfying the condition $\rho = -1$ in the plane of parameters e and μ.

A segment of the line is plotted in Fig.2.5. Use a point on cycle T_0 as the initial approximation of the Newton iteration procedure and try to find a cycle of period $2T_0$ (above the bifurcation line l in Fig.2.5). Calculations for $e = 0.2$ and $\mu = 2.9$ give the stable cycle of period $\simeq 2T_0$. By specifying the initial conditions on cycle T_0, which is saddle already now (above the bifurcation point $\rho = -1$), integrate the system (2.29) and find the unstable cycle T_0 for $\mu = 2.9$. The calculation results are presented at the top of Fig.2.5 in the form of the projections of corresponding phase trajectories onto xy-plane.

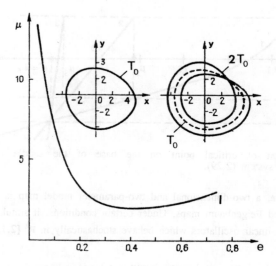

Fig.2.5. Period-doubling bifurcation line for the cycle in the system (2.29).

It is known that the sequence of Feigenbaum doubling is observed in system (2.29) by increasing μ for $e = b = 0.2$ which converges to the critical point $\mu \simeq 4.2$. The point could be found by calculating the sequence μ_k of the bifurcation points $\rho = -1$ of the period $2^k T_0$ cycle as $k = 1, 2, \ldots$ An alternative method can be proposed in terms of the universal value of unstable cycle multipliers at the critical point $\rho^* = -1.601\ldots$ Fig.2.6 shows the dependencies $\rho(\mu)$ for cycles $2T_0$, $4T_{0*}$ and $8T_0$. The trend for convergence of all the plotted lines to the point $\mu \simeq 4.2$, $\rho^* = -1.6$ is seen. Hence, the critical point $\mu^* = 4.2$ is found with the accuracy being practically enough. If there is a confidence in an infinite cascade of doublings, then the bifurcation line of critical values can be constructed in the plane of parameters e, μ which is defined as the line $\rho = -1.6$ for the cycles with periods $4T_0$ or $8T_0$.

In the present paragraph, the construction of algorithms was considered for numerically analyzing the periodic solutions of the system of differential equations (2.1). As a method, the technique of analyzing the stability of the Poincaré map's fixed points was chosen. Consequently, all the above algorithms can be applied to analyze immediately the fixed points of the discrete map when specifying the map operators in explicit form, with the exception of the procedure of numerical construction of the Poincaré map in the secant.

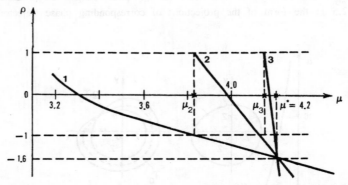

Fig.2.6. Calculation of critical point on the base of the saddle cycle multiplier universal value for system (2.29).

As an example, a two-dimensional and two-parameter model map is proposed in the form of two coupled Feigenbaum maps. Under certain conditions, it simulates a system of two interacting non-linear oscillators which behave stochastically in $\mathbb{R}^6$ [2.1,2.4,2.5]:

$$x_{n+1} = 1 - \alpha x_n^2 + \gamma(y_n - x_n),$$
$$y_{n+1} = 1 - \alpha y_n^2 + \gamma(x_n - y_n),$$
(2.30)

where $\alpha \in [-2,2]$, $\gamma \in [0,1]$ are the system parameters.

Analytical solution is assumed by system (2.30) for the cycles with period 1. It can be used for debugging a program which realizes algorithms on computer. The fixed points of the map (2.30) can be of two types, symmetrical (sm) (unstable point sm - 1 is not considered)

$$x = y = \frac{1}{2\alpha}(\sqrt{4\alpha + 1} - 1),$$
(2.31)

and asymmetrical (asm)

$$x_{1,2} = y_{1,2} = 0.5[-b(2\gamma + 1) \pm \sqrt{b}\sqrt{(b - 4b\gamma^2 + 4}],$$
(2.32)

where $b = \alpha^{-1}$. The multipliers of these fixed points are, respectively:

$$\rho_1^{(sm)} = 1 - \sqrt{1 + 4\alpha}, \quad \rho_2^{(sm)} = 1 - \sqrt{1 + 4\alpha} - 2\gamma, \quad (2.33)$$

and

$$\rho_{1,2}^{(asm)} = (1 + \gamma) \pm \sqrt{4\alpha - 3\gamma^2 + 1}. \quad (2.34)$$

Consider the bifurcations of the sm-1 point when increasing the parameter α as $\gamma \neq 0$ is fixed. The line on the plane of parameters α and γ, $\alpha = -0.25$, corresponds to the birth bifurcation of the sm-1 point ($\rho^{(sm)} = +1$). The multipliers of the sm-1 point decrease with the growth of parameter α. If the condition

$$\alpha = 0.25(4\gamma^2 - 8\gamma + 3) \quad (2.35)$$

is fulfilled, then the multiplier $\rho_2^{(sm)} = -1$ and $|\rho_1^{(sm)}| < 1$. A line on the plane for which condition (2.35) is met, is the bifurcation line of the asm-2 point birth (the asymmetrical fixed point of period 2). The fixed asm-2 point is a stable node because both its multipliers are real and smaller than a unity in modulus. The stable node is transformed into a stable focus as α is increased and, further, the Andronov-Hopf bifurcation for maps is realized, i.e., a complex-conjugate pair of multipliers emerges onto the unit circle. A closed invariant curve is born softly in the map on the x_n, y_n plane.

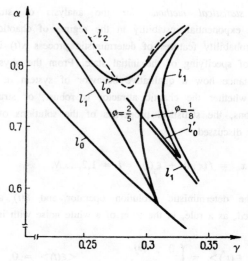

Fig.2.7. Bifurcation diagrams of system (2.30) for resonances 2:5 and 1:8.

The line l_0, where $\rho_{1,2}^{(asm)} = exp(\pm j2\pi\phi)$ corresponds to this bifurcation, is plotted on the diagram in Fig.2.7. *Winding number* ϕ varies continuously along the line and takes rational values in a discrete set of points corresponding to the *bifurcations of codimension 2*. The cycles of relevant periods are born at these points that lie on invariant circles. The regions of these cycles' existence are bounded by the lines of their multipliers turning to $+1$. They have the shape of *synchronization "tongues"* with the constant winding numbers.

It appears to be of interest to investigate the bifurcations of synchronous periodic points within the areas of their existence. The bifurcation lines $\rho = +1$ (l_1), $\rho = exp(j2\pi\phi)$ (l_0, l'_0), and $\rho = -1$ (l_2) within the synchronization regions with $\phi = 2:5$ and $\phi = 1:8$ are plotted on the diagram in Fig.2.7 (asm-10 and asm-16 are the periodic map points, respectively). As seen from the figure, the fixed points within the synchronization zones undergo the Andronov-Hopf bifurcation once again giving rise to a new system of the closed invariant curves. For flow systems, this corresponds to the soft birth bifurcations of two-dimensional tori from the resonant cycles on torus. As it will be elucidated further, the evolution of the above invariant curves with the growth of parameter d will lead to their destruction followed by formation of quasiattractors.

2.4. NUMERICAL ANALYSIS OF STATISTICAL PROPERTIES OF ATTRACTORS

The use of *statistical methods* for the analysis of strange attractors is substantiated by the exponential instability in the regime of chaotic oscillation, which gives rise to the probability features of deterministic process $x(t)$ in system (2.1) for the finite accuracy of specifying of the initial state. From the physical point of view, it is also of importance how a qualitative behavior of system in chaos holds under perturbations, i.e., whether the chaotic attractor is robust, or structurally stable. To answer these questions, the statistical properties of the solutions of relevant Langevin equations are usually discussed

$$\dot{x}_i = f_i(x_i,\mu) + \xi_i(t), \quad i = 1,2,...,N, \tag{2.36}$$

where $f_i(x_i,\mu)$ is the deterministic evolution operator and $\xi(t)$ are random system perturbations simulated, as a rule, in the form of a white noise with intensity D

$$<\xi_i(t) \cdot \xi_j(t')> = \begin{cases} 0, & i \neq j \\ D\,\delta\,(t-t'), & i=j, \end{cases} \quad <\xi(t)> = 0. \tag{2.37}$$

A noise source is introduced into the right hand parts of the equations to analyze the discrete systems:

$$x^i_{n+1} = P(x^i_n, \mu) + \xi^i(n). \qquad (2.38)$$

A complete information on the attractor's probability properties is provided by using the *distribution function* $P(x,t,\mu,D)$ which satisfies the appropriate Fokker-Plank equation. To solve this equation in multi-dimensional cases ($N \geq 3$) is far from being simple even with the help of modern fast computers. From pure physical observations, the distribution function $P(x,t,\mu,D)$ can be found much easier if the process $x(t)$ in the system is assumed to be *stationary* and *ergodic*. Due to the stationarity, the time dependence of distribution is eliminated and the ergodicity allows averaging over the ensemble to be substituted by averaging over the time. The required information can be obtained by numerically solving the stochastic equations (2.36) or (2.38) under such assumptions.

The distribution function $P_0(x,\mu,D)$ of the *stationary ergodic* process can be calculated as a limit of the relative residence time of the system trajectory $x(t)$ in the volume elements of the phase space ΔV_j corresponding to a certain discrete partition.

The system (2.36) is integrated by using any technique within a great time interval t_0 under particular partition, for example, in the form of N-dimensional boxes with equal edges Δ. With this, the data on the number of points k_j are stored which belong to the numbered elements of volume ΔV_j. If the storage is performed with equal time intervals specified, as a rule, by the integration step, then the number of points $k_j \in \Delta V_j$ referred to the total number of points in the file $n = t/\Delta t$, would define the probability of the volume element ΔV_j to be visited by a trajectory. The probability is equal to the relative time of the trajectory residence in the above element. An appropriate discrete law of the probability distribution is obtained by determining the probabilities for all $j = 1,2...$

Consider an example of numerically constructing the one-dimensional distribution function $p_0(x_i,\mu,D)$ where i is any number from 1 to N. Define the region of the coordinate $x_i(t)$ variation and divide it into equal intervals of the length

$$x_{j-1} \leq x^i_j \leq x^i_{j-1} + \Delta, \quad j = 1,2,... \qquad (2.39)$$

Calculate the number of points k_j of the realization $x_i(n \cdot \Delta t)$ which fall within each interval. If there is a rather great number of points on the average in each interval as defined by choosing t_0 and Δ, then the comparative numbers k_j/n would be close to the corresponding probabilities. We gain the discrete approximation of the distribution function $p_0(x_i,\mu,D)$ as a set of the probabilities

$$P_j(x^i_{j-1} \le x^i_j < x^i_{j-1} + \Delta) = k_j/n, \quad \sum_{j=1}^{j_m} P_j = 1, \qquad (2.40)$$

where number j_m is defined by the interval of values x^i:

$$j_m = (x_{i_{max}} - x_{i_{min}})/\Delta. \qquad (2.41)$$

The distribution law is graphically presented in the form of a histogram. But if Δ is small enough, then $\rho_i(x)$ can be shown approximately as a continuous dependence corresponding to the one-dimensional distribution $p_0(x,\mu,D)$.

Given distribution, the system's entropy can be calculated by using the discrete relation

$$H(\mu,D) = - \sum_j P_j \ln P_j. \qquad (2.42)$$

Its evolution can be studied, if wanted, depending on the noise intensity and parameters. If the distribution function $p_0(\mathbf{x},\mu,D)$ is known, then the moments such as mean values, mean-square values, dispersion and correlation function of the process $\mathbf{x}(t)$ in the system can be calculated. The procedure of calculating the distribution function moments can be remarkably simplified, however, in the assumption of the process stationarity by finding the corresponding quantities as tentative means. The calculation algorithms of the stationary ergodic process characteristics are well known. They are involved in standard computer programs and do not require any modifications as applied to dynamical chaos regimes.

The following moment of principle is to be noted. When integrating a system that is in the strange attractor regime, the specific form of $\mathbf{x}(t)$ realization will be quite sensitive to the initial condition changes, to the specification of integration accuracy and calculation step in time and will depend on a particular numerical integration technique. All this can come as a surprise first. The phenomenon is, however, just an evidence for the exponential trajectory instability in the attractor, i.e., it is inherent in the dynamical chaos principally. As an example, consider the results of integrating the following system of equations by using the Runge-Kutt fourth-order technique [2.1,2.4,2.5]:

$$\begin{aligned}\dot{x} &= mx + y - xz, \\ \dot{y} &= -x, \\ \dot{z} &= -gz + gI(x)x^2,\end{aligned} \qquad I(x) = \begin{cases} 1, & x > 0 \\ 0, & x \le 0 \end{cases} \qquad (2.43)$$

with the parameter values $m = 1.45$ and $g = 0.21$ corresponding to the strange attractor regime. The data of the $x(\tau)$ realization calculations are indicated in Fig.2.8(a) and 2.8(b) for different initial conditions at a fixed calculation accuracy ε(Fig.2.8,a) and for fixed initial data but with various integration accuracy ε (Fig.2.8,b). Four realizations differ from each other substantially; a similarity is seen, however, in the behavior of the time processes. The similarity is explained by all the trajectories belonging to one strange attractor what is confirmed by the statistical computer processing of all the four realizations. They are evaluated by using the same statistic data. Particularly, the positive LCE spectrum exponent is $\lambda_1 = 6.3 \cdot 10^{-2}$ in all the cases.

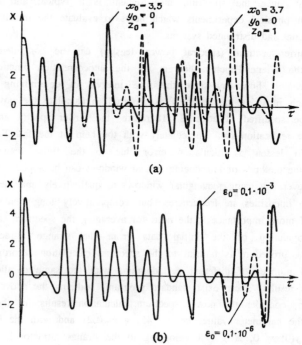

Fig.2.8. Dependence of the system (2.43) solutions on the initial data (*a*) and on the accuracy of integration (*b*).

The calculation of autocorrelation, entropy, LCE spectrum, dimension D_L and other attractor characteristics shows them to be independent of the factors affecting the computation of particular realizations. This fact is to be employed in experiments as a figure of merit for the time-calculation step and integration accuracy to be chosen correctly. Usually, the Runge-Kutt fourth-order technique is used. The accuracy from

10^{-4} to 10^{-8} is specified. The calculation step is defined by a particular system and is to be selected as compared to the least of its characteristic times. For example, the lowest characteristic time of the system is a quasiperiod of one trajectory rotation around the fixed point which is equal to 0.5-0.6 in the dimensionless variables for the popular Lorenz attractor. Owing to this, the optimal time integration step of 0.01 was chosen by Lorenz himself and was demonstrated experimentally.

Instructive information about the transition mechanisms and dynamical stochasticity features is gained by calculating the power spectrum of the realization process $x_i(t)$ in suitable regimes under the parameter variation. The calculations of the power spectrum are significant since the spectrum measurement is a typical and often the only information in physical experiments which helps to evaluate the transition to chaos in multi-dimensional and distributed systems.

The Fourier-spectrum (spectral power density) can be computed by using the relations of the Wiener-Hinchin theorem via the autocorrelation function. This is not exactly convenient, however, since the time interval of determining the correlation function, which provides the accuracy as required to calculate the power spectrum is difficult to be specified in advance. Therefore, the power spectra are computed by processing the realization data immediately with the help of the fast Fourier transform algorithm. To lessen a calculation error due to the finite realization of $x_i(t)$, $0 < t \le t_0$, the methods of introducing special windows can be applied. One can confine himself, however, by using rectangular windows to qualitatively analyze the distribution of oscillation intensities in frequencies, but comparatively long realizations must be processed. Of more importance is the need for averaging the spectrum calculation results over the population of the initial data from the strange attractor's basin of attraction. The procedure is faithful to the spectrum calculation according to a certain predetermined number of different periodograms with the same duration followed by averaging the results. There exist standard programs realizing the above algorithms. As an illustration, consider the power spectrum calculation results for the above system (2.43) with the parameter values $m = 1.45$, $g = 0.21$ and with the initial conditions $x(0) = 3.5$, $y(0) = 0$, $z(0) = 1$ belonging to the strange attractor basin of attraction [2.1, 2.4-2.5]. The spectrum was computed for realization $x(\tau)$ with the calculation step $\Delta\tau = 0.3$ by averaging n periodograms involving the arrays of 4096 points. Fig.2.9 shows the power spectra obtained by averaging over $n = 1$, 8, 16 and 48 periodograms. As seen from the figure, the spectra for $n = 16$ and 48 are almost indiscernible but they do not agree with the case as $n = 1$. The distinction is seen as well in the magnitude of the integral spectrum $J = \int S_x(\omega)d\omega$.

The power spectra of realizations $x_i(t)$ can be highly complicated in the course of experiments on the intricate bifurcational transitions to chaos because of, for example, the breakdown of the quasiperiodic regimes. Thereby, the diagnostics of the character of

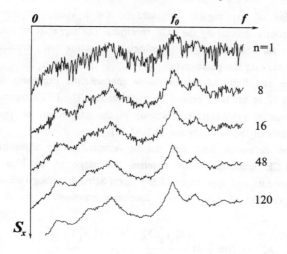

Fig.2.9. Spectrum of the time series $x(\tau)$ for system (2.43) depending on the number of periodograms averaged.

their rebuilding is hindered. In this case, they have recourse to constructing the spectra using the intersections of trajectory x_i^s with the specified secant S as an initial realization. I.e., the spectrum is calculated for the Poincaré map of the system. Being interpreted physically, the procedure can be in accordance with detecting an initial oscillation and analyzing the spectrum of the envelope of process $x_i(t)$. Machine time, required to set up the data of the map, is naturally increased. The quality of results obtained is, however, high enough for expenses to be justified.

2.5. ALGORITHMS FOR CALCULATING THE SPECTRUM OF LYAPUNOV CHARACTERISTIC EXPONENTS (LCE)

The knowledge of the total LCE spectrum for the attractor as a finite set in the system phase space allows one to make a number of important conclusions about its properties. So, for instance, the presence of positive exponents in the spectrum is a test for chaos and their sum in the LCE spectrum is related to the metric system entropy by the following inequality:

$$h_\mu \geq \sum_i \lambda_i^+. \tag{2.44}$$

The dimension of any regular attractor and the Lyapunov dimension of strange attractors are rigorously defined by the LCE spectrum. The signature of the LCE spectrum can be a key for revealing the topological structure of attractors and their bifurcations when varying the parameters.

The existence and finity of the Lyapunov characteristic exponents for any integral curve of equations (2.1) has been proved by mathematics. The LCEs represent the average rate of the exponential extension (contraction) of the projections of small perturbation onto the eigenvectors of the basis chosen [2.10-2.12].

However, the definition itself can not be accepted as an algorithm for rigorous calculating the LCE spectrum which requires the usage of an idea of generalized k-dimensional exponents and application of the Gram-Schmidt orthogonalization procedure.

Introduce a concept of *k-dimensional Lyapunov exponent* of the dynamical system (2.1)

$$\lambda_{\mathbf{x}_0}^k = \lim_{t \Rightarrow \infty} \frac{1}{t} \ln \frac{\| Y_{\mathbf{x}_0}^t \mathbf{e}_1^0 \wedge Y_{\mathbf{x}_0}^t \mathbf{e}_2^0 \wedge \ldots Y_{\mathbf{x}_0}^t \mathbf{e}_k^0 \|}{\| \mathbf{e}_1^0 \wedge \mathbf{e}_2^0 \wedge \ldots \mathbf{e}_k^0 \|}. \tag{2.45}$$

Here, $\{\mathbf{e}_i^0\}$, $\wedge$, $\|0\|$, and $Y_{\mathbf{x}_0}^t$ are the orthonormal basis of a k-dimensional subspace of the initial space $\mathbb{R}^N$ of the system (2.1) with the origin at point x_0; the external vector product; Euclidean norm, and the fundamental matrix of the solutions of the linearized system (2.1), respectively.

The vector $\mathbf{y}_{\mathbf{x}_0 i} = \mathbf{y}_{\mathbf{x}_0 i}^t \mathbf{e}_i^0$ is the solution of the following linear task

$$\dot{\mathbf{y}}_{\mathbf{x}_0} = A(\mathbf{x}) \mathbf{y}_{\mathbf{x}_0}, \quad \mathbf{y}_{\mathbf{x}_0}(0) = \mathbf{e}_i^0 \tag{2.46}$$

at the time moment t where $A(\mathbf{x})$ is the Jacoby matrix of system (2.1). Since the value of the external product of k N-dimensional vectors is equal to the volume of a k-dimensional parallelepiped which is rigorously constructed in the N-dimensional space on these vectors as on edges, then $\lambda_{\mathbf{x}_0}^k$ (i.e., the *k-dimensional Lyapunov exponent*) is the rate of variation of the parallelepiped volume along the trajectory emerging from the initial point x_0. The values of $\lambda_{\mathbf{x}_0}^k$ do not depend on the choice of basis $\{\mathbf{e}_i^0\}$ and on the norm definition method. They are only defined by the dimension of subspace k and the initial data $\mathbf{x}_0$. The existence and finity of limit (2.45) are rigorously proved [2.10 - 2.11].

Note the two important properties of k-dimensional exponents: a) $\lambda_{\mathbf{x}_0}^k$ can take at most C_N^k different values; b) k-dimensional exponents with the probability being close to a unity tend to the maximum value of all possible C_N^k under the arbitrary selection of the orthonormal basis $\{e_i^0\}$.

The LCE spectrum is defined, in the light of the above mentioned, by one-dimensional exponents put into order with respect to decrease:

$$\lambda_{x_{0,1}} \geq \lambda_{x_{0,2}} \geq \ldots \geq \lambda_{x_{0,N}}.$$

It is seen from the definition and the (b) property of the k-dimensional exponent (2.45) that only one maximum one-dimensional exponent $\lambda_{x_0}^1 = \lambda_{x_{0,1}}$ can be found as $k = 1$, only one maximum two-dimensional exponent $\lambda_{x_0}^2 = \lambda_{x_{0,1}} + \lambda_{x_{0,2}}$ can be determined as $k = 2$ and so on. Thus, all the one-dimensional exponents, i.e., the LCE spectrum, can be specified by calculating all the maximum k-dimensional exponents with the help of the following recurrence formula:

$$\lambda_{x_{0,1}} = \lambda_{x_0}^1, \quad \lambda_{x_{0,i}} = \lambda_{x_0}^i - \lambda_{x_0}^{i-1}, \quad i = 2,3,\ldots,N. \tag{2.47}$$

So, the task of finding the LCE spectrum is reduced to calculating the maximum k-dimensional exponents for $k = 1,2,\ldots,N$. But, their calculation by taking the definition (2.45) as an algorithm turned out to be impossible. This is due to the part of vectors growing in length rapidly along the "chaotic" trajectory and the angles between them becoming too small within the finite time for the volume of corresponding parallelepiped to be confidently calculated. This is the result of the fact that the solutions of the linearized system (2.46) obey the exponential law. This purely calculational difficulty can be bypassed using the following procedure. The vectors $\{y_i\}$ obtained from the initial basis $\{e_i^0\}$ are replaced in a certain fixed time τ by a new orthonormal system of vectors $\{u_i\}$ that are constructed with the help of the well known procedure of Gram-Schmidt orthogonalization [2.12]

$$\begin{aligned} v_1 &= y_1, \quad u_1 = v_1 / \|v_1\|, \\ v_{i+1} &= y_{i+1} - \sum_{k=1}^{i} (u_k \cdot y_{i+1}) u_k, \\ u_{i+1} &= v_{i+1} / \|v_{i+1}\|, \quad i = 1,2,\ldots,N \end{aligned} \tag{2.48}$$

The application of the above procedure of basis variation rests on the following feature of the external vector product. If two different basises $\{e_i\}$ and $\{u_i\}$ give rise to the same k-dimensional subspace, then the following relation is valid:

$$\frac{\|\bigwedge_{i=1}^{k} L\mathbf{e}_i\|}{\|\bigwedge_{i=1}^{k} \mathbf{e}_i\|} = \frac{\|\bigwedge_{i=1}^{k} L\mathbf{u}_i\|}{\|\bigwedge_{i=1}^{k} \mathbf{u}_i\|}, \quad L \in \mathbb{R}^k \times \mathbb{R}^k, \tag{2.49}$$

where $L \in \mathbb{R}^k \times \mathbb{R}^k$.

Let $\mathbf{x}_0 = \mathbf{x}(0)$ and $\mathbf{x}_n = \mathbf{x}(n\tau)$. Then, on account of the chain property of fundamental matrices, it can be written

$$Y_{\mathbf{x}_0}^{n\tau} = Y_{\mathbf{x}_0}^{\tau(n-1)+\tau} = Y_{\mathbf{x}_0}^{\tau(n-1)} \cdot Y_{\mathbf{x}_0}^{\tau}. \tag{2.50}$$

Denote a new basis, obtained from the system of vectors $\{Y_{\mathbf{x}_{n-1}}^{\tau} \cdot \mathbf{e}_i^{n-1}\}$ by using procedure (2.48), $\{\mathbf{e}_i^n\}$ and transform the expression under the logarithm sign in formula (2.45) to gain:

$$\frac{\|\bigwedge_{i=1}^{k} Y_{\mathbf{x}_0}^{n\tau} \mathbf{e}_i^0\|}{\|\bigwedge_{i=1}^{k} \mathbf{e}_i^0\|} = \frac{\|\bigwedge_{i=1}^{k} Y_{\mathbf{x}}^{\tau(n-1)}(Y_{\mathbf{x}_0}^{\tau}\mathbf{e}_i^0)\|}{\|\bigwedge_{i=1}^{k} Y_{\mathbf{x}_0}^{\tau} \cdot \mathbf{e}_i^0\|} \cdot \frac{\|\bigwedge_{i=1}^{k} Y_{\mathbf{x}_0}^{\tau} \cdot \mathbf{e}_i^0\|}{\|\bigwedge_{i=1}^{k} \mathbf{e}_i^0\|} =$$

$$\frac{\|\bigwedge_{i=1}^{k} Y_{\mathbf{x}_1}^{\tau(n-1)} \cdot \mathbf{e}_i^1\|}{\|\bigwedge_{i=1}^{k} \mathbf{e}_i^1\|} \cdot \frac{\|\bigwedge_{i=1}^{k} Y_{\mathbf{x}_0}^{\tau} \cdot \mathbf{e}_i^0\|}{\|\bigwedge_{i=1}^{k} \mathbf{e}_i^0\|} = \ldots = \prod_{j=0}^{n-1} \frac{\|\bigwedge_{i=1}^{k} Y_{\mathbf{x}_j}^{\tau} \cdot \mathbf{e}_i^j\|}{\|\bigwedge_{i=1}^{k} \mathbf{e}_i^j\|}. \tag{2.51}$$

For the orthonormal basis $\{\mathbf{e}_i\}$,

$$\|\bigwedge_{i=1}^{k} \mathbf{e}_i\| = 1 \tag{2.52}$$

is valid. Consequently, all the denominators in (2.51) are 1 for all $i = 0, 1, \ldots, n-1$ (the initial basis $\{\mathbf{e}_i^0\}$ is orthonormalized).

By substituting the relation (2.51) into the definition (2.45) taking into account (2.52), we obtain the final expression:

$$\lambda_{\mathbf{x}_0}^k = \lim_{n \to \infty} \frac{1}{n\tau} \sum_{j=0}^{n-1} \ln \|\bigwedge_{i=1}^{k} Y_{\mathbf{x}_j}^{\tau} \cdot \mathbf{e}_i^j\|. \tag{2.53}$$

Since the external products may be calculated immediately according to definition with no technical difficulties, expression (2.53) presents a straightforward algorithm for calculating the k-dimensional exponents and, thus, the entire LCE spectrum. As seen from (2.53), the system of $N \cdot (N+1)$ differential equations is to be solved to calculate N k-dimensional exponents. The first N equations comprise the starting nonlinear system (2.1) with the initial conditions $\mathbf{x}_0 = \mathbf{x}(0)$. Its solution is necessary to obtain a Jacoby matrix which defines the system of the linear approximation equations. The other N^2 equations appear due to the need of solving the linear equations (2.46) N times under N different initial conditions which vary within the time interval τ: $y_{\mathbf{x}_{0,i}}(j\tau) = e_i^j$, $i = 1,2,...,N$, $j = 1,2,...,N$. As $j = 0$, $\{e_i^0\}$ is a reference basis and as $j = 1,2,...,N$, it is the basis obtained as a result of procedure (2.48).

An algorithm for calculating the LCE spectrum is constructed identically with the same way applied to the N-dimensional maps

$$x_{n+1} = P(x_n). \qquad (2.54)$$

Denote the Jacoby matrix of map by $M_{\mathbf{x}_0}$, i.e.,

$$M_{\mathbf{x}_0} = \begin{bmatrix} \frac{\partial P_1}{\partial x_1} & \cdots & \frac{\partial P_1}{\partial x_N} \\ \vdots & & \vdots \\ \frac{\partial P_N}{\partial x_1} & \cdots & \frac{\partial P_N}{\partial x_N} \end{bmatrix}. \qquad (2.55)$$

Similarly to (2.45), define the *k-dimensional Lyapunov exponent for the map*

$$\lambda_{\mathbf{x}_0}^k = \lim_{n \to \infty} \frac{1}{n} \ln \frac{\| \bigwedge_{i=1}^{k} M_{\mathbf{x}_0}^n e_i^0 \|}{\| \bigwedge_{i=1}^{k} e_i^0 \|}. \qquad (2.56)$$

To prevent the interruptions of the order overflow and vanishing, as in case of a flow, apply the Gram-Schmidt orthogonalization procedure (2.48) at every step of the map iterating. Having transformed the expression (2.56) likewise as for the flow, obtain the formula for calculation of the LCE spectrum:

$$\lambda_{\mathbf{x}_0}^k = \lim_{n \to \infty} \frac{1}{n} \sum_{j=0}^{n-1} \ln \| \bigwedge_{i=1}^{k} M_{\mathbf{x}_j} e_i^j \|. \qquad (2.57)$$

It is seen from (2.45) and (2.57) that the k-dimensional exponents are the limits of a certain infinite sequence. Therefore, the calculation process must be interrupted when achieving a given accuracy in the convergence of the exponents to the limit values.

If there is no need of calculating all the N one-dimensional exponents of the spectrum, one can restrict himself with the number of $l < N$ largest exponents. This is enabled by the algorithm requiring the integration of $N(N + 1)$ equations and, thereby, saving the machine time.

In the simplest case, as $l = 1$, the maximum exponent of LCE spectrum is calculated. The renormalization of the only vector length by a unity corresponds to the orthogonalization procedure when orientation is maintained. Note, that the algorithm of the maximum Lyapunov exponent calculation for $l = 1$ was presented independently in [2.13].

2.6. METHOD OF NUMERICALLY CALCULATING THE SINGULAR SOLUTIONS

A number of practical tasks associated with defining the mechanisms of dynamical chaos evolution need analyzing *particular homoclinic and heteroclinic trajectories*. Some general conclusions about the motion in the vicinity of such trajectories can be made on the base of linear analysis in the vicinity of singular points. To find the singular trajectories and investigate the global solution features in their vicinity is possible only by using computation. Refer to the initial system of N non-linear differential equations (2.1) again. Let x_0 be a singular point of the system, i.e., $F(x_0) = 0$. The eigenvalues $\{s_1, s_2, s_3, \ldots, s_N\}$ of point x_0 can be divided into two groups that depend on their real part sign. It is known that, if condition $\text{Re } s_i > 0$ is fulfilled for m eigenvalues and $\text{Re } s_i < 0$ for other $N - m$ numbers, an m-dimensional unstable invariant manifold exists at point x_0 which is formed by trajectories entering x_0. The case as $\text{Re} s_i = 0$ is bifurcational since an abrupt invariant manifold rebuilding takes place with this. If system (2.1) has several equilibrium points, then each of them has stable and unstable invariant manifolds of appropriate dimensions. The interposition of all these manifolds in the phase space specifies the structure of its partition into regions with different regimes.

The most studied and interesting case is as $m = 1$, i.e., when there is the only positive eigenvalue at point x_0 that is conformed by the one-dimensional unstable separatrix and remaining eigenvalues give rise to the $(N - 1)$-dimensional invariant manifold. In such a situation, if the system depends on parameters, at some fixed values of them the unstable separatrix can form a *"loop"* : after having left the point x_0, it returns to it being tangent to the stable manifold. This is of interest since one of the mechanisms of cycle vanishing (or birth) is related to the loop [2.14]. Furthermore, complex phenomena, such as the birth of an infinite set of multiple cycles and multirotational loops, are observed under certain additional conditions in the vicinity of the loop that can lead to the onset of chaos [2.15, 2.16].

In this connection, one of important tasks of the numerical investigation of a system of type (2.1) is to construct a separatrix and, if the system's parameters are known, to find the bifurcational moment of the separatrix loop birth when varying one parameter, as well as to construct bifurcation curves which correspond to the loop occurrence in the plane of two parameters. Besides, the separatrixes are of interest, which go from one singular saddle point to another and for which the "loop" is a special case.

Let system (2.1) have two singular saddle points, x_0 and x_1, each of both having positive eigenvalues s_1^0 and s_1^1, respectively. Then, the one-dimensional unstable separatrix $\Gamma(x_0)$ emerges from x_0 and $\Gamma(x_1)$ does from x_1. $\Gamma(x_0)$ is a trajectory emerging from point x_0 toward the eigenvector v_1^0 which corresponds to the eigenvalue s_1^0. To construct $\Gamma(x_0)$, system (2.1) is to be simply integrated with the initial point lying on the vector v_1^0 at a very small distance from point x_0.

Assume the system (2.1) to depend on two parameters. It may be written in the form

$$\dot{x} = F(x, \mu_1, \mu_2).$$

Fix one of the parameters, for example, $\mu_2 = \mu_2^0$ and construct a separatrix $\Gamma(x_0, \mu_1)$ when varying the parameter μ_1. In general case, $\Gamma(x_0, \mu_1)$ having emerged from point x_0 passes by saddle x_1, for example, while bending up with respect to the stable manifold $W^s(x_1)$ of point x_1 (Fig.2.10,a). When varying μ_1, the situation can be altered and the separatrix will go below $W^s(x_1)$ (Fig.2.10,b). It is clear that, due to continuity, the separatrix $\Gamma_0(x_0, \mu^*)$ must fall on $W^s(x_1)$ exactly and come to point x_1 at a certain intermediate value $\mu_1 = \mu_1^*$ (Fig.2.10(c)). In case, when $x_0 = x_1$, the bifurcational value of parameter $\mu_1 = \mu_1^*$ will be answered by the *double-asymptotic trajectory* $\Gamma_0(x_0)$ of *separatrix loop type*. [2.17].

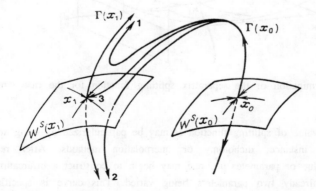

Fig.2.10. Possible behavior of unstable separatrix $r(x_0)$ in the vicinity of saddle x_0.

To particularize the bifurcational value μ_1^*, a certain function is to be assumed which tends to zero as $\mu_1 = \mu_1^*$. It can be determined as follows. Choose the system of the eigenvectors of the equilibrium point x_1 as a basis in the phase space $\mathbb{R}^N$. Let v_1^1 be the eigenvector corresponding to the positive eigenvalue s_1^1. Select from the negative eigenvalues the numbers having maximum real parts which are called the leading eigenvalues. There is available either one real leading number or a pair of complex leading numbers in general case. Call, by convention, point x_1 the saddle in the former case and the saddle-focus in the latter one. Assume, first, x_1 to be a saddle. Let the leading eigenvector, i.e., the vector corresponding to the leading eigenvalue, be denoted as v_2^1. Then lead a hypersurface S shifting slightly along the vector v_2^1, in parallel to an $(N - 1)$-dimensional subspace formed by all the other eigenvectors. But if x_1 is a saddle-focus, then draw the hypersurface S through the point x_0 in parallel to a certain vector lying in a leading invariant manifold. As the value of function H, take the distance between the point of intersection of $\Gamma(x_0)$ with hypersurface S and the projection of the intersection of S with the stable manifold $W^s(x_1)$ onto the $(N - 1)$-dimensional eigensubspace that corresponds to all the eigenvalues except v_1^1. The stable manifold $W^s(x_1)$ may be replaced with a hypersurface K being tangent to it at the saddle since the point of intersection of $\Gamma(x_0)$ with S lies close enough to x_1. Our considerations are visualized in Fig.2.11.

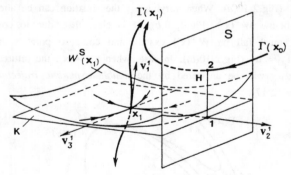

Fig.2.11. Determination of the separatrix splitting function for the case when x_1 is a saddle.

The zero value of splitting function H may be particularized by using any standard technique, for instance, dichotomy or interpolation methods. After refining the bifurcational value of parameter μ^*, one may begin to construct a bifurcation curve in the plane of already two parameters being varied. This curve is specified by the condition of the separatrix splitting function going to zero and is constructed like bifurcation curves that correspond to the loss of stability by the cycle.

Now, consider the periodic solution $x(t)$ of system (2.1). It is defined, as known, by multipliers $\rho_1, \ldots, \rho_{n-1}$. Like the equilibrium points, the periodic solutions (cycles) have stable and unstable invariant manifolds which are formed by trajectories "winding up" on the cycle and "winding off" from it. Let the cycle Γ have m multipliers with $|\rho_i| < 1$ and for other multipliers the condition $|\rho_i| > 1$ is fulfilled. Then, the dimension of stable manifold is $(m + 1)$ and that of the unstable one is $(N - m)$. So, for example, the saddle cycle Γ in a three-dimensional space has two-dimensional stable and unstable manifolds (Fig.2.12(a)). If there exist several cycles in the phase space, then their stable and unstable manifolds can intersect along the so-called *heteroclinic trajectories*. The stable and unstable manifolds of the same cycle can intersect along *homoclinic trajectories*. Complex phenomena can take place in the vicinities of these singular trajectories leading, in particular, to chaos. The problem of the numerical construction of homo- and heteroclinic trajectories is, however, a more complicated thing than that of constructing the equilibrium point separatrixes since the intersection lines of multidimensional surfaces are to be found. The dimension of the problem can be lessened if one passes to the corresponding Poincaré map. The point of intersection of the cycle and secant S is a fixed map point and the invariant manifold crossings with secant S form the invariant manifolds of the fixed point (Fig.2.12(b)). These manifolds are *one-dimensional* in case of the three-dimensional phase space and intersect at the points called *homoclinic*.

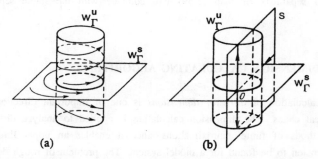

Fig.2.12. Two-dimensional stable W_s and unstable W_u manifolds of the saddle cycle Γ in $\mathbb{R}^3$ (a) and the corresponding one-dimensional separatrixes of the map (b).

So long as the task of construction of the homo- and heteroclinic trajectories of system (2.1) goes to finding the homo- and heteroclinic points of the Poincaré map, consider the following discrete map:

$$x_{n+1} = P(x_n, \mu), \qquad (2.58)$$

where $x_n = (x_n^1, x_n^2, \ldots, x_n^N)$. Let x_0 and $\rho_1, \rho_2, \ldots, \rho_N$ be the fixed point of map (2.58) and its multipliers, respectively. To illustrate, one of the multipliers (say, ρ_1) lies outside the unit circle and all the others are inside it. Then, multiplier ρ_1 corresponds to the one-dimensional unstable invariant curve which can be constructed as follows. Find the eigenvector corresponding to ρ_1 and lay off a segment of a very short length along this vector starting at point x_0. On this segment specify k points lying equally spaced from each other. Choosing them, in turn, as initial points, perform a sufficiently great number of the map (2.58) iterations from every point. As a result, a curve in the phase space is obtained. We have no information about its intersections with other invariant manifolds in such a way indeed, but, in return, we can judge about the degree of its smoothness being of great interest.

If the fixed point x_0 has a multiplier inside the unit circle and its other multipliers outside the unit circle, then the stable separatrix is one-dimensional. The one-dimensional stable separatrix can be constructed if the map (2.58) is reversible. Then, the following equation may be written

$$x_n = P^{-1}(x_{n+1}, \mu). \qquad (2.59)$$

Point x_0 will hold as the fixed point of map (2.59). But it will have one multiplier outside the unit circle and remaining multipliers inside it. The unstable one-dimensional separatrix of map (2.59) will coincide with the stable separatrix of map (2.58).

2.7. DIMENSION CALCULATING ALGORITHMS

The calculation of *attractor dimensions* is one of important directions in the study of dynamical chaos. The dimension calculating is applied to analyze distributed systems by the methods of finite approximations since it enables an upper limit of the phase space dimension to be found for a model system. The problem of rough description of the system dynamics with the help of discrete maps is connected as well with the attractor dimension.

The *fractal* (D_F) and *information* (D_I) dimensions are calculated more often as follows [2.18-2.20, 2.1-2.2, 2.4]:

$$D_F = \lim_{\varepsilon \to 0} \frac{lnM(\varepsilon)}{ln(1/\varepsilon)} ; \quad D_I = \lim_{\varepsilon \to 0} \frac{I(\varepsilon)}{ln(1/\varepsilon)} . \qquad (2.60)$$

where $M(\varepsilon)$ is the minimum number of n-dimensional boxes with ε size covering the attractor and $I(\varepsilon)$ is the Shennon entropy. The direct covering of available set of attractor points with N-dimensional boxes having the edge ε (N - the phase space dimension) is the simplest, although not always optimal, technique of calculating D_F and D_I. From the definition of (2.60) with finite ε, it follows that:

$$lnM(\varepsilon) = D_F ln(1/\varepsilon) + lnk,$$

where k is constant. The magnitude of D_F is determined by the slope of $M(\varepsilon)$ versus $1/\varepsilon$ in the double logarithmic scale.

When calculating the information dimension the probabilities p_i, that define the Shennon entropy in (2.60), are approximated by the relations n_i/n_0, where n_i and n_0 are the number of phase points in the i-th box having the edge ε and the total number of attractor points, respectively.

The method of direct covering proves to be efficient only for small-dimensional ($N \leq 3$) attractors and with a relatively small phase volume contraction degree in the vicinity of attractor, i.e., with the great values of the fractional part of dimension. Otherwise ($N > 3$, $d \ll 1$), very large data volume, small ε values and the enormous number of covering boxes $M(\varepsilon)$ are required.

A more effective algorithm of calculating the fractal dimension is based on the formula [2.21]:

$$D_F = N - \lim_{\varepsilon \to 0} [lnV(\varepsilon)/ln(1/\varepsilon)], \qquad (2.61)$$

where $V(\varepsilon)$ is the volume of a set consisting of all the points in the ε-vicinity of attractor.

To calculate the information dimension D_I, one may use effective algorithms realizing the methods of the most close neighbor, projection tracking methods, as well as special methods described in the literature [2.22].

An important representative of the probability dimension class is a *correlation dimension* D_s [2.23-2.25] defined by

$$D_c = \lim_{\varepsilon \to 0} \frac{\sum_{i=1}^{M(\varepsilon)} P_i^2}{ln\varepsilon} \qquad (2.62)$$

where P_i^2 is the probability for a pair of attractor points to belong to the i-th covering element. Note, that the correlation dimension D_c is a particular case of the so-called generalized dimension using the Renye entropy of the order of q instead of Shennon entropy in (2.60) [2.24]:

$$I_q(\varepsilon) = \frac{1}{1-q} \ln \left(\sum_{i=1}^{M(\varepsilon)} P_i^q \right). \tag{2.63}$$

The definition of the correlation dimension (2.62) corresponds to the case when $q = 2$. The correlation dimension is presented in the form:

$$D_c = \lim_{\varepsilon \to 0} \frac{\ln C(\varepsilon)}{\ln \varepsilon}, \tag{2.64}$$

$$C(\varepsilon) = \lim_{n_0 \to \infty} \frac{1}{n_0^2} \sum_{i,j=1}^{n_0} \theta[\varepsilon - |\mathbf{x}_i - \mathbf{x}_j|], \tag{2.65}$$

where

$$\theta[\varepsilon - |\mathbf{x}_i - \mathbf{x}_j|] = \begin{cases} 1, & |\mathbf{x}_i - \mathbf{x}_j| \leq \varepsilon, \\ 0, & |\mathbf{x}_i - \mathbf{x}_j| > \varepsilon, \end{cases} \tag{2.66}$$

and $\mathbf{x}_i$ is the vector of a point in the phase space. Consequently, dimension D_c is defined by the magnitude of the correlation integral $C(\varepsilon)$ describing the relative numbers of pairs of points $\mathbf{x}_i$ and $\mathbf{x}_j$ that lie at the distances $d_{ij} = |\mathbf{x}_i - \mathbf{x}_j| \leq \varepsilon$.

To calculate the correlation integral $C(\varepsilon)$ and, thus, the correlation dimension D_c, all the distances d_{ij} must be calculated and the number of pairs of points $n(\varepsilon)$ is to be determined for which $d_{i,j} \leq \varepsilon$. Then, the magnitude of the correlation integral will be evaluated as

$$C(\varepsilon) = n(\varepsilon)/n_0^2. \tag{2.67}$$

Distances d_{ij} are calculated faster with account of the representation of the numbers having floating points realized by computer [2.2].

The considered algorithms of calculating the fractal and information (correlation) dimensions of the attractor show that calculations require a set of points determined in the phase space of dimension N and belonging to the attractor. The number of points n_0 is naturally finite in calculation but it must be large enough. The problem of calculating the dimensions for cases, when the dynamical system is specified (either by the discrete operator of the map or by the differential equation system) presents no principal difficulties and is resolvable with the sufficient memory and fast-action of computer.

It is often necessary, however, to calculate the dimension of the attractor of a certain real system whose mathematical model is unknown. The dimension of its phase space is not known as a rule (even approximately). In this situation an experimenter has the information about the time behavior of a certain dynamical variable, for example, the time-dependence of current or voltage, pressure or rate or otherwise. Furthermore, the time interval of the experimental process realization is restricted. Is it possible to evaluate the attractor dimension under such conditions?

The way to solve this problem was proposed by F.Takens. In Ref. [2.26] it was proved that a new manifold may be constructed for almost all smooth dynamical systems with the help of the available time realization of one dynamical variable observed. Its main properties (particularly, the dimension) will be the same as for the initial manifold.

Let a certain dependence $x(t)$ be accessible for experimental measurement that is digitized with a regular finite time interval Δt. As a result, a sequence of numbers $x(t_i)$ is obtained, where $t_i = t_0 + i\Delta t$. It was offered by Takens to construct a set of m-dimensional vectors $\mathbf{v}_i \in \mathbb{R}^m$ on this sequence which are introduced as follows:

$$(x_i, x_{i+\tau}, x_{i+2\tau}, \ldots, x_{i+(m-1)\tau}); \quad x_i = x(t_i), \quad \tau = k\Delta t. \tag{2.68}$$

The discretization time Δt of initial realization, time τ in (2.68), and k are called the sampling rate, delay time, and embedding dimension, respectively. The Takens' main result is as follows. If the time realization $x(t)$ is represented as an infinite sequence of numbers $\{x_i\}$, then the infinite set of vectors $\mathbf{v}_i \in \mathbb{R}^m$, as $k = 1$, will specify the embedding of initial manifold almost for any chosen variable observed. This takes place if m is not smaller than the double dimension of the initial manifold.

The m-dimensional manifold found in such a way is the reconstruction of an initial manifold with the attractor of interest embedded into it. Anyone of dimensions introduced above may be calculated as applied to this reconstruction providing the evolution of dimension of the real attractor under study. Furthermore, the task of reconstructing the attractor or its projections becomes solvable [2.2, 2.22, 2.25].

As already mentioned, the time realization is presented in real experiment as a finite sequence of numbers and the dimension of the initial system phase space remains unknown. To minimize an error due to the finite set of experimental points calculations must be performed with a few different n_0 and m values. The dimension assessment obtained is sought to be independent of n_0 and m within a specified accuracy in numerical experiment. Since the values of x_1 and x_{i+1} will be close for small sampling times Δt, the proper choice of the delay time τ becomes of great importance. One has to seek for τ to be chosen so that correlation between x_1 and $x_{i+\tau}$ is as small as possible. Time τ can be roughly evaluated with the help of the autocorrelation function $\Psi_x(\tau)$ or by determining the minimum of mutual information between the measurements.

Note as well, that the dimension of the attractor can be evaluated, when calculating the LCE spectrum of a trajectory on the attractor, by using the Caplan-Yorke definition [2.27]. New algorithms are not required and the entire information is given by the results of calculating the LCE spectrum.

The question about the mutual compliance of all the above dimension definitions is natural and vital from the theoretical point of view. From the viewpoint of experiment, it would be sufficient at present to answer the question on a qualitative difference between the attractor dimension types shown. From the literature, the hypothesis on the mutual compliance of qualitative dimension magnitudes is as follows:

$$D_C \leq D_I \simeq D_L \leq D_F. \qquad (2.69)$$

Relation (2.69) is strengthened by the results reported in numerous papers [2.20, 2.28-2.30]. It is of significance that qualitative distinctions, while being observed, do not radically affect the main result, i.e., the definition of the effective number of degrees of freedom of a system running in its particular operating regime, at least, for typical low-dimensional attractors.

CHAPTER 3

INERTIAL NONLINEARITY OSCILLATOR. REGULAR ATTRACTOR BIFURCATIONS

3.1. GENERAL EQUATIONS OF ONE-AND-A-HALF-FREEDOM-DEGREE OSCILLATORS

Autonomous systems with a three-dimensional phase space are the most simple ones among the differential systems exhibiting chaotic behavior. This simplicity is relative and is to be understood in the sense that theoretical, numerical and experimental analyses of the dynamics of three-dimensional systems are simpler as compared to the analysis of multi-dimensional and, the more so, distributed systems. It is comparatively easy in the three-dimensional systems to find characteristic *homoclinic trajectories* and to study *typical bifurcations* in their vicinity by using theoretical and experimental methods. At the same time, the dynamics of systems, realizing one or another homoclinics type, obeys general regularities.

The *autonomous dynamical systems* that describe generation of chaotic oscillations in one-and-a-half-freedom-degree oscillators are of interest. Introduce the *most common equations* of such systems proceeding from the previously reported results as applied to the one-freedom-degree oscillator. In general form, self-oscillating systems on plane are described by the equation

$$\ddot{x} + \phi(x,\mu)\cdot\dot{x} + \Psi(x,\mu) = 0 \qquad (3.1)$$

where x is a variable oscillating periodically, $\mu = (\mu_1, \mu_2, ..., \mu_k)$ is a set of control parameters, $\phi(x,\mu)$ and $\Psi(x,\mu)$ are non-linear functions which define the action of forces providing self-oscillations.

Equation (3.1) can be generalized for a certain class of one-and-a-half-freedom-degree systems. Consider a radio engineering unit with the block diagram indicated in Fig. 3.1. The *main oscillator* is shown by the dashed line and consists of amplifier 1, selective element (e.g., oscillating circuit, resonator, or *Wien-bridge*), and a positive feedback circuit. When the suitable amplitude and phase conditions are fulfilled in such an oscillator self-oscillations described by equation (3.1) arise. Consider an additional feedback circuit which ensures the *inertial transformation* of affecting variable $x(t)$ into response $z(t)$ which controls the parameters of the main oscillator's amplifier and selective element. The equations for the total system (Fig.3.1) may be written as follows [3.1,3.2]:

$$\ddot{x} + F_1(x,z,\mu)\dot{x} + F_2(x,\dot{x},z,\mu) = 0,$$
$$\dot{z} = F_3(x,\dot{x},z,\mu). \tag{3.2}$$

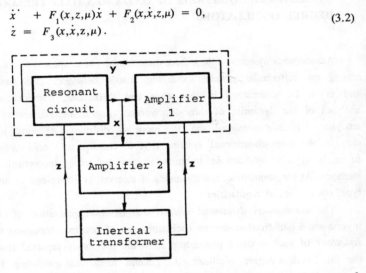

Fig.3.1. Scheme showing the inertial self-congruent effect on the main elements of classical oscillator.

Here, F_i are non-linear functions in general case. The phase variable $z(t)$ in (3.2) is related to the variable $x(t)$ by means of the first order differential operator. If the relation between the response $z(t)$ and the variable $x(t)$ is inertialless and is described by the following algebraic polynomial

$$z = G(x,\dot{x}), \tag{3.3}$$

then the equations (3.2) are reduced to the equation (3.1). If the variable z depends on x inertially and is specified by the first order differential equation, then the equations (3.2) describe oscillation processes in the three-dimensional phase space and present the generalization of equation (3.1) for this case.

The dynamical systems in $\mathbb{R}^3$, that model oscillations in the one-and-a-half-freedom-degree oscillators, admit the form of writing (3.2) by eliminating the third variable and, if required, by introducing a smooth substitution of the coordinate and time scales. This procedure is called the reduction to generalized equations (3.2) which constitute the mathematical model of an oscillator controlled automatically. Several examples are given below. The following model was suggested by O.E.Rössler [3.3]

$$\dot{x} = -(y+z), \quad \dot{z} = -cz + bx + xz,$$
$$\dot{y} = x + ay, \quad div F = a - c + x. \quad (3.4)$$

The differentiation of the first equation with respect to time and elimination of variable y from (3.4) will give:

$$\ddot{x} - a\dot{x} + [(1 + b + z)x - (a + c)z] = 0,$$
$$\dot{z} = -cz + bx + xz. \quad (3.5)$$

This system results from (3.2) when appropriately specifying the function F_i.

The equations of the relaxation oscillator which are studied in Refs [3.4] and are as follows

$$\dot{x} = 2hx + y - gz, \quad \dot{y} = -x, \quad \mu\dot{z} = x - f(z)$$
$$f(z) = z^3 - z, \quad div F = 2h - (2z^2 + 1)/\mu, \quad (3.6)$$

are also reduced by simple transformation to the form of (3.2)

$$\ddot{x} - 2h\dot{x} + (1 + g/\mu)x - gf(z)/\mu = 0,$$
$$\dot{z} = [x - f(z)]/\mu. \quad (3.7)$$

Consider the famous Lorenz system [3.5]

$$\dot{x} = \sigma(y - x),$$
$$\dot{y} = -y - xu + rz, \quad divF = -(\sigma + b + 1) \quad (3.8)$$
$$\dot{u} = -bu + xy.$$

Having eliminated the variable y from (3.8), we obtain the following equations[1]

$$\ddot{x} + (1 + \sigma)\dot{x} + \sigma(1 - r + u)x = 0,$$
$$\dot{u} = -bu + x^2 + x\dot{x}/\sigma, \quad (3.9)$$

which, by substituting the variables

---

[1]Equations (3.9) correspond to (3.2) in the form.

$$y = \frac{\varepsilon x}{\sqrt{2\sigma}}, \quad z = \frac{\varepsilon}{\sigma}(\sigma u - \frac{x^2}{2}), \quad \tau = \frac{\sqrt{\sigma} \cdot t}{\varepsilon}, \quad \varepsilon = (\sqrt{r} - 1)^{-1}, \tag{3.10}$$

are reduced to the system

$$\ddot{y} + \varepsilon h \dot{y} + y^3 + (z - 1)y = 0,$$
$$\dot{z} = -\varepsilon a z + \varepsilon \beta y^2, \tag{3.11}$$

where $h = (1 + \sigma)/\sqrt{\sigma}$, $a = b/\sqrt{\sigma}$, $\beta = (2\sigma - b) \cdot \sqrt{\sigma}$.

The substitution of variables (3.10) was suggested by V.I.Yudovich to investigate analytically the asymptotic behavior of Lorenz system at high values of r, corresponding to a small parameter ε in equations (3.11). The list of similar examples can be continued.

The general form (3.2) of writing the three-dimensional dynamical systems does not discover in detail possible, fundamental from the viewpoint of physics, distinctions of particular systems, such as the way of exciting oscillations, possibility to generate two-frequency and chaotic oscillations, etc. If required, some subclasses of the systems can be taken into consideration which satisfy eq. (3.2), but differ in certain criteria due to the specific restrictions for the explicit form of function F_i in (3.2). Particularly, one may distinguish between the two types of generating systems satisfying the equations (3.2) in general form, but differing in the mechanisms of exciting oscillations. In the first case, selfoscillations are excited due to the compensation of intrinsic losses by a negative friction under the positive feedback. The second type involves oscillators that represent a certain dissipative circuit forced parametrically because of the inertial influence of an amplified signal from the circuit on the elements of the circuit itself. An example of the second oscillator type is a Lorenz model in the form of a controlled pendulum (3.9), (3.11). In this pendulum, selfoscillations are excited owing to a parametric influence on the non-linear capacity of the dissipative circuit. The general block diagram of the second oscillator type is indicated in Fig.3.2. The differences from the block-diagram of Fig.3.1 are obvious. The second type of inertial oscillators covers many real oscillating systems, and namely, electromechanical vibrators, Helmholtz resonators, as well as some chemical and biological systems [3.6].

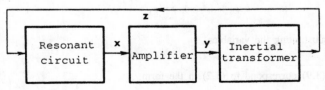

Fig.3.2. Inertial control of an oscillatory system in oscillators with parametric excitation.

The above partition of self-oscillating systems with respect to the oscillation self-excitation type may be of no principle, however, from the viewpoint of the general mechanisms of transition to chaos. Having appeared physically in different ways, self-oscillations can undergo identical bifurcational transitions to chaos with the growth of the control parameter. Therefore, the analysis of chaos birth calls for another system classification representing similarity and distinction just in the mechanisms of the strange attractor evolution in one-and-a-half-freedom-degree oscillators.

Refer to examples (3.4)-(3.11) and note that the *vector field divergence* depends in some models on phase coordinates and in others it is a constant negative quantity. For dynamical systems with three-dimensional phase space, two-frequency oscillations are encountered in the former case. In the latter case, they are impossible. The existence of a two-dimensional torus is not ruled out in the three-dimensional systems with variable divergence, that means the transition to chaos via quasiperiodic oscillations to be feasible. This mechanism is not realized in the three-dimensional systems with the constant negative divergence.

The systems may differ in the number of *singular points* where the vector of the flow phase velocity becomes zero. The *topology of limit sets* in the system phase space is defined by the singular points, and this critical circumstance is to be taken into account when classifying.

Unfortunately, there are no results available at present that permit to rigorously predict the presence of dynamical chaos and to indicate the mechanisms of its evolution and its structure even for the simplest case as $N = 3$.

In the following paragraph, one of the simplest autonomous systems with the three-dimensional phase space is introduced. Under certain conditions there exists a loop of saddle-focus separatrix in the system. As it will be shown, practically all instability hierarchies, leading to a typical quasiattractor in $\mathbb{R}^3$, are observed in the extended loop vicinity in the control parameter space of the above system.

3.2. STATEMENT OF EQUATIONS FOR A MODIFIED OSCILLATOR WITH INERTIAL NONLINEARITY

The classical oscillator with inertial non-linearity was proposed and described by Theodorchik [3.7]. Self-oscillations are provided in the system by introducing into an oscillatory circuit the thermoresistance $R(T)$ with the properties non-linearly and inertially depending on the current through it. The diagram of Theodorchik's oscillator with inertial non-linearity is presented in Fig.3.3. The equations for current $i(t)$ in the circuit have the form:

$$\frac{d^2i}{dt^2} + \left[\frac{R(T)}{L} - \frac{MS_0}{LC}\right]\frac{di}{dt} + \left[\frac{1}{LC} + \frac{1}{L}\frac{\partial R(T)}{\partial T}\frac{dT}{dt}\right]i = 0, \qquad (3.12)$$

where S_0 is the curve's slope of the amplifier which is assumed to be linear, M is the mutual induction of the feedback circuit, $R(T)$ is thermo-resistance depending on the absolute temperature T, L and C are the induction and capacity of the oscillatory circuit.

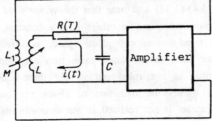

Fig.3.3. Classical scheme of the oscillator with inertial non-linearity.

Assuming the dependence $R(T)$ to be linear

$$R(T) = R_0 + LbT \qquad (3.13)$$

and setting the heat exchange process to obey the Newton's law

$$\rho q \frac{dT}{dt} + kT = R(T)i^2, \qquad (3.14)$$

where q and ρ are the specific heat capacity of the thermistor filament and its mass, respectively, we obtain a closed system of equations in the form

$$\begin{aligned}\frac{d^2i}{dt^2} + \omega_0^2 i &= (\mu - bT)\frac{di}{dt} - b\frac{dT}{dt}i, \\ \frac{dT}{dt} + \gamma T &= \alpha(T)i^2.\end{aligned} \qquad (3.15)$$

Here, the following notation is used:

$$\begin{aligned}\mu &= \omega_0^2 S_0 M - R_0/L, \quad \omega_0^2 = 1/LC, \quad \gamma = k/\rho q, \\ \alpha(T) &= \alpha_0 + bLT/\rho q, \quad \alpha_0 = R_0/\rho q.\end{aligned} \qquad (3.16)$$

Inertial Nonlinearity Oscillator. Regular Attractor Bifurcations

The use of the dimensionless variables as follows:

$$x = ai, \quad \dot{y} = -x, \quad z = bT/\omega_0, \quad \tau = \omega_0 t, \quad a = \sqrt{abpq/\omega_0 k} \qquad (3.17)$$

provides equations (3.15) in the form:

$$\begin{aligned}
\dot{x} &= mx + y - xz, & m &= \mu/\omega_0 = \omega_0 S_0 M - R_0/\omega_0 L, \\
\dot{y} &= -x, & g &= \gamma/\omega_0, \\
\dot{z} &= -gz + gx^2, & \dot{x} &= \frac{dx}{d\tau}.
\end{aligned} \qquad (3.18)$$

In the three-dimensional two-parametric system (3.18), parameter m is proportional to the difference between energies introduced and scattered in the circuit, and g is a parameter defining the relative time of the thermistor relaxation. Parameters m and g will be called further the oscillator excitation and inertia parameters, respectively.

Consider the diagram shown in Fig.3.4. The depicted oscillatory circuit has no non-linear elements in contrast to the classical case (Fig.3.3). Amplifier 1 is controlled by the additional feedback circuit comprising a linear amplifier and an inertial transformer. The differential equations for the oscillator may be explicitly written by particularizing the slope dependence of amplifier 1, $S(x,V)$, and specifying the equations of the inertial transformation $V(x)$.

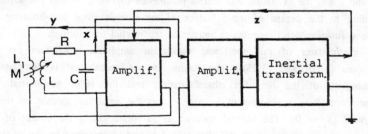

Fig.3.4. Modified scheme of the oscillator with inertial non-linearity.

Fit the function $S^1(x)$, i.e., the slope of amplifier 1 without reference to the effect of the additional feedback, by the following polynomial:

$$S^1(x) = S_0 - S_1 x^2, \qquad (3.19)$$

where x is the voltage at the input of amplifier 1 and S_0 and S_1 are the fixed positive coefficients. Suppose the mechanism of the inertial feedback circuit influence to obey

the regularity

$$S = S^1(x) - bV = S_0 - S_1 x^2 - bV, \qquad (3.20)$$

where $V = V(x)$ is voltage at the output of the inertial transformer and b is a parameter. Let the inertial transformation be performed according to the equation

$$\dot{V} = -\gamma V + \phi(x). \qquad (3.21)$$

The following equation for current in the oscillator circuit (Fig.3.4)

$$L\frac{di}{dt} + Ri + \frac{1}{C}\int (i - MS\frac{di}{dt})dt = 0, \qquad (3.22)$$

along with equations (3.20) and (3.21), give a closed system which is reduced, in the dimensionless variables, to the form

$$\begin{aligned}\dot{x} &= mx + y - xz - dx^3, \\ \dot{y} &= -x, \\ \dot{z} &= -gz + g\Phi(x),\end{aligned} \qquad (3.23)$$

where m and g are the excitation and inertia parameters (3.18), $d = d(S_1)$ is a parameter corresponding to the degree of the influence of the curve's slope non-linearity (3.19), and $\Phi(x)$ is a function describing the properties of the inertial transformer.

Two mechanisms of the non-linear oscillation amplitude restriction act in the oscillator (see (3.20) and (3.23)). The former is non-inertial and is associated with the non-linearity of the amplifier characteristics, and the latter is inertial and is due to the slope S as a function of voltage V. Let the amplifier operate on the linear slope segment ($S_1 = 0$). The inertial transformer is connected in the circuit of a two-half-wave square-law detector with RC-filter (Fig.3.5) and is described by the equation:

$$\dot{z} = -gz + gx^2. \qquad (3.24)$$

Fig.3.5. A scheme of a version of practical realization of inertial transformer.

The inertia parameter g is equal to the ratio of the circuit oscillation period T_0 to filter time constant $\tau_\Phi = R_\Phi \cdot C_\Phi$.

With the above assumptions, equations (3.23) are transformed into the classical equations of the oscillator with inertial non-linearity (3.18). Thus, if amplifier 1 is linear and if the inertial transformer satisfies eq. (3.24), then the mathematical oscillator models in Figs 3.3 and 3.4 are identical. The circuit comprising an inertial detector is more convenient for leading the experiment since it allows the inertial properties of the oscillator to be changed by controlling the time constant of the filter what is impracticable when using a thermistor.

The form of equations (3.23) holds if an RC-chain in the form of Wien-bridge is employed as a selective element. To provide for oscillation conditions in this case, two amplifier stages must be used, as it is shown in Fig.3.6. For a symmetrical Wien-bridge, the control parameters m and g in equations (3.23) are represented simply and experimentally more convenient via the circuit parameters

$$m = K_0 - 3, \quad g = R_0 C_0 / \tau_\Phi, \tag{3.25}$$

where K_0 is the amplification factor of a two-stage amplifier and $R_0 C_0$ and τ_Φ are the time constants of the Wien-bridge and the detector filter, respectively. Parameters m and g are easy to vary and to measure by varying the amplification factor and the time constant of the filter.

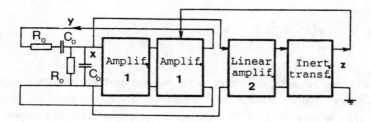

Fig.3.6. Scheme of RC-oscillator with inertial non-linearity.

The dynamics of system (3.23) depends in much on the specific form of function $\Phi(x)$ defining the inertial transformer's properties. A more abundant pattern of oscillation regimes, including chaotic ones, and of their bifurcations is observed in the case when a one-half-period detector is used [3.2]. Function $\Phi(x)$ may have the form

$$\Phi(x) = \exp(x) - 1 \tag{3.26}$$

or, what is convenient for numerical investigation,

$$\Phi(x) = I(x)x^2, \quad I(x) = \begin{cases} 1, & x > 0, \\ 0, & x \leq 0. \end{cases} \tag{3.27}$$

Defining the function $\Phi(x)$ according to (3.27), we obtain from (3.23) the *equations of a modified oscillator with inertial non-linearity*. They represent the following three-dimensional three-parametric non-linear dissipative system:

$$\begin{aligned} \dot{x} &= mx + y - xz - dx^3, \\ \dot{y} &= -x, \\ \dot{z} &= -gz + gI(x)x^2, \end{aligned} \tag{3.28}$$

which will be the main object under study in the present book.

By eliminating the variable y, the equations for the oscillator with inertial non-linearity (3.28) are reduced to the form (3.2)

$$\begin{aligned} \ddot{x} - (m - z - 3dx^3)\dot{x} + [1 - gz + g\Phi(x)]x &= 0, \\ \dot{z} &= -gz + g\Phi(x). \end{aligned} \tag{3.29}$$

An automatically controlled non-linear oscillator (3.29) is defined by the inertial dependence of dissipation and frequency on the variable x. In case of a strong system inertia ($\tau_\Phi \gg T_0$), as $g \to 0$, the initial system (3.28) degenerates into the two-dimensional one:

$$\begin{aligned} \ddot{x} - a(1 - bx^2)\dot{x} + x &= 0, \\ a = m - z_0, \quad b &= 3d/(m - z_0), \quad z_0 = z(0) \end{aligned} \tag{3.30}$$

and coincides in the form of writing with the Van-der-Pole oscillator's equations regardless of the type of function $\Phi(x)$. Parameters a and b depend on the initial conditions $z(0)$, as seen from (3.30). Alternative asymptotical case is presented by a non-inertial oscillator that corresponds to the magnitude of parameter g tending to infinity. Under this condition, the algebraic relation between the variables z and x follows from the third equation of system (3.23) thus reducing the initial system to the form

$$\ddot{x} - [m - \Phi(x) - 3d^2x^2]\dot{x} + x = 0. \tag{3.31}$$

The identity with the Van-der-Pole equations is achieved in this limit case under the condition $\Phi(x) = x^2$. In a real oscillator with inertial non-linearity, the region of the inertial parameter g values, where the system behaves like the three-dimensional one, is bounded by a certain interval $g_1 \leq g \leq g_2$. Beyond it, the considered asymptotical equations in the phase plane can serve as an approximate description.

3.3. PERIODIC OSCILLATION REGIMES IN THE OSCILLATOR AND THEIR BIFURCATIONS UNDER VARIATION OF THE PARAMETERS

Formulate in outline the purpose of bifurcational investigation and algorithms for its performing. Make an attempt to describe characteristic (from a set of possible ones) oscillation regimes in the system and their rebuilding under variation of the parameters. To do this, by using computation, ascertain the structure of the parameter plane partition into the regions of qualitatively different motion types, indicate their phase portraits and particularize the types of regime bifurcations at the region boundaries. For two-parametric systems, the overall algorithm for constructing bifurcation diagrams consists of the following items:

1. To find the *singular points* of the system, to study their stability, and to reveal characteristic bifurcations, particularly, the bifurcation of periodic motion (cycle) birth.

2. To examine the *character* of the cycle birth *bifurcation* that defines its stability.

3. To study one-parametrically the evolution of cycles along different parameters and to find the *points of characteristic bifurcations*.

4. To perform the two-parametric investigation of cycles which comprises the construction of bifurcation lines corresponding to different types of codimension one bifurcations. The *points of codimension two bifurcations* are to be found on these lines.

At this point, the qualitative study of two-parametric systems is completed.

In the systems with three parameters, the two-parametric analysis is repeated for the selected values of the third parameter and the bifurcation situations of a higher codimension are investigated.

The mathematical model of the modified oscillator with inertial non-linearity (3.23) is a non-linear three-dimensional dissipative system having three independent parameters which specify the flow in $\mathbb{R}^3$,

$$-\infty < x < \infty, \quad -\infty < y < \infty, \quad 0 \le z < \infty,$$

where variable z is defined on a positive half-axis since, from the physical point of view, it presents a detected voltage $x(t)$ at the output of the filter. The divergence of the flow velocity vector field (3.23) depends on the parameters and phase coordinates as follows:

$$\mathrm{div} F = (m - g) - 3d^2 x^2 - z. \qquad (3.32)$$

The study within a quasi-linear approximation $m < g \ll 1$ shows that the system is globally dissipative and that, for any initial data from the domain of definition of the phase variables, the following is always valid

$$divF < 0.$$

In a quasi-linear approximation the variable $z \simeq m$ and the divergence is negative independently of the magnitude of coordinate x. With the increase of the parameter $m > g$, g is finite (the most interesting region of generation of non-linear oscillations), the sign of divergence depends on coordinates. The condition of dissipation used here is:

$$m - g < z + 3dx^2. \quad (3.33)$$

This condition is always fulfilled for selfoscillations as $d \neq 0$. In this sense, parameter d defines the inertialess dissipative non-linearity of the system. If the amplifier operates in the linear curve segment and the non-linear restriction of the amplitude, due to inertia, sets in earlier than the values of variable x go into the non-linear segment of curve $S(x)$, expression (3.33) takes the form:

$$m - g < z(\tau). \quad (3.34)$$

The phase space of system (3.23) is separated by the above inequality into two regions with the plane $z = z^0 = m - g$. For $z > z^0$, the system is dissipative, and for $z < z^0$ the phase volume in the vicinity of any system trajectory is increased. Stationary self-oscillation regimes are realized when the energy pumping-up and its consumption are compensated in average over time. This is possible under the condition

$$m - g < \bar{z}, \quad (3.35)$$

where $\bar{z}$ is the mean-over-time value of variable $z(\tau)$. For sufficiently great $m > 1$, inequality (3.35) may not be fulfilled, and the system trajectories will go to infinity if the dissipative non-linearity is absent ($d = 0$).

3.3.1. Andronov-Hopf bifurcation. System (3.23) is featured by the only one singular point at the origin. If the function $\Phi(x)$ has no terms linear by x, then linearizing the system at the singular point gives the characteristic polynomial [3.2]

$$(g + s)(s^2 - ms + 1) = 0, \quad (3.36)$$

with the eigenvalues

$$s_{1,2} = \frac{m}{2} \pm \frac{i}{2}\sqrt{4 - m^2}, \quad s_3 = -g. \quad (3.37)$$

The real parts of all the eigenvalues are negative and the singular point is stable in the region of the parameter plane $g > 0 - 2 < m < 0$. From the physical point of view, parameter g is always positive being the ratio of the characteristic times of the system (of the oscillation period to the relaxation time of the filter). Parameter m may be both smaller (oscillator is not excited) and greater than zero (in the generation regimes). The singular point in the interval of $0 < m < 2$ is a saddle-focus with two-dimensional unstable and one-dimensional stable manifolds (3.37). The line $m = 2$ is bifurcational and corresponds to the substitution of the saddle-focus by a saddle-node.

As evident from (3.37), the eigenvalues $s_{1,2}$ cross the imaginary axis at the bifurcation point $m = 0$ with the non-zero velocity

$$\frac{\partial \text{Re } s_{1,2}(m)}{\partial m}\bigg|_{m=0} = \frac{1}{2}.$$

The third eigenvalue $s_3 = -g$ is separated from the imaginary axis. A *classic Andronov-Hopf bifurcation* is realized, the bifurcation of a cycle birth from the saddle-focus. The linear analysis of the cycle birth bifurcation is insensitive to both the presence of dissipative non-linearity (the eigenvalues are independent of the coefficient d) and the form of the function $\Phi(x)$ which must only not involve the term linear by x. Hence, the line $g > 0$, $m = 0$ is a bifurcation line of the cycle birth in the region of the control parameters of the system that is physically realizable and presents a positive quadrant of the plane $m \geq 0$, $g > 0$.

At first, examine the system (3.38) for the case when $d = 0$ [3.8]:

$$\begin{aligned} \dot{x} &= mx + y - xz, \\ \dot{y} &= -x, \\ \dot{z} &= -gz + gI(x)x^2, \end{aligned} \qquad (3.38)$$

restricting ourselves to the two-parametric analysis. The effect of the inertialless dissipative non-linearity will be considered in its own right.

The calculation of the fixed point stability within the linear approximation is practically the only task which is possible to be solved analytically for the system under study. Further numerical investigations will be carried out by using computation and experiments will be performed on a radiophysical oscillator.

To solve the problem on the stability of a cycle being born, analyze the type of Andronov-Hopf bifurcation. Numerical calculations have shown the first Lyapunov quantity $L_1(g)$ to be negative at the singular point everywhere along the line of cycle birth. The system's limit cycle being born is stable (*supercritical bifurcation*). The first Lyapunov quantity may be approximately found by calculating analytically with the use of

Bautin's algorithm [3.9][2]. Approximate analytical and numerical results agree qualitatively. Thus, in systems (3.38) and (3.23) on the line $m = 0$, $g > 0$ a stable limit cycle is born with the radius growing in proportion to $\sqrt{m}$ and with the period, according to the theorem, being equal to

$$T_0 \simeq \frac{2\pi}{|s_{1,2}(0)|} = 2\pi \qquad (3.39)$$

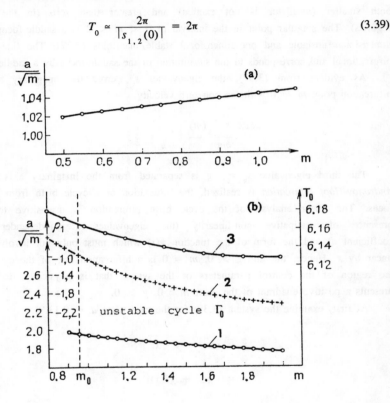

Fig.3.7. Normalized amplitude of stable cycle $\Gamma_0(m)$ (a), the amplitude $a/\sqrt{m}$, the largest multiplier ρ_1 and period T_0 of cycle $\Gamma_0(m)$ as a function of parameter m when passing through the bifurcation point of period doubling (b).

It was found by numerically integrating the system (3.38) that, for values $0 < m < 2$ and $0 < g < 2$, the stable periodic oscillation with the amplitude $\simeq \sqrt{m}$ and

---

[2] Problems with using the above algorithm appear due to a discontinuity of the second derivative $\Phi(x) = I(x)x^2$ in (3.38). If one approximates the $\Phi(x)$ by the exponential function $\exp(x) - 1$ and restricts himself to the first three terms of its Taylor's expansion, then calculation can be completed and can be shown that $L_1(g) < 0$ for any $g > 0$.

period $T_0 \approx 2\pi$ is the solution of the Cauchy task with the initial conditions close to the singular point at zero. For the values of $m < 0.5$, the agreement to the theorem of cycle birth, within the accuracy of calculation, is virtually adequate. With the increase of $m > 0.5$, small deviations appear in the amplitude and the period of the cycle $\Gamma_0(m)$ dependencies from theoretical predictions. In fig.3.7(a) the data are indicated which were calculated for the size of the stable limit cycle $\Gamma_0(m)$ divided by $\sqrt{m}$ (normalized) for $g = 0.097$. The deviation of the dependence from the constant is visible in the region of $0.5 < m < 1.1$ and is not higher than 2.5-3.5%. The normalized size and period and the greatest in modulus cycle multiplier are given in Fig.3.7(b) as a function of parameter m for $g = 0.2$ when the parameter passes the point of period doubling bifurcation ($m^* = 0.966...$). As calculations demonstrate, the limit cycle evolution is described well enough by Andronov-Hopf bifurcation theorem not only in the small vicinity of the parameter value corresponding to the birth and not only of stable, but also of saddle cycle Γ_0. The deviation of the oscillation period from 2π and that of cycle size from $\approx \sqrt{m}$ are here ± 3%, though the exceeding over the generation threshold is essential and corresponds to the regimes of non-linear oscillations. The last conclusion is confirmed by realizing period doubling bifurcation.

3.3.2. Limit cycle bifurcations. Let us treat one-parametrically the evolution of the born cycle family $\Gamma_0(m,g)$ to find the points of typical bifurcations. Survey different sections of the plane of parameters m and g by recording the former and calculating the cycle and its multipliers under variation of the latter. Calculations show the cycle family $\Gamma_0(m,g)$ to have the following bifurcations: 1) the greatest in modulus multiplier ρ_1 becomes -1 at the bifurcation point what corresponds to the oscillation period doubling bifurcation; 2) the multiplier ρ_1 of cycle Γ_0 takes the value $+1$ what corresponds to the merging and vanishing (or the birth) of stable and unstable cycles; 3) there are situations when the cycle multipliers' product satisfies the condition $|\rho_1 \cdot \rho_2| = 1$ when varying the parameter. The above condition will be called the *neutrality condition* since it corresponds to the case when the sum of Lyapunov exponents of cycle Γ_0 becomes zero: $\lambda_1 + \lambda_2 = 0$. If the multipliers are complex-conjugate with this, then the bifurcation of a two-dimensional torus birth is realized. For the real multipliers ρ_1 and ρ_2 there is no bifurcation situation available, the cycle is saddle. For the cycle family $\Gamma_0(m,g)$, both cases were observed, i.e., the bifurcation of torus birth occurs in system (3.38)!

Having determined the values of parameters corresponding to the bifurcations of cycles $\Gamma_0(m,g)$ of our interest, we proceed to a two-parametric analysis - the construction of bifurcation lines in the parameter space. The bifurcation diagram is indicated in Fig.3.8 for the family of period-one cycles Γ_0 being born as a result of Andronov-Hopf bifurcation. One of the cycle multipliers becomes -1 on line l_{01} (the *period doubling bifurcation line*). The cycle Γ_0 is saddle inside the region bounded by line l_{01} and is stable outside this region, since both of its multipliers belong to the inside of the unit circle. On line l_{02}, $\rho_1 = +1$. Here, the merging and the subsequent vanishing of stable and saddle cycles take place. Or, if one moves along parameters in the opposite direction, a pair of cycles is born from the closeness of trajectories. Line l_{02} will be called further the *line of multiple cycles*, or the *multiplicity line*. Inside the region bounded by the multiplicity line, three period-one cycles always exist, the saddle one Γ_0'' and two cycles Γ_0 and Γ_0' which can be stable or saddle.

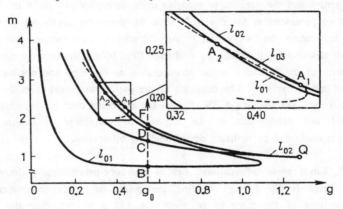

Fig.3.8. Bifurcation diagram for the family of cycles Γ_0 of system (3.38).

The situation is made clear by Fig.3.9 where the calculated dependence is qualitatively shown for the largest multiplier of cycle ρ_1 under the variation of the parameter m for the fixed g_0 as indicated in Fig.3.8. A pair of cycles Γ_0 and Γ_0'' is born at point C, the above cycles merge and vanish at point F, cycle Γ_0 undergoes the period doubling bifurcation at point B, becomes saddle, and finds its stability again at point D[3]. The situation is described for the increase of the parameter m. The typical points of peculiarities B, C, D and F are also marked in Fig.3.8.

The multiplicity line l_{02} forms a characteristic angle with the vertex at point Q where all the three cycles Γ_0, Γ_0', and Γ_0'' merge in one cycle. Point Q is bifurcational and has codimension two. In the theory of catastrophes, this point is called the *point of cusp*. The presence of a cusp corresponds to the simplest and more frequently available catastrophe in general multi-parametric systems [3.10]. The relation between the cusp catastrophe and the dynamics of the system under study will be considered

---

[3] Such calculations, requiring transition to unstable cycles at points C and F, are non-trivial, but possible with the suitable modification of calculation algorithms for the cycle multipliers. The normal practice of iteration leads to the loss of cycle and to the abrupt change of regimes here.

below.

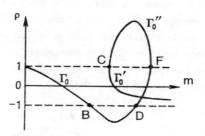

Fig.3.9. Qualitative view of dependence $\rho_1(m)$ for cycle Γ_0 in the section $g=g_0$. Points B,C,D and F correspond to those shown in Fig.3.8.

Fig.3.8 shows a segment of bifurcation line l_{03} where the condition of the cycle Γ_0 neutrality is fulfilled, $|\rho_1 \cdot \rho_2| = 1$. Line l_{02} is strictly bifurcational in the segment from point A_1 to A_2 where the cycle multipliers are complex-conjugate and emerge to the unit circle. At points A_1 and A_2, both multipliers are either -1 (point A_2) or $+1$ (point A_1) what corresponds to the resonances $1:1$ (A_1) and $1:2$ (A_2). Like point Q, the bifurcation points A_1 and A_2 have codimension two. The resonance conditions are satisfied at these points besides the emerging of a pair of multipliers to the unit circle. As illustrated by calculations, the bifurcation of torus birth from cycle Γ_0 is subcritical. The stable quasiperiodic oscillations in the autonomous system (3.38) were not found. The two-parametric analysis of the stability character of the period-one cycles of the system may be completed since typical bifurcations have been defined and the corresponding diagram has been constructed in the plane (Fig.3.8).

Cycle Γ_0 is stable not in the small (the first Lyapunov quantity is strictly negative at the singular point) when approaching the line of doubling l_{01} from below. It means that the crossing of the line l_{01} will lead to the soft birth of a stable cycle Γ_1 with a period being twice greater ($T_1 \simeq 2T_0$) within the linear approximation. Taking as an initial approximation a point on cycle Γ_0 close to the point of doubling bifurcation, find the cycle Γ_1 defined by a period-two fixed point in the Poincaré map. Having shifted along the parameter behind the bifurcation line l_{01}, determine cycle Γ_1 by means of numerical integration. It is really stable, has a period close to the doubled one, and goes twice round the period-one cycle, that has lost its stability, in the phase space.

Fig. 3.10 illustrates the above for the value $g = 0.2$. Below the bifurcation point $m^* = 0.966...$ cycle Γ_0, shown in the figure for $m = 0.5$, is stable in the system. Upwards in parameter ($m > m^0$), cycle Γ_1 with doubled period is stable, cycle Γ_0 becomes saddle as indicated by the dashed line in Fig.3.10.

Investigate, in a similar manner, a family of period-two cycles $\Gamma_1(m,g)$ first one-parametrically and then two-parametrically. The bifurcation analysis shows the character of bifurcations and the structure of mutual location of the corresponding bifurcation lines in the parameter plane for cycles Γ_1 with doubled period to repeat completely the picture for period-one cycles (Fig.3.9). The only difference is that two independent families of period-two cycles Γ_1^1 and Γ_1^2 appear [3.11].

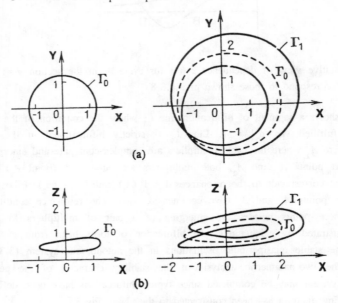

Fig.3.10. Period doubling of cycle Γ_0.

In Fig.3.11, the bifurcation diagram of one of the families of cycles $\Gamma_1^1(m,g)$ is presented confirming the above. The presence of two families of period-two cycles is answered by the dependence of multiplier on parameter m in the form of two loops in contrast to one loop for the cycles with period T_0. Calculations demonstrate that there are two independent families of period-four cycles inside each region bounded by the bifurcation lines of period doubling of cycles $\Gamma_1^1(m,g)$ and $\Gamma_1^2(m,g)$ [3.11]. The hierarchy of multiplication of cycle families may be assumed to extend to infinity. Their bifurcation diagrams constitute a system of *topologically equivalent embedded structures* that correspond to universal properties which generalize the regularities of Feigenbaum type similarity for the case of two parameters. This was verified experimentally and confirmed qualitatively for cycles Γ_k with periods $T_k \simeq 2^k T_0$, $k = 0,1,2$ and partially for $k = 3$. Quantitative regularities are difficult to ascertain since there is a need of analyzing numerically the cycles with sufficiently great periods to a high degree of accuracy. This problem is more convenient to be considered as applied to two-parametric, two-dimensional model maps.

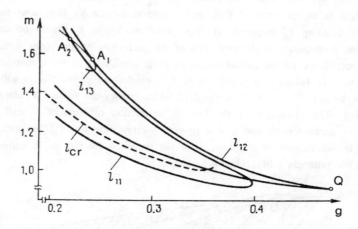

Fig.3.11. Bifurcation diagram for the family of cycles $\Gamma_1(m,g)$; l_{cr} is the critical accumulation line of period doubling bifurcations.

The complexity of phase space partition into different trajectory types is visualized in the relatively simple system (3.38) by geometrically representing the obtained results. We introduce into consideration a combined three-dimensional space, where the dependence $\xi = \xi(m,g)$ is graphically represented, with ξ taken as one of the fixed point coordinates in the Poincaré section for the cycle. For example, if the secant plane $x = 0$ is introduced in $\mathbb{R}^3$ of system (3.38), then ξ may be taken as the coordinate z of a two-dimensional map in the secant plane. A complicated two-dimensional surface S^0 in the above space indicated in Fig.3.12 is a locus of points corresponding to stable and unstable cycles $\Gamma_0(m,g)$ of the system. Surface S^0 emerges from the line of cycle Γ_0 birth and has *two folds* and a *cusp* as $\xi \geq 0$. For clearness, the section of surface S^0 by the plane $g = g_0$ is given in the figure. Three sheets of surface corresponding to cycles Γ_0 (upper sheet), Γ_0' (lower sheet), and Γ_0'' (internal sheet) are available inside the region between points C and F. The projection of the complicated surface onto the plane of parameters m and g, due to the presence of folds, provides a peculiarity line l_{02} consisting of upper and lower branches which intersect at point Q where the folds and cusp vanish. A hysteresis phenomenon, caused by "jumps" at points F (when decreasing m) and C (when increasing m) is inevitably connected with the variation of the parameter m ($g = g_0$). These abrupt switchings are well known in the theory of catastrophes and form, as a matter of fact, the effect of the cusp catastrophe itself. Besides the fold and cusp, there can not be any other peculiarity under projection of the surface onto the plane in two-parametric families.

Thus, the bifurcation multiplicity line l_{02} in the diagram shown in Fig.3.8 owes its origin to the presence of folds and a cusp at surface S^0. How does the bifurcation line of doubling l_{01} originate? If only period-one cycles Γ_0 are to be analyzed, then this line corresponds to the projection of the corresponding line onto surface S^0 which is in accordance to the critical values of cycle amplitudes, when a multiplier takes the value of - 1. Taking into account cycle Γ_1 with doubled period that is softly born with this, a surface S_0^1 is marked in Fig.3.12 which crosses the surface S^0 along the line of doubling. The projection of the line of intersection of surfaces S_0^1 and S^0 onto the plane of parameters m and g will provide a bifurcation line l_{01}. No catastrophe takes place in the vicinity of this line, since the transition from a stable regime to another stable one proceeds softly, without hysteresis.

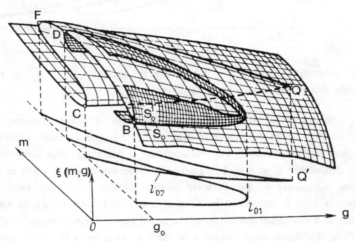

Fig.3.12. Geometric representation of critical phenomena in system (3.38). The peculiarity points are the same as in Figs.3.8 and 3.9.

The presence of doubling bifurcations complicates the general pattern of dynamic properties of the system with variation of the parameters, because only those transitions are really observed which are followed by the change of stable regimes. However, just the above difficulty allows one to mentally imagine the possible versions of regime changes in the real oscillator when varying the parameter. So, if one varies the parameter m from zero in the plane $g = g_0$, the period-one cycle is lost at point B, the system passes to the regime of periodi-two oscillations what is corresponded by going out onto surface S_0^1 in Fig.3.12. There will be no hysteresis for cycle Γ_0, since the doubling will occur earlier. The mutual location of bifurcation points B, F, C and D in different sections along parameter g will vary, and due to this the particular sequence of oscillation regime's change and the hysteresis may depend on the choice of the section. The geometric representation as in Fig.3.12 becomes practically hindered in case when a cascade of accumulated doubling bifurcations is realized. The construction of the dependence of stable cycles' amplitudes on parameter is more evident. It is the so-called "bifurcation tree" of branching of amplitudes which is obtained by calculating the stable cycles' amplitudes as the functions of only one parameter. The evolution of the oscillatory regimes' amplitudes is noticeable in Fig.3.12 for the first step of doubling in the section $g = g_0$.

CHAPTER 4

AUTONOMOUS OSCILLATION REGIMES IN OSCILLATOR

4.1. TWO-PARAMETRIC ANALYSIS OF TRANSITION TO CHAOS VIA THE CASCADE OF PERIOD DOUBLING BIFURCATIONS

The bifurcation diagram for the period-one cycles (Fig.3.8) comprises three characteristic bifurcation lines corresponding to a certain type of stability loss. Consider the contribution of these bifurcations into the overall picture of changing oscillatory regimes of oscillator (3.38) and interconnection between the type of stability loss, on the one hand, and the mechanisms of birth and properties of the system's attractors, on the other. We begin with ascertaining the role of doubling bifurcations [4.1, 4.2].

As mentioned, the analysis of the nesting structure of the bifurcation diagrams of cycles with periods $2^k T_0$[1] shows that the number of cycle families grows twice with every doubling. Analyze the period doubling bifurcations as applied to a family of cycles whose bifurcation diagrams lye close to the line $m = 0$ of cycle Γ_0 birth. In this case, a cascade of period doubling bifurcations can be expected. Carry out a numerical experiment on studying one-parametrically the sequence of bifurcational values of parameter μ_k corresponding to the points of period doubling bifurcations of cycles $2^k T_0$. Choose two directions of motion in parameter plane: the fixed $g = 0.3$ provides the variation of parameter m and, as $m = 1.45$, the variation of parameter g. The bifurcational values of parameters μ_k are to be calculated for which the multiplier of the cycle with period $2^k T_0$ becomes -1. The values of parameters m_k and g_k were defined with the accuracy of $\varepsilon = 10^{-6}$. They are indicated in Table 4.1 for $k \leq 4$.

---

[1]The magnitude of period is not to be understood literally, since $T_0 = T_0(m,g)$. For differential systems, the equality $T_k = 2^k T_0$ is not fulfilled in general case in contrast to the maps and it mainly represents the topology of the trajectories.

Table 4.1

	δ = 0.3			m = 1.45		
	m_k	δ_k	m^*	g_k	δ_k	g^*
0	0.7700	–	–	0.1200	–	–
1	1.0200	–	1.0880	0.16898	–	0.18233
2	1.0713	4.873	1.0853	0.18162	3.876	0.18506
3	1.08216	4.724	1.08511	0.18438	4.582	0.18513
4	1.08449	4.66896	1.08512	0.18497	4.66836	0.18513

Bifurcational values of parameters m_k, g_k, Feigenbaum constant δ_k and critical points under period doublings in system (3.38)

The magnitude of Feigenbaum's constant δ_k and values of critical points m^*, g^* (Table 4.1) were calculated by using the following asymptotic relations:

$$\delta_k = (\mu_{k+1} - \mu_k)/(\mu_{k+2} - \mu_{k+1}), \quad \mu^* = (\mu_{k+1} \cdot \delta - \mu_k)/(\delta - 1).$$

The data obtained are the evidence that a converging sequence of period doubling bifurcations, obeying Feigenbaum's law, takes place when moving in the parameter plane, both along m and g. For two-parametric systems, the bifurcation line of critical values having codimension one must correspond to the transition to a strange attractor via doublings in the parameter plane. This line may be constructed by applying the above one-parametric approach. The alternative method, saving machine time substantially, consists of calculating a line in the parameter plane that corresponds to the critical multiplier value of cycle $2^k T_0$, $k \gg 1$, $\rho^* = -1.60119...$ [see Chapt.2]. To provide the accuracy needed for practical use, $k = 2$ is enough to be taken, i.e., one may confine himself to the calculation of the multiplier of a period-four cycle[2]. The line of critical parameter values, calculated in such a way, is shown in Fig.3.11 by dashed line.

The analysis of the system (3.38) dynamics, carried out during its evolution to stochasticity via a series of period doubling bifurcations of the main family of cycles, demonstrates a comparatively rare possibility to verify the conclusions, obtained numerically, in experiment not only qualitatively, but also quantitatively. Really, cycle Γ_0 is the only limit cycle of system (3.38) being softly born from a singular point at zero with small m and finite g. Every subsequent cycle with doubled period is

---

[2]With this, the confidence is needed in a doubling chain to be infinite, and the control is obligatory with the help of one-parametric analysis at selected points.

softly born from the preceding one at the bifurcation points m_k, g_k and is stable. If the intrinsic noise of the system is small enough, then the evolution of limit cycles can be followed in a radiophysical oscillator with the excitation parameter growth up to the critical point of the onset of chaos and of small threshold exceedings. At this point the tangency bifurcations of cycles with comparatively small periods do not influence the oscillation regimes (their multiplicity lines lie over the line of critical parameter values). Such an experiment is not related with an exact specification of initial conditions that is generally difficult to provide in a physical system. The physical experiment on quantitatively investigating the doubling bifurcations in inertial non-linearity oscillator becomes possible.

The RC-oscillator for a low radio-frequency range $(f_0 = 1/T_0 \simeq 7\text{kHz})$ was developed and fabricated for full-scale experiments according to Fig.3.6 [4.3]. The basic oscillator involved two amplifier stages having a sufficiently large linear segment of the curve and a selective element in the form of Wien-bridge. The complementary chain of negative inertial feedback consisted of the anode-like half-wave square-law detector and RC-filter. The highest possible correspondence between real and mathematical system models was experimentally achieved by thoroughly choosing the oscillator element parameters and characteristics of detector, amplifiers and filter [4.1-4.3]. A measuring set allowed one to independently vary and to quantitatively evaluate the magnitude of excitation parameter m by measuring the amplification factor of the basic amplifier K_0 and the inertia parameter g controlled by the magnitude of the filter capacity C_Φ. The accuracy of the measurement of control parameters m and g was better than $\pm$ 1 %. Note an extreme feature of system (3.38) and thus of the oscillator, and namely: the independent variation of parameters m and g makes it feasible to control separately the degrees of contraction $(s_3 = s_3(g))$ and expansion $(s_{1,2} = s_{1,2}(m))$ in the vicinity of a saddle-focus in the course of experiments, i.e., to *independently* control the properties of *stable and unstable manifolds of a singular point* (see relations (3.37)).

The oscillation regimes and their bifurcations were studied with the help of a system of oscillographes, spectrum analyzer and selective microvoltmeter involved in the measuring set.

We follow the evolution of oscillation regimes when increasing the parameter m for a number of discrete parameter g values. The experimental diagram of oscillation regimes in a portion of the parameter plane is indicated in Fig.4.1. As it was expected, when varying m in experiment, the accumulation of bifurcation doubling lines l_{k1} towards the line of critical values of parameters l_{cr} is observed. Doubling bifurcations are reliably recorded for $k = 0,1,2$, i.e., up to the transition $4T_0 \to 8T_0$. The transition $8T_0 \to 16T_0$ was observed experimentally as well, but it was unstable in some regions of the parameter plane due to natural and technical setting noise. The corresponding bifurcation line l_{31} is not shown in Fig.4.1.

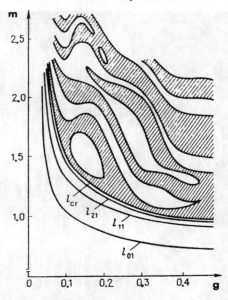

Fig.4.1. The bifurcation diagram of experimental oscillation regimes in oscillator. Hatched regions correspond to chaos.

The transition to stochasticity, including a small threshold exceeding, proceeded *without hysteresis*. The above bifurcation lines l_{k1} and l_{cr} coincided within the accuracy of measurement when both increasing and decreasing the parameter m. A noticeable threshold exceeding resulted in abrupt transitions to the regimes of oscillations with period $p \cdot 2^k T_0$ ($p = 1,2,3..., k = 1,2,...$) which were alternated with the regions of strange attractors exhibiting another structure. In this region the hysteresis phenomenon was typical and the data, shown in Fig.4.1, present the picture of regime change under increase of the parameter. The properties of the system under the considerable exceeding of the stochasticity threshold will be considered in the following paragraphs. Here, our attention focuses on the transition to chaos via a *series of doubling bifurcations*.

Fig.4.2 shows a number of oscillation regime photographs illustrating doubling bifurcations $2T_0 \rightarrow 4T_0 \rightarrow 8T_0$ and the form of a strange attractor near the critical point. The experimental results correspond to the value $g = 0.2$ and to different m's. The period-one cycle, preceding the first doubling, is not shown in the figure. The photographs display the evolution of time realization $x(t)$ and the projections of phase trajectories onto the plane of variables x, y for different m's. The qualitative picture of transitions is quite obvious. To ascertain qualitative regularities, we present the findings of numerical and physical experiments below.

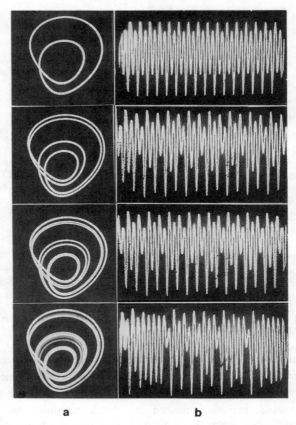

Fig.4.2. Phase trajectories' projections onto xy-plane and the corresponding parts of time series $x(t)$ for the transition to chaos via the sequence of period doubling bifurcations.

We fix the parameter value $g = 0.2$ and construct, by using computation, a *bifurcation tree* of branching of the amplitudes of stable cycles when varying the parameter m. To do this, introduce in the system's phase space a secant plane $y = 0$ and record the coordinate x on the secant under variation of the parameter m. The calculation results are indicated by solid curves in Fig.4.3. Having set the value $g = 0.2$ in the course of physical experiment, measure the oscillation amplitudes of variable $x(t)$. The normalized results are shown for comparison in Fig.4.3 by dashed curves.

The comparison of numerical and experimental data gives a good qualitative agreement. The *calculated* and *measured* parameter values m_k *do not coincide*, however, at doubling bifurcation points! The reason turned out to be hidden in the following. It is apparent from the model equations that the detector characteristic $\Phi(x)$ obeys the square law (3.27). The real characteristic $\Phi(x)$ can differ somewhat from approximation (3.27) in the oscillator, but it allows correction by varying the detector anode voltage of the inertial cascade of oscillator.

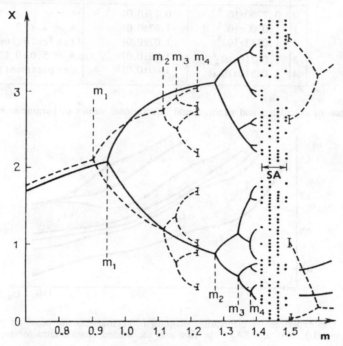

Fig.4.3. The bifurcation tree of amplitude branching during doubling (solid curves: calculation, dashed-and-doted curves: experiment).

In Fig.4.4, the bifurcation diagram of oscillation regimes is presented which was obtained experimentally in the plane of parameters m and U_a, the latter being the detector anode voltage for the value $g = 0.3$. Notation in the figure is the same as in Fig.4.1. Special measurements have shown the best agreement between the experimental characteristics $\Phi(x)$ and theoretical approximation (3.27) to be achieved as $U_a^0 = 120$ V. For $U_a = U_a^0$, the bifurcational values of parameter m_k ($k = 0,1,2$ and 3) and m_{cr} were measured. The results obtained are indicated in Table 4.2.

A good correlation of results within the accuracy of experiment is observed for the first three bifurcations. The experimental and numerical values of δ_2 are close, as well as the critical point values. The data obtained testify that the transition to stochasticity via the sequence of doubling bifurcations obeys Feigenbaum's law in the system under study, the similarity constant to be $\delta = 4.669...$.

Table 4.2

k	m_k (calculation)	m_k (experiment)	δ_2
0	0.7700 ± 10^{-6}	0.77 ± 0.01	
1	1.0200 ± 10^{-6}	1.02 ± 0.01	$\delta_2 = 4.873$
2	1.0713 ± 10^{-6}	1.07 ± 0.01	(calculation
3	1.08216 ± 10^{-6}	1.08 ± 0.01	$\delta_2 = 5.0\pm0.12$
4	1.08516 ± 10^{-6}	1.09 ± 0.01	(experiment)

Comparison of calculated and experimental bifurcational values of parameter m for $g=0.3$.

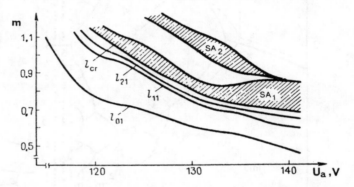

Fig.4.4. The experimental bifurcation diagram of regimes on the plane of parameters m, U_a for $g = 0.3$. SA_1 and SA_2 - the first and the second chaotic attractor regions.

4.2. THE POINCARE MAP

The appearance of a positive exponent in the LCE spectrum of the system is a criterion of stochasticity. For systems in $\mathbb{R}^3$ the signature of LCE spectrum may be, in general case, as follows: "-", "-", "-" for a stable steady-state point, "0", "-", "-" for a limit cycle, and "+", "0", "-" for a strange attractor. Calculate the whole LCE spectrum of system (3.38) while varying the parameters in those sections where the bifurcational values of m_k and g_k were calculated for doublings ($m = 1.45$ and $g = 0.3$). The LCE spectrum signature is insensitive to regime rebuilding due to doublings up to the critical point where a positive exponent appears in the spectrum indicative of the birth of a strange attractor. The LCE spectrum bifurcation point must belong to the line of critical parameter values. The possibility is given to independently determine the critical parameter values and to compare these values to the data obtained by calculating the cycle multipliers with doublings.

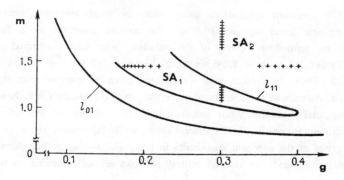

Fig.4.5. Bifurcation diagram portion with positive Kolmogorov entropy points.

Fig.4.5 shows a portion of the plane of parameters m and g comprising the bifurcation lines of doublings l_{01} and l_{11}. Here, the plane points corresponding to the presence of a positive exponent in the LCE spectrum are marked with crosses. The LCE spectrum is of the type "0", "−", "−" at other points of the straight lines indicated, testifying the presence of periodic oscillations in the system. When varying both m and g, two localized stochasticity regions are found. We call them the first and the second regions, SA_1 and SA_2. Examination has shown every of the regions to be a complicated region where different types of both regular and strange attractors are realized replacing each other under variation of parameters and initial conditions. The calculation results for $\lambda_1(m)$ and $\lambda_1(g)$ are given in Fig.4.6.

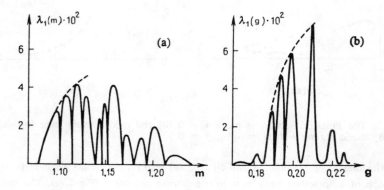

Fig.4.6. The results of $\lambda_1(m)$ and $\lambda_1(g)$ calculation; a: $g = 0.3$, b: $m = 1.45$.

The LCE spectrum calculations using relatively great averaging time intervals ($\tau \geq 10^4$) and the linear approximation of the positive exponent as a function of parameter in the immediate vicinity of the transition point have confirmed the results indicated in Table 4.1, i.e., $m^* = 1.0853 \pm 5 \cdot 10^{-4}$ for $g = 0.3$ and $g^* = 0.1854 \pm 5 \cdot 10^{-4}$ for $m = 1.45$. Thus, the *critical points*, within the given accuracy of calculation of the LCE spectrum, *correspond* to those found numerically by *Feigenbaum's law*. Note that one must be particularly careful when calculating in the vicinity of the critical point, where the maximum exponent of LCE spectrum undergoes the bifurcation "0" → "+", due to a slow convergence of the exponent magnitudes in time. The averaging time decrease under determination of λ_1 leads to lowered critical point values as compared to the precise ones.

The knowledge of the whole LCE spectrum of the strange attractor of the system enables one to determine its *Lyapunov dimension*. In the two above stochasticity regions, $D_L = 2 + d$, as $0 < d < 0.4$, which decreases with the growth of g. The dimension is 2.341 and 2.187 at the points of developed stochasticity $m = 1.45$, $g = 0.21$ and $m = 1.16$, $g = 0.3$, respectively. The fractional part of d near the critical line tends to zero, thus substantiating the approximate description of the transition dynamics with the help of one-dimensional map [4.4]. Numerical experiments give evidence in favor of such an

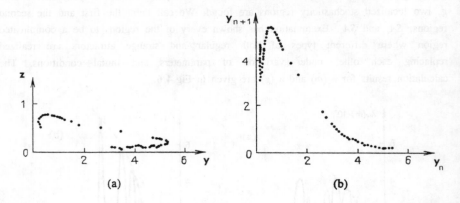

Fig.4.7. The Poincaré section in plane $x = 0$ (*a*) and model return map (*b*), calculated for the chaotic regime with $m = 1.16$, $g = 0.3$.

approximation to be possible. A two-dimensional map on the secant plane yz ($x = 0$) has the shape of a horseshoe being near to one-dimensional curve (Fig.4.7a). The model map $y_{n+1} = \phi(y_n)$ (Fig.4.7b), constructed numerically for variable y, is presented as a function belonging to the class of Feigenbaum map (discontinuities are absent, and there is one smooth maximum). *Cantor structure* is not visible in the transverse section of the almost one-dimensional map in the scale of the figure (Fig.4.7a) due to comparatively strong dissipation ($g = 0.3$). However, if the value of parameter $g \leq 0.1$ is decreased, then the system map on the secant would appear as a system of nested, nearly one-dimensional curves of horseshoe-type demonstrating the Cantor structure features even without varying the pattern scale [4.5].

The *Poincaré section* can be *visualized* in a physical experiment [4.6]. To accomplish this, the electron beam of the oscillograph is to be intensified at the time moments of a phase trajectory travelling through a given cross-section. For example, short intensifying pulses can be formed when realization $x(t)$ passes through zero values. In this case, the projection of the Poincaré section by the plane $x = 0$ will be displayed by brighter points on oscillograph screen simultaneously with the attractor projection onto the phase plane. The photographs of two types of chaotic attractor are presented in Fig.4.8 along with the corresponding Poincaré sections [4.2]. The oscillation regime shown in Fig.4.8a corresponds to the parameter values for which the section and return functions shown in Fig.4.7 were calculated.

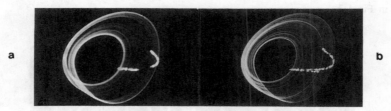

Fig.4.8. The Poincaré section in physical experiments.

The qualitative agreement of full-scale and numerical experiment results is obvious. Notice that a period-two band of an attractor with the Poincaré map, comprising two noncrossing chaotic zones, is realized in the situations presented in Figs 4.7 and 4.8. In case of the developed chaos, these zones merge and the Poincaré section takes the characteristic shape very similar to the one-dimensional square parabola (Fig.4.8b).

The joint results of numerical and physical experiments performed attest that the *universal Feigenbaum's law* holds in the differential system (3.38) under transition to chaos via doublings. Let us verify quantitatively the regularities in the character of the positive LCE spectrum as a function of the supercriticality level. As seen from Fig.4.6, the non-linear dependencies $\lambda_1(m)$ and $\lambda_1(g)$ are true near the threshold interrupted by sudden depressions when varying the parameter. The broken character of curves $\lambda_1(m)$ and $\lambda_1(g)$ corresponds to the appearance of periodic motion stability *"windows"* due to tangency bifurcations. The envelope of plots $\lambda_1(m)$, $\lambda_1(g)$ near the threshold allows approximation by using the following relations:

$$\lambda_1(m) = c_1(m - m^*)^\gamma, \quad g = 0.3;$$
$$\lambda_1(g) = c_2(g - g^*)^\gamma, \quad m = 1.45,$$

where $c_1 = 0.181$, $m^* = 1.085$, $c_2 = 0.395$, $g^* = 0.185$, and $\gamma = \ln 2/\ln \delta \simeq 0.449$. The approximating curves are drawn in Fig.4.6 by dashed line. They corroborate quantitatively the theoretical dependence $\lambda_1(\mu)$ ascertained for Feigenbaum's transition on the base of one-dimensional maps [4.7].

Consider the evolution of the oscillation power spectrum on the way to chaos via a series of period doubling bifurcations. The sequence in discrete spectrum enrichment with *subharmonics* correspond in a one-to-one manner to a set of cycles with periods, being doubled while passing the bifurcation points. Subharmonic amplitudes satisfy the quantitative regularity proceeding from the scale-invariant properties of branching of the cycle amplitudes. In Fig.4.9, the experimental power spectra $S_x(f)$ of oscillations are shown. The corresponding time dependencies $x(t)$ and phase trajectory projections are given in Fig.4.2.

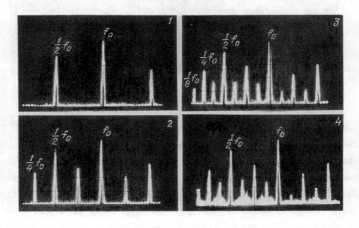

Fig.4.9. Oscillation power spectra $S_x(f)$ evolution for the transition to chaos via period doubling bifurcations.

The spectrum of period-one oscillations (not indicated in the figure) consists of the basic line $f_0 = 1/T_0$ and its harmonics. In the spectrum of period-two oscillations, the harmonics with half-frequencies $nf_0/2$ appear. Their amplitudes grow when increasing the depth of modulation and become saturated at the instant when the subsequent doubling bifurcation proceeds. The components of spectrum $nf_0/4$ appear and at the bifurcation point smoothly increase with the parameter growth corresponding to the soft birth of a cycle with period $4T_0$. The picture is the same for the transition $4T_0 \to 8T_0$. Thorough measurements with the help of the spectrum analyzer, comprising a memory, have demonstrated the ratio of the subharmonic intensity $f_0/2^k$ to $f_0/2^{k+1}$ to be 10-12 dB for $k = 0$ and 1. As $k = 2$, it reaches the value of 13.0 ± 0.4 dB. The calculations of the power spectrum of realization $x(\tau)$, corresponding to the cycle with period $16T_0$ ($m = 1.084$, $g = 0.3$), provide a close result of 13.03 ± 0.3 dB. The calculations were performed by using the algorithm of Fast Fourier-transform for subharmonic intensity ratios $f_0/4$ to $f_0/8$. The theoretical value of $\simeq 13.5$ dB is assumed to be reached with sufficiently large k.

The process of spectrum enrichment by subharmonics on the way to chaos via doublings is defined by the *scale-invariant properties*. If the spectrum of period-two cycle is to be considered, then it comprises two lines with a lower intensity $f_0 \pm f_0/2$ in the vicinity of frequency f_0. $\Delta f_2 = f_0/4$ is the scale unit for a period-four cycle. In the vicinity of line $f_0/2$, the spectrum has as well two lateral components $f_0/2 \pm f_0/4$. Within the limit of accumulation of the period doubling bifurcations ($k \gg 1$) the *scale-invariant structure of the spectrum* becomes valid at the bifurcation points.

When achieving the critical point, the number of preceding doublings tends to infinity, and the frequency interval between the nearest subharmonics $\Delta f_k \to 0$. The spectrum becomes continuous. The intensity of subharmonics descends, however, quickly with growing k and tends to zero as $k \to \infty$. Because of this, the real experimental spectrum will be discrete under the finite recording accuracy at the critical point. With exceeding the stochasticity threshold, the process of broadening the spectrum lines of the highest subharmonics sets in that leads to a higher intensity of continuous noise component in the spectrum. The continuous spectrum is really recorded in experiments starting at a certain parameter value (*always over the critical level*). Usually, it is a slightly smeared spectrum of period-(8-16) cycle.

Fig.4.9,*4* illustrates the spectrum of a strange attractor under a small threshold exceeding which corresponds to the smeared spectrum of period-eight cycle. Such an attractor presents an period-eight band in the phase space (Fig.4.2).

With the growth of supercriticality, the successive merging of multi-periodic attractor bands to period-one band proceeds forming, as a result, a developed period-one band. Consider the process of the merging of a period-two band to the period-one band shown in Fig.4.10. At the instant where a period-four band merges to the period-two one (Fig.4.10,*1*), the subharmonics lines $f_0/2^k$ ($k \geq 3$) are entirely noisy in the oscillation spectrum. It is seen, how the spectrum lines $nf_0/4$ become smeared. With this, the spectrum lines nf_0 and $nf_0/2$ are virtually not smeared. The developed period-two attractor band (Fig.4.10,*2*) is formed under the parameter growth. Here, the oscillation spectrum is free from peaks in the harmonics of frequency $f_0/4$. The spectrum lines $nf_0/2$

become smeared rapidly (in parameter), and the spectrum of the developed period-one band is featured by a continuous noise component having the peaks only in the basic frequency f_0 and its harmonics (Fig.4.10,3).

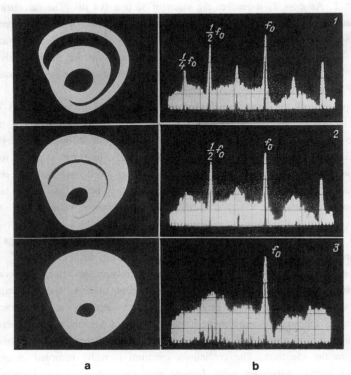

Fig.4.10. Phase portraits and oscillation power spectra while passing through the attractor bifurcation point.

The mechanism of the *merging of multi-periodic bands* of an attractor with the growth of supercriticality, followed by successively smearing spectral lines of subharmonics $nf_0/2^k$ ($k = \infty,...,4,3,2$ and 1) is due to homoclinic effects[3]). A dynamical system is characterized by the presence of the infinite number of saddle cycles with periods $2^k T_0$ at the critical point of transition. The parameter growth results in the intersection of two-dimensional stable and unstable cycle manifolds followed by robust homoclinic curves. In the Poincaré map on the secant plane, the stable and unstable separatrixes of a saddle point, that make the prototypes of appropriate cycle manifolds, intersect robustly providing for the birth of a homoclinic point and, thus, of a

---

[3]The bifurcation analysis of the chaotic attractor band merging effect is given in detail below.

homoclinic structure. Clear succession in the appearance of such structures is observed being inverse to the regularity of the doubling bifurcation accumulation process. Under the supercriticality growth, the homoclinic structures of saddle cycles with periods $2^k T_0$ appear in the inverse order in k. In connection with this, the attractor bifurcations in a supercritical region are called sometimes the *inverse, or connectivity, bifurcations*.

Universality in the frequency distribution of oscillation energies is, on the face of it, at variance with the process of the amplitude scale splitting in the cascade of Feigenbaum doubling bifurcations. We apply to a bifurcation tree of branching of the amplitudes (Fig.4.3). The amplitudes of unstable cycles with periods $2^k T_0$ disappear with every subsequent doubling making place for the amplitudes of stable cycles with periods $2^{k+1} T_0$. The pattern of energy distribution in the spectrum is different, and, namely, the spectrum lines of unstable cycles hold after every doubling, and new subharmonics are added having the amplitudes that are in the limit lower by 13.5 dB. The eigen frequencies of periodic motions, that have lost their stabilities, are really present in the spectrum of oscillation! The phenomenon is easy to understand by considering the onset of chaos under doublings as a process of initial oscillation self-modulation that becomes more and more complicated. With the first doubling bifurcation, the lateral components $f_0 \pm f_0/2$ are added to the spectrum of frequency f_0 self-oscillations which correspond to the regime of resonant quasiperiodic oscillations with the modulation frequency $f_0/2$. The total energy of a signal incorporates carrier energy and that of the modulation signal. The loss of stability by the initial periodic motion does not mean the carrier frequency disappearance thus providing the basic line in the oscillation spectrum to be maintained. The cascade of doubling bifurcations, leading to the successive amplitude decrease of modulating components of lower frequencies, results in a remarkable part of the attractor energy to be concentrated at the carrier frequency and first subharmonics. The contribution of higher subharmonics is highly small due to the successively decreased modulation energy at frequencies close to zero. The above properties of the oscillation spectrum in the Feigenbaum attractor regime are typical for small supercriticality levels. They can serve as a distinct criterion of this quasiattractor type.

The totality of experimental data presented allows one to conclude about the possibility to approximately describe the transition to chaos in a modified oscillator with inertial non-linearity via the sequence of doubling bifurcations with the help of one-dimensional map of square parabola type. Such a presentation is valid at least for situations as $0.15 \leq g \leq 0.35$ and the parameters slightly exceeding the critical values' line [4.1,4.2].

4.3. SYSTEM DYNAMICS IN THE SUPERCRITICAL RANGE OF PARAMETER VALUES. HYSTERESIS AND TRANSITION TO CHAOS VIA INTERMITTENCY INDUCED BY FLUCTUATIONS

We return to discussion of the results of calculating the maximum LCE spectrum exponent $\lambda_1(m)$ (Fig.4.6). With exceeding the threshold, dependence $\lambda_1(m)$ involves discrete regions in parameter m where the positive exponent becomes zero. Initially, $\lambda_1 = 0$ in the narrow parameter value intervals. But the interval widths increase with the growth of supercriticality. This is a sufficient demonstration of the presence of stable periodic regimes which may be observed in the course of an experiment. The appearance of stability windows above the critical point is caused by two reasons. First, the stable and unstable cycles of different families are born from the trajectories' closeness as a result of bifurcations $\rho = +1$. Secondly, initial cycles with periods $2T_0$, which are all saddle at the critical point, acquire their stabilities via inverse doubling bifurcations (the multiplier enters the unit circle through -1). The width of regions of stable periodic regime existence above the critical point can be defined by the corresponding cycle multiplier as a function of parameter shown qualitatively in Fig.3.9 and, thus, by the location of multiplicity line in the diagrams (Fig.3.8 and 3.11). As an example, consider the evolution of period-two cycle Γ_1 being born from cycle Γ_0 as a result of doubling. The analysis will be performed in the section $g = 0.3$ by varying the parameter m. At the bifurcation point $m = 0.77$, cycle Γ_1 is softly born and is stable further. Fig.4.11 presents the maximum multiplier ρ_1 of cycle Γ_1 as a function of parameter m obtained by calculation. Curve $\rho_1(m)$ is not simple and has the shape of a loop with peculiarities at points B, D, F, C and E where the multiplier modulus is a unity. At point B ($m_B \simeq 1.02$), cycle Γ_1 undergoes the doubling bifurcation and further, up to point D ($m_D \simeq 1.257$), remains saddle. At point D, the multiplier enters the unit circle through -1, and cycle Γ_1 acquires its stability again. At point F ($m_F \simeq 1.267$) cycle Γ_1 vanishes by merging with the unstable cycle Γ_1'' as a result of bifurcation $\rho_1 = +1$. A window of cycle Γ_1 stability appears which has the width in parameter $m_F - m_D \simeq 0.01$. Then, at point C ($m_C \simeq 1.101$), if one increases the parameter m, a pair of cycles is born, stable Γ_1' and unstable Γ_1''. The second window of stability appears since cycle Γ_1' is stable within the interval in parameter $m_E - m_C \simeq 0.22$. Then, cycle Γ_1' loses its stability at point E ($m_E \simeq 1.32$) via doubling, and when increasing m, the cascade of Feigenbaum doubling bifurcations becomes possible as applied to the family of cycles Γ_1'[4]. If we calculate the dependence $\rho_1(m)$ for cycle Γ_1', then it will qualitatively follow the loop shown in Fig.4.11.

---

[4]The investigation of doubling bifurcations of cycle Γ_1' in different sections along the parameter g has shown the realization of the cascade of Feigenbaum doulings to start at a certain $g < g_0$. The number of doublings is finite for $g > g_0$.

Abrupt changes of oscillation regimes (catastrophes) and intricate hysteresis phenomena are observed at the boundaries of stability windows. Let, for example, the initial conditions correspond to cycle Γ_1 in the region of its stability between points D and F. When increasing m, the cycle disappears at the point F, and the jump of a representative point onto the stable cycle Γ_1' would be probable. With decreasing parameter m, the stable cycle Γ_1' exists everywhere up to the critical point C. Due to abrupt bifurcation it is difficult to predict what new regime will appear when the old one vanishes behind point C. The realization of one of the multi-periodic cycles of the system or of a strange attractor is not improbable since there are no stable cycles with periods of $\simeq 2 \cdot T_0$ below the point C in the parameter m.

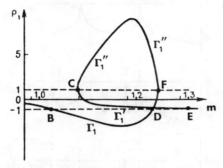

Fig.4.11. $2T_0$-cycle's largest multiplier ρ_1 as a function of parameter m at $g = 0.3$.

Investigate the effect of *abrupt bifurcations* near points C and F (Fig.4.11) in detail. At the point $m_C = 1.10112$, system (3.38) was integrated for times $\tau_0 \geq 10^3$, and the power spectrum, autocorrelation and distribution functions, the entire LCE spectrum and attractor dimension were calculated. Then, parameter m was slightly decreased while retaining a point on cycle Γ_1' as initial conditions, and the above calculations were performed again. The following was revealed. With a small offset from point C ($m = 1.1010 < m_C$), a long *laminar oscillation phase* was observed nearly repeating the motion on cycle Γ_1'. However, the system has gone over to the regime of a strange attractor localized in the system's phase space in such a manner that it comprised no vicinity of the fixed point of cycle Γ_1' in the Poincaré map. The long-duration of the laminar phase was decreased when moving along the parameter away from point C, however, the phenomenon of intermittency, i.e., that of an accidentally repeating return of a trajectory to the vicinity of cycle Γ_1', was not recorded! The non-stationary character of the prolonged process of transition to chaos nearby the point C lead to smearing the spectrum lines of initial cycle Γ_1', and a continuous spectrum of the Feigenbaum attractor appeared when moving away from the point of tangency bifurcation. The calculation results are illustrated in Fig.4.12. The spectrum lines nf_1' of stable cycle

Γ'_1 are gradually broadened under motion away from point C. As a consequence, new lines nf_1 appear which rigorously correspond to Γ_1, and the transition to an attractor is performed that appears as a result of doubling sequence at the critical point $m^* = 1.085...$ In the vicinity of point F, the picture is similar in the whole, but the switching is performed from the vanishing cycle Γ_1 to the stable cycle Γ'_1. Only abrupt transitions either to chaos or to the stable periodic motion proceed at the bifurcation points C and F with no intermittency. This is a radical finding which must be rigorously explained.

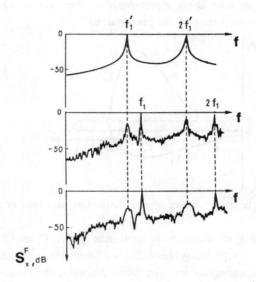

Fig.4.12. Oscillation power spectra with departing along the parameter from the point of tangency bifurcation.

How does a true intermittency appear? In the Poincaré map, the fixed points of saddle and node (or focus) types correspond to the unstable cycle Γ''_1 and the stable cycle Γ_1, respectively. Before their merging unstable saddle separatrixes may not intersect with the stable ones. The intermittency is realized if the instant of saddle-node bifurcation is preceded by formation of a homoclinics in the vicinity of the above points. As the saddle-node is merging and vanishing, the homoclinic structure becomes attractive, and a representing trajectory in map returns randomly in time to the vicinity of disappearing stable point. In the dynamical system (3.38), the above bifurcational mechanism is not realized! This is verified by numerical experiments. We apply to Fig.4.6a again and center on the parameter values' region close to $m = 1.15$ where a stability window exists. Here, depending on initial conditions, stable periodic regimes with different structures are conceivable.

Choose a period-four cycle and denote it Γ. Cycle Γ is stable at the point of the parameter plane for which $m = 1.1520$ and $g = 0.3$. Its multipliers are $\rho_1 = 0.3890$ and $\rho_2 = 0.0228$. The projection of cycle Γ onto xy-plane is shown in Fig.4.13. At the point with $m = 1.15231$, $g = 0.3$, multiplier ρ_1 becomes $+1$. This point is identical to point F in Fig.4.11, cycle Γ disappears at it by merging with the unstable cycle. Vary parameter

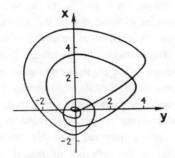

Fig.4.13. Stable $4T_0$-cycle of system (3.38) in the overcritical region of parameter values ($m = 1.152$, $g = 0.3$).

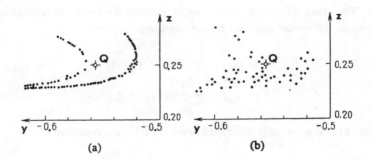

Fig.4.14. The Poincaré sections at the abrupt switching to a chaotic attractor resulting from bifurcation of $+1$; a: fluctuations are absent, b: with fluctuation intensity $D = 10^{-3}$.

m up to the point with $m = 1.155$ and integrate the system (3.38) with the initial conditions on cycle Γ, eliminate the long relaxation time $\tau \geq 10^3$, and construct the Poincaré section in the plane $x = 0$ for the regime established. The results are indicated in Fig.4.14a. They are indicative of an abrupt switching from cycle Γ to the strange attractor. There is no intermittency as demonstrated by the relative position of the attractor and the fixed point of cycle Γ denoted by Q. Point Q is isolated, i.e., its vicinity is not involved in the attractor. If one constructs the one-dimensional return function $y_{n+4} = \Phi(y_n)$ using the section calculation results (Fig.4.14a), then the plot of a model map turns out to be typical for tangency bifurcation in the local vicinity of point Q. The non-local map properties do not provide, however, the trajectory return to the point Q vicinity thereby eliminating the intermittency!

The effects of switching from one regime to another, that exhibit the character of a catastrophe, are manifested, under computation, with a high accuracy degree in the intervals of the parameter values $\Delta m \simeq 10^{-3}$. In a radiophysical oscillator, the controlled variation of parameter m is feasible with the accuracy of 1%, i.e., rougher by an order. Thus, the detailed investigation of switching effects is obviously excluded in the full-scale experiment. Nevertheless, the switching, hysteresis and intermittency phenomena are confidently observed in the physical experiment. How is this to explain?

There are at least two reasons of possible disagreement between experimental and calculated data. The former resides in mathematical model's incompatibility with a real oscillator at the parameter values lying in the supercritical region. The numerical and full-scale experiments reject such a proposal and this is illustrated by the whole following description. The latter is the influence of natural and technical fluctuations on oscillation processes in the oscillator. It is of principle and deserves particular attention[5].

The effect of natural fluctuation is modeled by using the appropriate system of *stochastic differential equations* for the oscillator:

$$\begin{aligned} \dot{x} &= mx + y - xz + \xi_1(\tau), \\ \dot{y} &= -x + \xi_2(\tau), \\ \dot{z} &= -gz + gI(x)x^2 + \xi_3(\tau), \end{aligned} \qquad (4.1)$$

where $\xi_i(\tau)$ is a normally distributed noise satisfying the conditions

$$<\xi_i(\tau)> = 0, \quad <\xi_i(\tau)\xi_j(\tau')> = \begin{cases} 0, i \neq j, \\ D\delta(\tau - \tau'), i = j \end{cases} \qquad (4.2)$$

and D is the noise intensity.

We investigate oscillation processes in the system in the presence of additive noise on the assumption the external influence to be a small perturbation of initial equations (3.38). The dependence $\lambda_i(m)$ for $g = 0.3$ is to be calculated by varying noise intensity D. Fig.4.6 corresponds to the case of $D = 0$. The growth of the noise intensity leads to the successive vanishing of the smallest in parameter stability windows of multi-periodic cycles. As $D \simeq 10^{-2}$, the system attractor becomes *as if hyperbolic*, i.e., it comprises no stable periodic trajectories in a certain finite region of

---

[5]The problems of fluctuation role in the chaotic system's dynamics will be considered in detail in Chapters 10, 11.

parameter values Δm. The critical point of stochasticity birth is shifted towards lower parameter values due to the noise influence on cycles with period $2^k T_0$ starting at a certain $k > k_0$.

The system's response to the *additive* external normal *noise* was found to be principally non-linear in regimes being close in parameters to the points of tangency bifurcations. It depends on intensity D and on the distance in parameter from the bifurcation point. Near the critical point, the duration of nearly periodic oscillations (the laminar phase) increases with the growth of intensity D, and the quantity λ_1 is decreased. The movement away from the critical point gives an opposite result, i.e., the positive exponent λ_1 grows with the increase of noise intensity D, and the time of switching to another stable regime decreases. The results of calculating the influence of additive fluctuations on the system dynamics are illustrated in Fig.4.15. Thus, the effect of external noise on system (3.38) near the point of tangency bifurcation is manifested in that the characteristic relaxation times for a new stable regime depend complexly on the fluctuation intensity and parameter values. The intermittent stochasticity phenomenon is not induced, however, *by the external additive noise* [4.1,4.2].

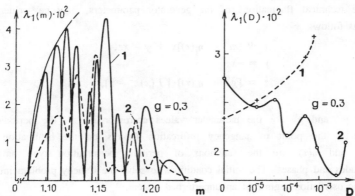

Fig.4.15. Noise effect on motion chaoticity degree; a - $\lambda_1(m)$ for $D = 0$ (curve 1) and $D = 10^{-2}$ (curve 2), b - $\lambda_1(D)$ for $m = 1.1005$ (curve 1) and $m = 1.1010$ (curve 2); $m^* = 1.10112$.

We refer to Fig.4.14b where the Poincaré section is shown for the parameter values given in Fig.4.14a, but as the additive noise with an intensity of $D = 10^{-3}$ is introduced. Being close to the one-dimensional band in the absence of noise, the attractor band "swells" as affected by fluctuations, becomes principally two-dimensional. But the representative points fill the region near an unperturbed attractor essentially with the equal probability and do not concentrate in the vicinity of point Q.

Cycle Γ (Fig.4.13) is concerned again. For the parameter values $m \leq m^0 = 1.15231$, the cycle is stable. The increase of parameter $m \geq m^0$ implies the abrupt system's switching to the SA regime (Fig.4.14a). The system will generate either periodic or stochastic oscillations if parameter m is slowly varied as compared with the basic oscillation period of the oscillator $T_0 \simeq 2\pi$ within the small limits of $\pm (0.005 \div 0.010)$. Under the random parameter variations, the oscillations will exhibit the laminar phases (motion on the limit cycle Γ) that are irregularly interrupted in time by the turbulent phases (motion on the SA with the parameter values in excess of the critical one). The time realization of the oscillation process $x(\tau)$ will not differ from that being typical for intermittency. However, this process is not self-oscillating, but *a modulating process, induced by the system parameter fluctuations*! Unlike the systems of Morse-Smale type, where noise leads to a broader spectrum line of oscillation, the intermittent chaos can be induced by the multiplicative noise in the systems which exhibit stochastic behavior near the multiplicity lines!

The mathematical model, describing the process of *modulation intermittency* induced by the technical fluctuations of the generator parameters, may be written in general form as follows

$$\dot{x} = [m^0 + \eta_1(\tau)]x + y - xz,$$
$$\dot{y} = -x, \qquad (4.3)$$
$$\dot{z} = [g^0 + \eta_2(\tau)] \cdot [I(x)x^2 - z],$$

where m^0 and g^0 are the parameter values corresponding to the periodic oscillation regime at the point of tangency bifurcation (points on the bifurcation multiplicity lines), and $\eta(\tau)$ are the functions of periodic perturbation type having random amplitudes and phases. The initial conditions to solve the Cauchy problem must belong to periodic oscillation regime of an unperturbed system.

Carry out the numerical experiment with system (4.3) by specifying $\eta_1(\tau) = \xi(\tau) \cdot sin(p\tau)$, $\eta_2(\tau) \equiv 0$, considering $\xi(\tau)$ to be a white noise with intensity D, and assuming $p = 0.01$. Fig.4.16 presents the calculation data for the Poincaré section and power spectrum of oscillation regime under the above conditions. It is seen that the representative points in the section are concentrated in the vicinity of point Q testifying to the trajectory return (laminar oscillation phases). At the same time, in the Poincaré section, the representative points are present in the vicinity of chaotic band (see Fig. 4.14a as $D = 0$) testifying to the *"turbulent" motion phases*. The spectrum of a regime, being discrete in the absence of noise, acquires the characteristic form which represents the process of modulation intermittency.

Now, look at the physical experiment. The stable periodic regime in the vicinity of the bifurcation point $\rho_1 = +1$, $m \simeq 1.15$ and $g = 0.3$ is shown in Fig. 4.17 (photo 1). In photo 3, a chaotic attractor is presented to which the system is switched as parameter m increases to a value of $\simeq 1.16$. In the second photo, the phase trajectory projection onto xy-plane is presented in the regime of intermittency realized by specifying the intermediate parameter value of $1.15 < m < 1.16$. As seen from the comparison of calculated and experimental data, just the modulation intermittency is realized in the physical experiment with the regularities in the character of the process spectra being typical.

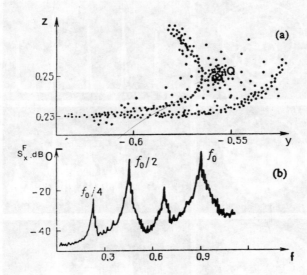

Fig.4.16. The Poincaré section (*a*) and oscillation power spectra (*b*) for by modulation intermittency ($m = 1.1522$, $g = 0.3$, $D = 0.01$, $p = 0.01$).

In the course of experiments, the phenomenon of modulation intermittency near the multiplicity line of a multi-periodic cycle was confidently recorded for the period-three, -four, and -five cycles. The analysis of truncated realizations of time processes by using a storage oscillograph has proved the random character of appearance and duration of the laminar phases, of oscillations interrupted by turbulent motion on the chaotic attractor.

The presence of folds at surface S^0 (Fig. 3.12) whose number increases with the growth of the "periodicity" of periodic oscillations, as well as the period doubling bifurcation, lead to complex, unpredictable hysteresis phenomena whose basic reason is the cusp catastrophe. The narrowness of our imagination, being unable to clearly fancy the picture of the infinite branching of the infinite number of periodic solutions, results at first in the sensation of hopelessness and doubt about our capability to probe the mysteries of the dynamics of the simplest system in R^3 having two parameters. What is to anticipate in case of the increased phase space dimension and the number of the control parameters of a dynamical system? A. Poincaré found himself in similar situation at the beginning of century when he had come to recognize that a homoclinic structure appearing due to the perturbation influence on the separatrix loop is unable to be visualized. Such moods must be rigorously rejected if one aims at the study of systems exhibiting stochastic behavior for which the above phenomena are typical.

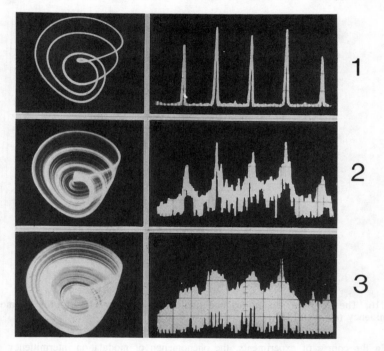

Fig.4.17. Modulation intermittency in physical experiment. 1 - stable $4T_0$- cycle near the point $\rho = +1$, 2 - intermittency, 3 - chaotic attractor.

4.4. INTERACTION OF CHAOTIC ATTRACTORS. INTERMITTENCY OF "CHAOS-CHAOS" TYPE

The infinite number of attractors, with different structures is realized in *quasihyperbolic systems* with the chaotic dynamics depending on parameters and initial conditions. The recording of a control parameter does not change the picture: the infinite number of attractors holds and their basins of attraction are separated in the phase space by separatrix surfaces. The problem of describing the dynamics of a quasihyperbolic system becomes highly complicated due to many regular regimes coexisting along with the chaotic attractors in the phase space.

The modulation intermittency is a typical phenomenon for quasihyperbolic systems considered in section 4.3. Conditions for its realization are, first, the coexistence of only two stable (regular and chaotic) regimes close in parameters and in the phase space, and, second, parameter fluctuations. In the present paragraph, consider just one more phenomenon, namely, the interaction between chaotic attractors leading to the *intermittency of "chaos-chaos" type* [4.8-4.10]. A distinct feature of this mechanism is its purely dynamical character unrelated to the external system's perturbation.

We perform a numerical experiment using the dynamical system (3.38) in the section of the parameter plane $g = 0.097$ to study the oscillation regimes when varying the parameter m. The basic (initial) limit cycle of the system with period T_0 evolves, as a result of accumulation of period doubling bifurcations, to a strange attractor (denote it SA_1) that is softly born at the critical point $m^* = 2.39$. This attractor was considered in detail in sections $g = 0.2$ and 0.3 above. In the supercritical region $m \geq 2.39$, the attractor SA_1 evolves according to the regularities discussed earlier.

At the bifurcation point $m = 2.31$, a pair of period-three cycles, saddle and stable ones, is born abruptly as a result of bifurcation $\rho = +1^6$. The stable period-three cycle undergoes as well the cascade of period doubling bifurcations with the growth of parameter m which is completed by the birth of attractor SA_3 at the critical point $m_3^* = 2.35$. While the parameter m grew to the values being lower than a certain critical one $2.44 < m_0^* < 2.45$, both SA_3 and SA_1 demonstrate typical regularities, including the appearance of the stability windows of multi-periodic cycles of different families. Both strange attractors, SA_1 and SA_3, exist independently in the interval of parameter $2.40 < m < m_0^*$ with the basins of attraction in the phase space separated by a complicated two-dimensional separatrix surface.

The calculation results for the Poincaré section of system (3.38) in plane $y=0$ and for the appropriate spectra are presented in Fig.4.18. Sections (a) and (b) correspond to the value of parameter $m=2.41$ with the initial conditions on attractors SA_1 and SA_3, respectively. It is seen from the location of points on the Poincaré section that there exist two independent attractors localized in different phase space regions of system (3.38). Attractors SA_1 and SA_3 maintain their structures and statistical properties under a small variation of control parameters and the influence of small noise.

The attractors begin to interact as parameter m grows. The separatrix surface,

---

[6]Depending on m and initial data, several types of period-three oscillations appear abruptly in the section $g = 0.097$. The first family of period-three cycles is born as $m = 1.25$, and evolves to chaos via the doublings in the critical point m* = 1.413. A chaotic attractor based on period-three cycles and a stable cycle with period T_0 exist simultaneously in different phase space regions in the interval of $1.4 < m < 1.5$.

separating SA_1 and SA_3 upon reaching a certain threshold m_0^*, is collapsed and the attractors merge. A merged attractor SA_0 is formed that comprises both SA_1 and SA_3. In Fig.4.18,c, a stochastic set is shown in the secant plane for the parameter value of $m = 2.45 > m_0^*$ involving attractors $SA_1(a)$ and $SA_3(b)$ if one takes into account their minor evolution with the growth of parameter m. The merged attractor is realized independently of whether the initial conditions belong to the SA_1 or SA_3 basins of attraction, or not, showing the separatrix surface to collapse. The merging of the SA_1 and SA_3 basins of attraction seems to take place under formation of the merged attractor SA_0.

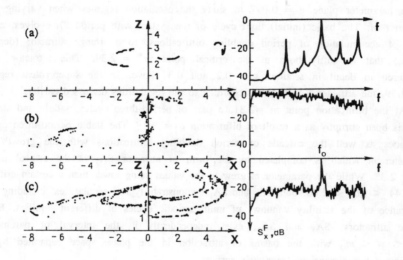

Fig.4.18. The Poincaré section (a) and power spectra of intermittency "chaos-chaos".

The detailed study of time realizations, autocorrelation, power spectrum and integral intensity of oscillations in the SA_1, SA_3 and SA_0 regimes shows that, whilst the parameter m passes the critical point of merging of attractors m_0^*, the system first spends a longer time on SA_1 continuously. The switching to SA_3 proceeds irregularly in time and has the character of *"turbulent"* splashes on the background of relatively more prolonged "laminar" motion phase on SA_1 with the energy being remarkably lower than that of SA_3. The switching to attractor SA_3 can be determined by the time realization $x(\tau)$ when a negative splash appears in dependence $x(\tau)$ with the amplitude exceeding the maximum possible ones when moving on SA_1. The statistical analysis of the relative residence time of the system in the states of "turbulent" and "laminar" phases in the merged attractor regime has shown that for small threshold exceeding, the dependence of the mean residence time of the system on SA_1 is evaluated approximately in the form of a ratio being valid for the "cycle-chaos" intermittency [4.10]:

$$\langle \tau_1 \rangle = c \, (m - m_0^*)^{-1/2}, \quad c = 18.0, \quad m_0^* = 2.445 \, . \tag{4.4}$$

The duration of laminar oscillation phase (here, it corresponds to a chaos regime which is structurally close to SA_1) decreases with the growth of supercriticality m having the critical index of -1/2 while the mean duration of turbulent splashes (motion on SA_3) increases, respectively. An obvious analogy to the "cycle-chaos" intermittency regime allows one to call the phenomenon described the *intermittency of "chaos-chaos" type*.

As supercriticality grows, turbulent splashes occur more often leading to the increase of SA_0 mean energy. The intensity of selfoscillation is remarkably built up and its power spectrum becomes smoother evolving simultaneously to lower frequencies. The autocorrelation function of oscillation process seeks to be δ-shaped demonstrating a strong intermixing.

The averaged power spectrum and autocorrelation function of the merged attractor SA_0 are indicated in Fig.4.19. They were calculated using realization $x(\tau)$ for $m = 2.55$. The turbulent splashes proceed within relatively larger time intervals on the average. This is the reason for the power spectrum to evolve to the region of lower frequencies.

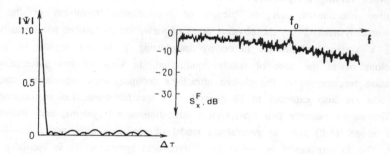

Fig.4.19. Autocorrelation function and power spectrum of the developed attractor SA_0 ($m = 2.55$, $g = 0.097$).

The merged attractor does not collapse as affected by an additive δ-correlated noise. The noise effect resulted, under the numerical simulation of a process by stochastic equations (4.1), in variation of the particular time moments of the system's switching from SA_1 to SA_3. However, the mean statistical characteristics of SA_0 were practically not altered. A more intricate type of "chaos - cycle - chaos" intermittency can be induced depending on the intensity of noise influence on SA_0. Such a selfoscillation process randomly includes the realization segments belonging to SA_1, SA_3

and to one or several multi-periodic cycles.

The results presented here may be treated, particularly, as an experimental proof of the fact that the branching of chaotic solutions of a system of non-linear differential equations is possible when varying the control parameters.

4.5. DISSIPATIVE NONLINEARITY INFLUENCE ON ATTRACTOR BIFURCATIONS

In low dimensional systems chaotic oscillations take place, particularly, in the models that have dissipative terms in the equations defining the mode damping. The self-oscillations of such systems are due, as a rule, to parametric excitation and to the non-linear inertial energy transfer into the damping mode. The refinement of models for real systems to be described more correctly often requires consideration of *non-linear dissipative components* in the right-hand parts of initial differential equations[7]. As an example the system of equations (3.28) can serve which presents the generalization of oscillator equations (3.38) for the case of a non-linear slope of the basis oscillator.

As it was already discussed, the study of the mechanisms of stochasticity evolution in dynamical systems requires the bifurcation phenomena to be investigated that take place when varying the parameter. It seems to be of interest to define the influence of dissipative non-linearity on the picture of bifurcational transitions in the three-dimensional systems. From the physical point of view, the dissipative non-linearity must act as a system stabilization factor by transferring it to the regime of periodic oscillations or to the state of saddle equilibrium. In view of this assumption, the bifurcation mechanisms of the chaotic attractor's collapse when introducing non-linear dissipation are also expected to be general in a specific sense. Let us illustrate this by taking as an example two most typical three-dimensional systems, and, namely, the Lorenz model (3.8) and the generalized model of oscillator with inertial non-linearity (3.28). The Lorenz model is typical for a dynamical system which is inertially driven [4.11] where the strange attractor regime may abruptly arise. The model of oscillator with inertial non-linearity is an example of a system with a soft excitation of quasihyperbolic chaos as a result of the cascade of period doubling bifurcations.

The classical Lorenz model (3.8), describing the three-mode convection of viscous fluid, does not correspond to the real convection with the great Prandtl number ($\sigma > 1+b$), but remains of interest for physicists (laser equations), mechanics and radio-physicists (inertial, parametrically driven, non-linear oscillator (3.11)), as well as from the viewpoint of rigorous mathematics, as a system with a structurally unstable

---

[7]The non-linearities, favoring the additional contraction of a phase flow, will be called dissipative.

hyperbolic attractor. Leave aside the detailed discussion of the consistency between the mathematical model and the real system and consider the following equation:

$$\begin{aligned}\dot{x} &= -\sigma(x - y), \\ \dot{y} &= rx - y - xz, \\ \dot{z} &= -bz - dz^3 + xy.\end{aligned} \quad (4.5)$$

In the third equation of system (4.5), the term $-dz^3$ is added regarding, in the first approximation, the non-linear character of dissipation that can be due, for example, to the non-linear thermal conductivity. The divergence of the velocity vector field (4.5) depends on coordinates

$$divF = -(\sigma + 1 + b + 3dz^2) < 0. \quad (4.6)$$

The classical model (3.8) bifurcations have been much studied [4.12]. For the traditional values $\sigma = 10$, $b = 8/3$ as $1 < r < r_1 = 13.92$, there is a saddle point at the origin and two stable foci O_1 and O_2. The separatrixes become double-asymptotic to the saddle at the origin as $r = r_1$, and the transition through the bifurcation point $r = r_1$ is accompanied by the birth of a pair of saddle periodic motions. The only attracting set of the system are foci O_1 and O_2 in the interval of $r_1 < r < r_2 = 24.06$. The Lorenz attractor appears in the interval of $r_2 < r < r_3 = 24.74$, but the stable foci still remain. Three attractors are realized in this region of parameter r values depending on the initial conditions, but the only one, the Lorenz attractor, holds as $r \geq r_3$.

The results of calculating the typical bifurcation lines on the plane of parameters r and d of system (4.5) are presented in Fig.4.20 for fixed $\sigma = 10$ and $b = 8/3$ [4.13]. Line l_1 was calculated using the condition of the separatrix return into the saddle at the origin. System (4.5) was integrated with recording the coordinates x and y in the secant plane specified by the equation $\dot{z} = 0$ with the initial conditions on the separatrix determined from the solution of the equations linearized in the vicinity of zero. With the parameter having reached the bifurcational value, the separatrix returned to the saddle. This was illustrated by coordinates x and y simultaneously changing their signs due to the separatrix outgoing into the second focus' region. The influence of non-linear dissipation on this bifurcation seems practically to escape for values of d in the interval $0 < d < 3 \cdot 10^{-4}$. This is explained by the coordinate z exceeding the quantity $(r - 1)$ only slightly under a bifurcation, and the term $3dz^2$ in (4.6) being negligible.

The effect becomes detectable for $d > 10^{-3}$. The bifurcation line l_1, corresponding to the regime of separatrix "winding" around the unstable cycle, is featured by the complicated non-linear dependence $r(d)$, as it is seen from Fig.4.20. Calculation was performed by determining the values of parameters d_1 and d_2 (or r_1 and r_2) that lye at different sides from line l_2. With this, the separatrixes behave differently in principle, either randomly or converging into foci. Then, the difference Δd (or Δr) was smoothly decreased to a small value depending on the calculation accuracy of the bifurcation line l_2.

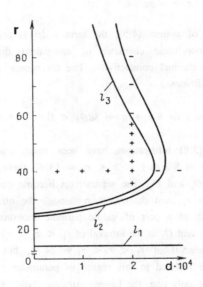

Fig.4.20. Bifurcation diagram on the parameter r and d plane of system (4.5) for $\sigma = 10$, $b = 8/3$.

Line l_3 is the bifurcation line of the unstable cycle vanishing when only the Lorenz attractor remains in the system. This line was computed by linearizing the vector field at the equilibrium point O_1 this way:

$$x = y = [b(r-1) + d(r-1)^3]^{1/2}, \quad z = r - 1 \qquad (4.7)$$

and by finding the conditions under which the following characteristic polynomial

$$\begin{aligned}&3d^2(r-1)^5 + 9(\sigma+1)d^2(r-1)^4 + d(1-\sigma+4b)(r-1)^3 + \\&+ 3(\sigma+1)(\sigma+2b+1)d(r-1)^2 + b(1+\sigma+b)(r-1) + \\&+ (\sigma+1)(1+\sigma+b)b = 0\end{aligned} \qquad (4.8)$$

has two pure imaginary roots.

The algebraic equation (4.8) specifies the desired dependence $r(d)$ at fixed σ and

b, i.e., the bifurcation line l_3. The behavior of the system at the boundary of the stability region requires consideration of the non-linear terms of equations (4.5), calculation and definition of the signs of Lyapunov quantities. The algorithm for calculating the first Lyapunov quantity L_1 was developed for three-dimensional systems by N.N. Bautin and is described in Ref. [1.15]. The first Lyapunov quantity L_1 calculation for system (4.5) has shown L_1 to be positive at the stability region boundary (on line l_3). Thereby, the subcritical character of Andronov-Hopf bifurcation being found for the classical model (3.8), is proved to hold.

The subcritical character of bifurcation does not demonstrate yet the onset of chaos although the system's robustness relative to the introduction into equation (3.8) of a non-linear dissipative component, enables one to assume a strange attractor to be present within the region of parameter values bounded by the bifurcation line l_3 (Fig.4.20). The presence of a positive exponent in the LCE spectrum of the system and, as a consequence, the fractional dimension of attracting set is the chaos criterion. To determine the dimension of the attractor, the whole LCE spectrum of system (4.5) was calculated along the straight lines in the plane of parameters $r = 40$ and $d = 2 \cdot 10^{-4}$ (Fig.4.20). Inside the region bounded by the bifurcation line l_3, the type of the LCE spectrum corresponds to the attractor with the Lyapunov dimension $D_L = 2.065 \pm 5 \cdot 10^{-3}$ and with the value of the positive Lyapunov exponent $0.98 < \lambda_1 < 1.06$. At the point where $r = 40$ and $d = 2 \cdot 10^{-4}$, the LCE spectrum is $+ 1.032, 0.000, - 15.497$ corresponding to the dimension $D_L = 2.0666$.

The exit from the chaos region (transition through the line l_3) is accompanied by bifurcation in the LCE spectrum consisting now of three negative exponents. This is due to the change of the stability character of equilibrium points, and outside the bifurcation line l_3, the system trajectories are attracted to one of them O_1 or O_2. In Fig.4.20, the points with positive metric entropies are marked with the sign "+" ($\lambda_1 > 0$) and those corresponding to stable equilibrium points with "-".

The location and character of bifurcation lines in Fig.4.20 demonstrate visually that the region of the Lorenz attractor existence in the system's parameter space decreases gradually with the growth of dissipative non-linearity d. As a result, chaos disappears making room for a regular attractor being a stable equilibrium. It should be noted, that the Lorenz attractor suffers the collapse while varying the parameter d as a result of a sequence of typical bifurcations that are realized in inverse order to the sequence of critical phenomena leading to an abrupt Lorenz attractor birth in the classical model (3.8).

We consider the influence of dissipative non-linearity on bifurcations in system (3.28). Numerical experiments were performed as follows. For the parameter values $d = 0$, $g = 0.3$, and $m = 0.2$, one was looking for the initial (basic) limit cycle of the system Γ_0 with period $\simeq 2\pi$ and following its multipliers with the growth of parameter d. The above technique was used for the one- and two- parametric analysis of the stability of a family of period-one cycles Γ_0. The calculation results are presented in Fig.4.21. The bifurcation line l_{01} corresponds to the doubling of cycle Γ_0 (the cycle multiplier on line l_{01} is -1), and line l_{02} is the bifurcation multiplicity line. Two independent families of period-two cycles Γ_1 and Γ'_1 exist inside the region bounded by line l_{01} whose bifurcation diagrams quantitatively follow the picture for the period-one cycle Γ_0. The bifurcation pattern is similar to that discussed when two-parametrically analyzing the critical phenomena in system (3.38) in the plane of parameters m and g.

The sequence of doubling bifurcations for each of cycle families leads to the birth of an attractor according to Feigenbaum's law. In Fig.4.21, the line of critical values of parameter l_{cr} is marked, above which one of attractors is born due to the accumulated doubling bifurcations of initial (basic) limit cycle family.

If one moves from the region of stochasticity in the parameter plane (Fig.4.21) along the straight line $m = const$ in the direction of increase of d, then the bifurcation lines of doubling of cycles with periods $2^k T_0$ intersect successively for $k = \infty,...,n,..., 3, 2,$ and 1. A sequence of *bifurcations of "period halving"* takes place. With every bifurcation, the cycle period becomes half as much, and the process is completed by setting the regular oscillations with period T_0 in the system. The increase of the dissipative non-linearity d leads to the chaos collapse in the system during which the sequence of critical phenomena rigorously corresponds to the inverse sequence of bifurcations resulting in the birth of a strange attractor. This is proved by the results of calculating the rate of accumulation of period doubling bifurcations while varying the parameter d which show Feigenbaum's law to be obeyed with the constant $\sigma = 4.66920....$.

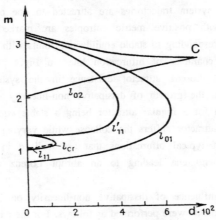

Fig.4.21. Bifurcation diagram on the plane of parameters m and d of system (3.28) for $g = 0.3$.

The chaos collapse with the increase of dissipative non-linearity parameter can be observed in the course of a physical experiment. Due to the non-linearity of the amplifier's characteristics, the degree of dissipative non-linearity influence depends on the location of a point on the curve which can be moved by the externally controlling the fixed shift. The degree of influence of the non-linear segment of curve $S(x)$ can be smoothly varied by changing the shift level and by maintaining the amplification factor of the main oscillator amplifier in a special manner in the physical experiment. This corresponds to the variation of parameter d of the mathematical model (3.28). The experiments have shown a chaotic attractor to collapse with decreasing the negative shift level (increasing the degree of dissipative non-linearity influence). First, the periodicity of the attractor band increases. Then, a period-eight cycle is found followed by the recorded sequence of period "halving" bifurcations which results in the transition to the period-one oscillation regime. The experimental findings are completely similar to those shown in Fig.4.2 if they are considered in the direction of decrease of the parameter m.

The investigation of the dissipative non-linearity influence on the dynamics of particular systems, which exhibit different chaos birth mechanisms, show the regime of chaotic attractor to be collapsed by the dissipative non-linearity. In the Lorenz model, this leads to the transition to a stationary state, and in the oscillator with inertial non-linearity this results in stable period-one oscillations. The bifurcation mechanism of the chaos collapse in the models discussed corresponds rigorously to the mechanisms providing its onset. The above mechanisms are realized, however, in inverse order with the growth of dissipative non-linearity parameter.

CHAPTER 5

QUASIATTRACTOR STRUCTURE AND PROPERTIES AND HOMOCLINIC TRAJECTORIES OF AUTONOMOUS OSCILLATOR

5.1 OSCILLATOR DYNAMICS IN THE VICINITY OF A HOMOCLINIC TRAJECTORY OF SADDLE-FOCUS SEPARATRIX LOOP TYPE

One of the main mathematical difficulties in investigating transitions to chaos in finite-dimensional dynamical systems is a *non-local nature* of these phenomena. The latter gives a principally nonlinear form for evolution equations which are not amenable to exhaustive analysis by traditional analytical methods. The problem can be overcome in different ways. One of them is to reduce the basic aspects of global dynamics to a local task. It is known that transitions to chaos, being global when considered one-parametrically, can be studied as the local ones using the multi-parametric analysis. The above method has its origin in the study of high-codimension (two or more) bifurcational phenomena. The second possibility involves the theoretical analysis of the mechanisms of transitions to chaos as regularities in a complicated behavior of a particular class of systems as a whole. This program is though more attractive, but difficult to perform. Nevertheless, the finding of universal similarity laws, as applied to some classes of one- and two- dimensional maps, raises optimistic hopes.

Homoclinic trajectories, arising in the vicinity of saddle periodic motions or saddle equilibrium points, play a basically important role in understanding the complexity of dynamic phenomena in a system exhibiting quasihyperbolic properties, as it was already mentioned. The homoclinic trajectories (points), as a result of the rough intersection of stable and unstable saddle cycle manifolds (the stable and unstable separatrixes of fixed saddle points), are, in their way a forerunner of a complicated aperiodic motion of the system. This phenomenon was discovered and studied by A.Poincaré, G.Birkhoff and S.Smale. The presence of a denumerable set of stable and unstable periodic trajectories with different periods, including a continuum of trajectories stable after Poisson, in the vicinity of the homoclinic trajectories follows from their availability with some additional assumptions. One would not be largely mistaken to conclude a homoclinic trajectory to inevitably give rise to quasiattractors in its vicinity in the parameter and phase space. Therefore, the demonstration of homoclinic points and trajectories in dynamical systems is certainly a fundamental step in chaos investigation and may be considered as its criterion. Rigorous justification for the problem of homoclinic trajectories existing in general dynamical systems and, thereby, the proof for the presence of denumerable number of cycles is the task not solved for the present. Due to this, we resort to the help of computer simulation more often in particular cases.

A versatile experimental analysis of the origin mechanisms and topological structure of chaotic attracting sets in a modified oscillator with inertial nonlinearity results in an idea about a homoclinic trajectory of the type of *equilibrium state separatrix loop* to exist in an autonomous dynamical system. It is not difficult, however, to show the saddle-focus separatrix loop *not to be realized* at zero of coordinates in system (3.38).

Really, a singular point of the system is defined by two-dimensional unstable and one-dimensional stable manifolds. Let us reverse the time in (3.38) and specify the initial conditions $x(0) = y(0) = 0$, $z(0) > 0$ for the unstable one-dimensional manifold. System integration will confirm the drift of trajectory into infinity along the z-axis. It follows from equations (3.38) that $z(\tau) = z(0) \cdot exp(g\tau)$. As $\tau \to \infty$, the trajectory *does not return* to the singular point!

We carry out another experiment where the system (3.38) is integrated in direct time with initial conditions on the chaotic attractor and move in the plane of parameters to the region of the most developed chaos while decreasing parameter g and increasing m (Fig.3.8). When integrating for greater times, the quantity R_{min} is recorded, i.e., the shortest distance of the phase point from the stable one-dimensional manifold of the system (z-axis). Calculation results are as follows [5.1]:

$m = 1.16$, $g = 0.3 - R_{min} = 0.13$;
$m = 1.45$, $g = 0.21 - R_{min} = 0.53 \cdot 10^{-1}$;
$m = 2.45$, $g = 0.1 - R_{min} = 0.28 \cdot 10^{-1}$, and finally
$m = 2.8$, $g = 0.06 - R_{min} = 0.46 \cdot 10^{-2}$;

A phase trajectory in the regime of developed chaos, that is rigorously not tangent to the z-axis, comes to it very closely, however. With this, time-nonregular return of the trajectory to the close vicinity of singular point at zero is observed. There is no saddle-focus loop here, but the situation is highly similar!

A hypothesis naturally arises: the saddle-focus *separatrix loop exists* in a *certain perturbed* system and causes it to behave chaotically. With no perturbation the loop vanishes itself, but the structure of the phase space partition into trajectories is maintained. To strengthen these considerations, one has to demonstrate the presence of the saddle-focus separatrix loop in the system in order to ascertain the structures of attractors, and to study their evolution as perturbation is removed. The solution of the above task is not unique. but the particular form of small perturbation is to be of little importance, due to the robustness.

Adding to the second equation of system (3.38) a constant positive term γ for the system perturbed [5.1] yields:

$$\begin{aligned}\dot{x} &= mx + y - xz, \\ \dot{y} &= -x + \gamma, \\ \dot{z} &= -gz + gI(x)x^2.\end{aligned} \qquad (5.1)$$

A singular point of the flow (5.1) is the only one as before. It is slightly shifted relative to the origin and represents the saddle-focus. Its coordinates are $x^0 = \gamma$, $y^0 = \gamma(\gamma^2 - m)$, $z^0 = \gamma^2$. Like the initial system, the perturbed system (5.1), for $m > 0$, is defined by two-dimensional unstable and one-dimensional stable manifolds. To find the loop Γ_0^1 in the system equations, we reverse the time and solve repeatedly the Cauchy's problem with initial conditions on the unstable one-dimensional manifold for the fixed $g = 0.3$ and various m and γ. Having selected a small value of $\gamma = 0.1$, we find the bifurcation point $m^* = 1.176...$, where the *one-rotational saddle-focus loop* Γ_0^1 *is realized*. The three-dimensional map of a double-asymptotic trajectory Γ_0^1 and of its projection onto two planes are indicated in Fig.5.1. The trajectory Γ_0^1 is structurally

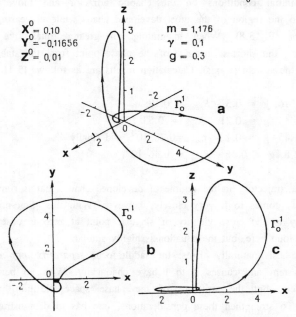

Fig.5.1. Separatrix loop of saddle-focus in the system (5.1) - *a*, its projections on selected planes - *b,c*.

unstable and the loop is destroyed towards A as $m > m^*$ and towards B as $m < m^*$ according to designations in Fig.5.1.

Let us perform numerically the two-parametric analysis of bifurcational phenomena in the system (5.1) having constructed the basic bifurcation lines of codimension one in the plane of parameters m and g for the fixed value of $\gamma = 0.1$. The bifurcation diagram of the system is shown in Fig.5.2 where the Andronov-Hopf bifurcation line l_{00}, period doubling and multiplicity lines l_{01} and l_{02} are presented for the initial (basic) family of cycles. The location of the above bifurcation lines in the parameter plane does not differ qualitatively from the appropriate diagram of system (3.38). There are distinctions in the deflection of the birth bifurcation line of cycles l_{00} from the straight line $m = 0$, in a certain increase of bifurcational values of parameter m with doubling and widening of the region of doublings in parameter g (cf. the diagram in Fig.3.8). The relative location of line of critical values of parameter l_{cr} is also maintained on the diagram.

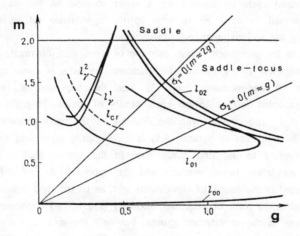

Fig.5.2. Bifurcation diagram of the perturbed system (5.1) for $\gamma = 0.1$.

In addition, a system of bifurcation lines is placed on the diagram that characterizes the critical phenomena with respect to equilibrium state. The line l_γ^1 corresponds to the condition of existence of double-asymptotic trajectory Γ_0^1 in (5.1); the line $m \cong 2$ separates the regions where the singular point of the system is a saddle-focus ($m < 2$) and a saddle ($m > 2$); the lines where the saddle equilibrium state quantities σ_1 and σ_2 (1.54) become zero are also indicated. Calculations show that, on the bifurcation line l_γ^1, the saddle value σ_1 of singular point of system (5.1), fixed in reverse time according to *Shilnikov's theorem*, differs from zero and is negative. The loop is *not dangerous*! Hence, the only stable cycle is to be born under its destruction towards A which corresponds to crossing the line l_γ^1 while increasing the parameter m. This is proved by calculations. The cycle is born with very small parameter deviations from the line l_γ^1 towards A and undergoes doubling bifurcations in a narrow region of parameters. In the initial system (5.1) (in direct time), the cycle, born from the separatrix loop, is *absolutely unstable* and evolves in parameter into a *saddle cycle with doubled period*. Fig.5.2 comprises one more bifurcation line l_γ^2 corresponding to the presence of a more complicated *two-rotational separatrix loop of saddle-focus* Γ_0^2. On this line, the unstable manifold of singular point, before it closes, makes two rotations in the vicinity of the main loop Γ_0^1. Similar phenomena are met more and more often when solving specific problems. The bifurcation line l_γ^2 begins at a bifurcation point Q of codimension two and goes to the region where the singular point of the system is an *unstable saddle-focus*. It is extremely difficult to analyze numerically the problem of multi-rotational separatrix Γ_0^k loops and critical phenomena in their vicinity due to an infinite set of bifurcations realized in a narrow region of parameter values. Abrupt birth (death) cycle bifurcations are a major obstacle on this way. Some set of loops Γ_0^k is assumed to exist giving birth to the appropriate set of multi-rotational cycles undergoing the various cascades of bifurcations.

According to the parameters' space partition of the system obtained, let us examine the oscillation regimes. If we increase the parameter m as $g = const$ while remaining below the line of one-rotational loop l_γ^1, then the sequence of period doubling bifurcations is recorded relative to the basic family of cycles. Feigenbaum attractor is born on the line l_{cr} and evolves with the decrease of the band periodicity, crisis and intermittency. The dynamics of system (5.1) is topologically equivalent to the dynamics considered as applied to the first chaos zone in the system (3.38). Moreover, the appearance of oscillation phase portraits and the spectrum do not differ practically from those obtained in the course of experiments with an unperturbed system. Two chaotic zones were found in the plane of parameters m and g (Fig.4.5) by concerning with chaotic oscillation regimes in the unperturbed system, but only the structures and properties of attractors in the former (lower) zone were analyzed in detail. This was done deliberately since it was impossible to assess the mechanisms of the onset of chaos in the latter zone without studying the role of homoclinic trajectories.

Now we discuss the dynamics of system (5.1) in the vicinity of separatrix loop Γ_0^1 being destroyed. Under a small deviation of parameter values from the line l_γ^1 downwards, integration with initial conditions on an unstable cycle, being born from the loop, gives the evidence on an abrupt system switching to the regime of period-two band of Feigenbaum's attractor. No specific attracting hyperbolic set of trajectories was revealed in the vicinity of loop Γ_0^1. In Fig.5.3a, the typical one-dimensional return map of such an attractor is presented that has the well-known form of a smooth square-law parabola. As the value of parameter m increases (removing upwards from the line l_γ^1), attractors appear which have somewhat different structure. The form of attracting sets themselves becomes complicated and the phase trajectories are less regular. Random oscillation phase failures (the random nature of a time sequence for trajectories' return to the secant plane) are recorded besides the chaotic amplitude modulation. The smoothness of model return maps is disturbed and discontinuities appear. As a rule, stronger mixing and Lyapunov dimension growth are observed.

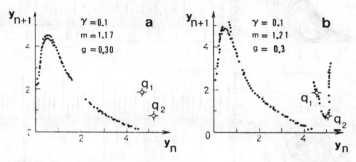

Fig.5.3. Maps modeling the quasiattractor of system (5.1), which corresponds to the vicinity of loop Γ_0^1 in the parameter regions below the line $l_\gamma^1(a)$ and above it (b).

In Fig.5.3b, the typical attractor map, which looks as a one-dimensional map, is shown for the specified region of parameter values. A characteristic discontinuities in map is seen at points q_1 and q_2, belonging to the local vicinity of a loop that is involved in the attractor unlike the case indicated in Fig.5.3a. Thereby, the fact of time-chaotic phase trajectory return to the local vicinity of the destroyed saddle-focus separatrix loop is proved. In the vicinity of the loop, the representative point slows down its motion when approaching the equilibrium state (the velocity near the saddle on the loop tends to zero). This results in a failure of the phase of oscillation process; then, a relatively uniform rotation follows (motion on a spiral path with amplitude modulation) which undergoes a sudden slow-down when approaching the singular point again, and so on.

If we further vary the parameter m starting at the loop birth line l_γ^1, then an infinite set of regular and chaotic attractors is realized depending on initial conditions.

The above bifurcation of attractors in the vicinity of the destroyed saddle-focus loop was pointed out in a number of papers dealing with computer simulation and was analyzed as the transition from spiral-type chaos to the chaos of a screw-type (the attractor having the random amplitude and phase oscillation modulation, or the Shilnikov's attractor) [5.2-5.4]. Computer simulation and full-scale experiments have satisfactorily bolstered all the above bifurcations and attractor properties of the second band to be maintained with γ tending to zero, as the homoclinic trajectory Γ_0^1

vanishes and the system (5.1) goes to (3.38). As an example, in Fig.5.4 and Fig.5.5, the

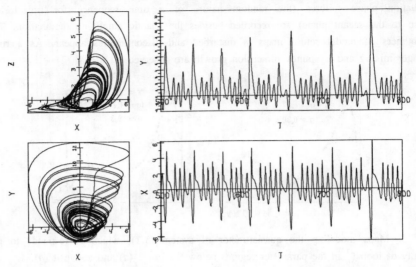

Fig.5.4. The attractor of the second zone of chaos in the system (3.38) for the parameter values $m = 1.45$ and $g = 0.3$.

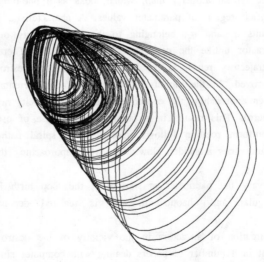

Fig.5.5. "Three-dimensional" representation of Shilnikov's attractor (Fig.5.4).

phase trajectory projections, temporal realization intervals and the general view of Shilnikov's attractor in the system (3.38) are shown which were obtained numerically for the parameter values from the second chaos band. The time-nonregular phase failures are seen on the background of chaotic amplitude modulation caused by the slow-down in the vicinity of singular point. Note, that the screw-type attractor can be easily assumed as intermittency during full-scale experiments, when there is no way of analyzing the structurally unstable homoclinic trajectories, since the motion on the "spiral" oscillation phase looks often as a laminar phase under tangency bifurcation (Fig.5.4).

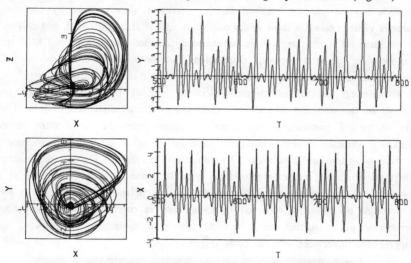

Fig.5.6. Two-dimensional projections and time series of processes for the attractor of the first chaos zone in the system (3.38) for $m = 1.5$ and $g = 0.2$.

Fig.5.7. "Three-dimensional" representation of the attractor (Fig.5.6).

To make sure that the typical oscillation regimes in the systems (3.38) and (5.1) are topologically equivalent, we refer to the results of computer simulation and full-scale experiments. In Fig.5.6, the findings of system (3.38) integration are presented which illustrate the developed chaos in the first zone. The general view of the chaotic attractor is shown in Fig.5.7 represented three-dimensionally. An essentially similar oscillation regime can be realized in the perturbed system (5.1) by choosing the parameter values under the same initial conditions $x(0) = 3.5$, $y(0) = 0$, and $z(0) = 1.0$. The similarity of motion regimes is confirmed by detailed calculations: the LCE spectra, autocorrelation, integral power, distribution functions, and Fourier-spectra of the above regimes are almost indiscernible.

Experiments on a radiophysical oscillator enabled us to observe the chaotic attractors of the first and second zone and, under selected directions of motion in the plane of control parameters, the smooth transition from the spiral attractor to the Shilnikov's attractor. Their topological structure is in a qualitative agreement with the data of computer simulation. To illustrate the above, Fig.5.8 demonstrates a photograph of chaotic attractor projections onto the plane of variables xy and xz taken from the screen of oscillograph. The fundamental role of saddle-focus separatrix loop in the structure of attractors observed in its vicinity is corroborated by comparing the experimental and calculated results with the forms of homoclinic trajectory Γ_0^1 and of its appropriate projections onto the plane (Fig.5.1).

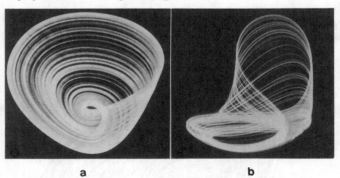

Fig.5.8. The attractor projections onto the plane xy (*a*) and xz (*b*). Experiment for $m = 1.5$, $g = 0.2$.

One of the most probable reasons for further variation of the oscillation's nature is the presence of multi-rotational loops of separatrixes Γ_0^k and, as a consequence, a more complicated structure of attracting sets in the system's phase space in their vicinity.

How are the obtained results related to the conclusions of the theory of dynamics regularities for systems having structurally unstable homoclinic curves of saddle-focus?

Computer simulation with the Rössler's system and with a number of other ones are referred to the classical situation where the saddle-focus loop was dangerous ($\sigma_1 > 0$) [5.2-5.3]. The appearance of the screw chaos was naturally connected with this by the authors of the above papers. No serious problems were caused by the fact that no chaos was recorded locally in the vicinity of the destroyed loop during our experiments: the Shilnikov's theorem does not prove a hyperbolic subset near the loop to be attracting. As a consequence, the structure and properties of nonattracting hyperbolic subset in the vicinity of the dangerous loop were not specially studied.

It is by far clear that the homoclinic trajectories of separatrix loop type play a fundamental role in the global mechanisms of quasiattractor birth with no dependence on the sign of saddle magnitude.

The constructive theoretical results, being capable to describe, in more general cases, the dynamical properties of systems having the homoclinic trajectories of saddle-focus loop type, would be driven when developing the Shilnikov's classical approach for the case of several parameters, i.e., when investigating the bifurcational situations of codimensions two and higher.

5.2 ROLE OF HOMOCLINIC SADDLE CYCLE TRAJECTORIES IN THE CHAOTIC ATTRACTOR BIFURCATIONS

As the threshold of the quasiattractor birth is exceeded, a further supercriticality increase leads, as it was already mentioned, to *bifurcations in chaos*, particularly, to a cascade of merging attractor bands. This phenomenon is typical for systems with period doubling and is attributable to the properties of homoclinic trajectories of saddle periodic motions.

Consider the Poincaré section and oscillation power spectra in the system (3.38) with the parameter values indicated in Fig.5.9. In case of a, the section involves four separated chaotic sets and, in case of b only two of them. According to this, a- and b-type attractors will be called the period-four and -two attractors, respectively. The structure of oscillation power spectrum $S_x(f)$ is uniquely connected with the attractor's band periodicity. In case of a, the spectrum comprises resonance lines at frequencies nf_0, $nf_0/2$ and $nf_0/4$ and, in case of b, the lines at frequencies $nf_0/4$ ($n = 1, 3, 5...$) are smeared.

Within the interval of parameter variation $1.09 < m < 1.10$, the chaotic attractor undergoes the merging of period-4 band to the period-two band. In Fig.5.9, points $q_{1,2,4}$ are marked in the Poincaré section which correspond to the intersections of the appropriate saddle cycles Γ_1, Γ_2, and Γ_4, having lost their stabilities via period doubling bifurcation, with a secant. As seen from Fig.5.9a, the points q_4 of the period-four cycle are involved in the chaotic set and points $q_{1,2}$ are not. Due to the band merging bifurcation, the period-two cycle points q_2 turned out to be involved in the attractor (Fig.5.9b).

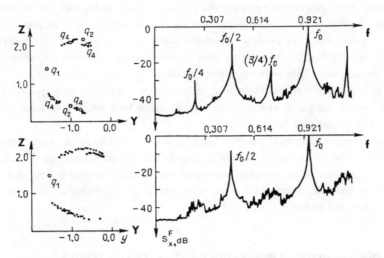

Fig.5.9. Evolution of Poincaré sections and power spectra as the the parameter m passes through the point of merging bifurcation: a - $m = 1.09, g = 0.3$, b - $m = 1.10, g = 0.3$.

As it was established, the internal bifurcations of attractor band merging are caused by variations in the nature of unstable saddle cycle manifolds. Let us analyze an example of the bifurcation of period-four attractor band merging to the period-two band.

The saddle cycle Γ_2 has two-dimensional stable and unstable invariant manifolds. They cross the Poincaré secant $x = 0$ along one-dimensional curves representing the stable and unstable separatrixes of saddle fixed point q_2. The above separatrixes can be constructed numerically with the results indicated in Fig.5.10. With $m = 1.09$ (a), the separatrixes do not intersect and as $m = 1.10$ (b), they intersect and form homoclinic points[1]. Before these points appear, the separatrix surfaces of saddle cycle Γ_2 separate the regions of existence of the phase trajectory flows in the phase space of system (3.38). The structure of the phase space partition into trajectories is rebuilt under the birth of a homoclinic trajectory. Particularly, the bifurcation of chaotic attractor band merging is realized through the destruction of corresponding separatrix surfaces. A newly formed homoclinic structure is "hooked up" to the attractor leading to the growth of the positive LCE spectrum exponent. The exponent increases from the value $\lambda_1 = 0.016$ ($m = 1.09$) to $\lambda_1 = 0.026$ ($m = 1.10$) as a result of the bifurcation considered. The

---

[1]In Fig.5.10, calculation results are given for the initial branches of unstable separatrices leading to the appearance of the first homoclinic points.

phenomenon of hooking up the new homoclinic structure manifests itself qualitatively in one of the period-two attractor band branches in the Poincaré section that repeats the behavior of the unstable separatrix of saddle point q_2 (cf. Figs 5.9 and 5.10). The vicinity of the unstable manifold becomes attracting under such bifurcation.

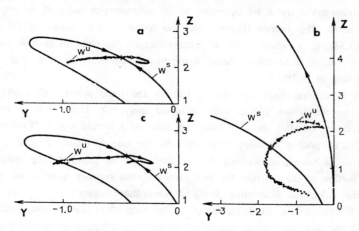

Fig.5.10. The separatrixes of saddle points in the Poincaré section of system (3.38): $a - m = 1.09$ (cycle $2T_0$), $b - m = 1.10$ (cycle $2T_0$), $c - m = 1.10$ (cycle T_0).

A homoclinic trajectory near the saddle cycle Γ_4 is responsible for the period-four attractor band existence, as $m = 1.09$. The absence of period-one band (if our considerations are valid) must be due to the absence of the homoclinic trajectory at period-one cycle Γ_1. This is confirmed by calculations. As seen from Fig.5.10c, the stable and unstable separatrixes of point q_1, as $m = 1.10$, do not intersect yet. At the same time, the form of the unstable separatrix of period-one point q_1 allows one to predict the form of the Poincaré section realized after a period-two band merging to the period-one band. This was substantiated by experimentation as well [5.1, 5.5].

The bifurcation diagram of system (3.38) is desired to be constructed in the plane of parameters to represent a more complete picture of bifurcation phenomena caused by homoclinic trajectories. However, this is extremely difficult because of intricate automated calculation. In any case, the computation will require a greater machine time. We choose an alternative approach based on the following qualitative considerations. As stated above, the dynamics of transition to chaos via a series of period doubling bifurcations admits an approximate description with the help of one-dimensional square-law map. The transition picture will be simulated more detailed by a two-dimensional map. Such a map can be, as experiments have demonstrated, the classical dissipative Henon map.

The equations for the Henon model are as follows:

$$x_{n+1} = 1 - ax_n^2 + y_n, \quad y_{n+1} = bx_n, \quad (5.2)$$

where parameters a and b are equivalent to the parameters m and g of the oscillator in a physical sense. The discrete Henon model was found to be an analog of the differential system (3.38), in a certain sense. It demonstrates on the plane of control parameters a and b a set of bifurcations being qualitatively equivalent to that realized in the Poincaré map of (3.38).

The bifurcation lines for stability lost by the fixed points with multiplicities of 1,2,...,6 for the Henon system were calculated analytically [5.6]. It is shown that two fixed points q_0 (saddle) and q_1 (stable) are born on a certain line l_0. Furthermore, the point q_1 undergoes a doubling bifurcation on the line l_1 (stable 2-cycle q_2 is softly born). Then, a cascade of doublings follows that obeys the universal Feigenbaum's law.

A portion of the diagram for such Henon system regimes is presented in Fig.5.11 where the calculated *bifurcation lines of homoclinic tangency* $l^h_{0,1,2,4}$ are shown as applied to the saddle fixed points $q_{0,1,2,4}$ of map (5.2) along with doubling lines $l_{2,4,8}$ and the line of critical parameter values l_{cr} corresponding to the birth of the Henon attractor [5.5].

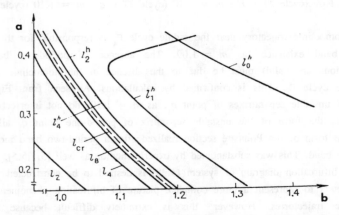

Fig.5.11. A portion of the bifurcation diagram of Henon map on the plane of parameters a and b.

The bifurcation of fixed points q_i of the map corresponds qualitatively to the sequence of the limit cycle bifurcations of system (3.38). To the right of any of lines l^h_i, a rough intersection of stable and unstable separatrixes of saddle points q_i takes place that is equivalent to formation of the robust homoclinic trajectories of the system's (3.38) saddle periodic trajectories. This is proved by calculating the separatrixes of saddle points q_i for various parameter a and b values. The separatrixes behave in a qualitatively equivalent manner as compared with the behavior shown in Fig.5.10. For all the saddle points q_i of map (5.2), the conditions of theorem [5.7] are fulfilled (as well as for the saddle cycles of system (3.38)). Thus, the quasihyperbolicity of Henon attractor and its adequacy to the chaotic attractors of system (3.38) in this sense are confirmed.

Let us fix the parameter value $b = 0.3$ and consider a number of the parameter a values placed sequentially between the bifurcation lines l_{cr} and l_i^h. With the growth of parameter a, a sequence of *attractor band merging bifurcations* is realized, followed by formation of period-one bands as $a = 1.17$. Thus, the lines of homoclinic tangency l_i^h are the bifurcation lines on which multi-periodic bands of the Henon map attractors merge. In Fig.5.12, the calculation results are shown for the period-four attractor band merging to the period-two attractor while the parameter a crosses the bifurcation line l_2^h. As for the oscillator, the chaotic sets of attractors in the Henon model are concentrated in the vicinity of unstable saddle point separatrixes having homoclinic structure. The results of calculation of the stable and unstable separatrixes of saddle points q_0 and q_1 are presented in Fig.5.13 which give evidence to their intersection. As seen from the plots, the form of a chaotic set of the developed Henon attractor is defined by the geometry of the unstable separatrix.

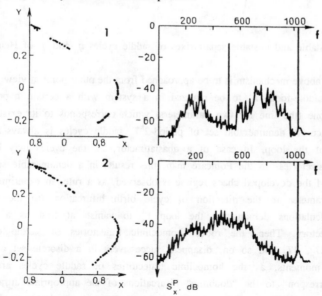

Fig.5.12. Phase portraits and power spectra of Henon map as the the parameter a passes through the point of merging bifurcation: I - $b = 0.3$, $a = 1.07$, II - $b = 0.3$, $a = 1.10$.

Our experiments showed the period doubling mechanism to be more typical in simple chaotic systems having singular points of saddle-focus type. The motion in the parameter space in the direction, being transverse to the hypersurface of Andronov-Hopf bifurcation, leads to the cascade of doublings, birth, and evolution of chaotic attractors.

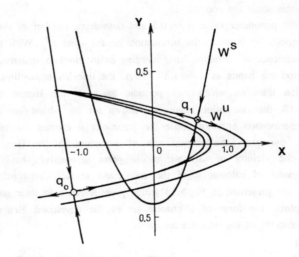

Fig.5.13. Stable and unstable separatrixes of saddle cycles q_0 and q_1 of Henon map.

Now the above mechanism is to be approached from the other point of view. We assume the loop of saddle-focus Γ_0^1 to be realized in a system with a certain hypersurface of codimension one (the line for two parameters) which corresponds to its existence in the parameter space. A denumerable set of period-$2^k T_0$ saddle cycles is always available in the vicinity of the loop. In case of a quasiattractor, all the cycles may have robust homoclinic trajectories. In the Poincaré map, this results in a denumerable set of Smale horseshoes and the developed chaos regime is observed, as a rule, in experiments. Let us vary the parameter in the direction of cycle birth bifurcation (to the line l_{00} in Fig.5.2). Calculations demonstrate the loop Γ_0^1 to vanish at first as a structurally unstable trajectory. Then, the robust homoclinic trajectories of saddle cycles with periods T_0, $2T_0$, $4T_0$ and so on, disappear successively in a discrete set of parameter values. The moments, as the homoclinic trajectories of saddle cycles are vanishing rigorously correspond to the "doubling" bifurcations of the appropriate attractor bands. For example, a bifurcational transition from period-one to period-two chaotic band corresponds to the disappearance of homoclinic points at the saddle cycle with period T_0. The band periodicity corresponds rigorously to the presence of a homoclinic trajectory at the saddle cycle with period $2^k T_0$, $k = 0,1,...,\infty$. For $k = 0$, the band has period one, for $k = 1$, it has period-two and so on. The band periodicity tends to infinity in the close vicinity of the bifurcational supersurface of critical parameter values from above. Thus, a sequence of quasiperiod doubling bifurcations of the attractor take place being similar to the cycle doubling sequence when approaching the critical value of parameter from the other side. The structure of attractors being realized, if it is studied in the Poincaré section, demonstrates clearly its relation with the form of unstable saddle cycle manifolds having the homoclinic trajectories. In essence, the attractors in the section "visualize" the unstable manifolds of the entire generality of saddle cycles. Since the above manifolds in the Poincaré section form the Smale horseshoes, it is a good practice to call the appropriate attractors "Smale attractors" [5.8]. A denumerable set of saddle cycles near the loop Γ_0^1 results in the denumerable set of horseshoes. When varying the parameters away from the hypersurface of the loop existence, the number of horseshoes is finite and decreases further.

Here a circumstance should be noted that is of importance for applications. When analyzing the positive Lyapunov exponent of the attractor as combined with the autocorrelation function of the process, the following is revealed. For the developed period-one attractor band, as the homoclinic orbits of the whole generality of period $2^k T_0$ saddle cycles exist, the following fundamental ratio is valid

$$\lambda_1 = (\tau_k)^{-1}, \qquad (5.3)$$

where τ_k is the time of autocorrelation function decrease by a factor of $e = 2.718...$, i.e., the correlation time. The decrease of the rate of process time correlations is defined in this case rigorously by the Kolmogorov entropy. The above relation is not directly fulfilled for the attractor's multi-periodic bands. The thing is that the periodic motion components appear. Mixing is proved by the Kolmogorov entropy being equal to the positive exponent of the LCE spectrum. But the mixing occurs for the multi-periodic attractor bands in a certain vicinity of the saddle cycle of the corresponding periodicity and is not associated with the presence of periodic components. Autocorrelation, as characteristic of the temporal process, is uniquely related with the form of time dependencies and reflects, naturally, the occurrence of the periodic components of the process. The equality (5.3) is not fulfilled, in general, for the multi-periodic attractor bands. It turns out to be valid, however, if one analyzes the mixing (λ_1) and the rate of correlation decrease (τ_k^{-1}) by removing from consideration the appropriate periodic component. For example, a chaotic set of points in the Poincaré section is to be analyzed which were obtained by recording every second point of the phase trajectory intersection with the secant if the attractor band has period-two. The results of computer simulation performed for the system (3.38) and two-dimensional Henon map speak for the above and will be considered below.

Similar picture is proposed for the destroyed multi-rotational loops of separatrix Γ_0^k.

5.3. PHYSICAL INTERPRETATION OF EXCITING THE NONPERIODIC OSCILLATIONS IN OSCILLATOR WITH INERTIAL NONLINEARITY

The analysis of homoclinic trajectories in a perturbed model of the oscillator with inertial nonlinearity allows us to explain the findings of numerous performed experiments from the physical and qualitative points of view. As it was ascertained, the system (3.38) is structurally stable relative to small flow perturbations. Topologically equivalent oscillation regimes are maintained by introducing the dissipative nonlinearity (3.28), by deforming the function $\Phi(x)$ that describes the characteristics of the inertial cascade detector, and by perturbing the second equation via adding a constant term (5.1). What is the mechanism of maintaining the oscillation regime in the oscillator with inertial nonlinearity? With exceeding the generation threshold, $m > 0$, small deviations from the equilibrium point lead to the increase of variable x. The energy of this mode is transferred in a nonlinear manner into the decaying mode z which starts growing with the decay due to noninertia. As certain values of variables x and z are reached, the influence of nonlinear restriction of the rate of x growth is observed, the process that is due to inertial and inertialless mechanisms is general. As a result, the rate of variable x increase becomes negative which leads to decay. The decay process, being complicated by the system's inertia and nonlinearity, yields, as a consequence, lower values of variable x. The influence of nonlinearity in the first and third equations of the system becomes negligible and x begins growing again. The ratio of the characteristic times of the system, defined by excitation and inertia parameters, on the one hand, and the degree of influence of nonlinear energy transfer and the mechanisms of oscillation amplitude growth confinement, on the other hand, define, a certain fundamental time T of the system. The latter is the time of energy exchange between dissipative oscillation elements and the source energizing the circuit via the positive feedback. With low nonlinearities, T is small and defines the period of quasiharmonic oscillations. With their growth, the period T increases due to complicated relaxation processes hindering the synchronization of characteristic times of the system. The period T tends to infinity in the structurally unstable bifurcation case when a homoclinic loop of the equilibrium point separatrix is feasible in the system. Synchronization to periodic regime is excluded here. The time of T is finite, in general, in the vicinity of the loop (and this is already a robust, realizable case), but quite high. Periodic oscillations are theoretically probable, but there is a denumerable set of them in the vicinity of the loop, their period is very high, and the existence and attraction basins are small and are easily covered by small fluctuations. A quasiattractor is realized in this region and chaotic oscillations are generated by the real system both in full-scale experiments and computer simulation.

We write the system (3.38) in the form of the oscillator with inertial nonlinearity as follows (3.2):

$$\ddot{x} + (z - m)\dot{x} + [1 - gz + gI(x)x^2]x = 0,$$
$$\dot{z} = -gz + gI(x)x^2.$$
(5.4)

The generation of stationary self-oscillations is possible only when the coefficient at the first derivative $(z - m)$ becomes zero in the mean for time T. This coefficient would be a function that reverses its sign in time. *Effective delay* in the oscillatory system is *negative* in the region of $z < m$. The system is unstable and goes out from the equilibrium point when slightly perturbed. Because of the nonlinear energy transfer, the level of a negative feedback signal begins to increase, as does the magnitude of variable $z(\tau)$. The system will be transformed into a *nonlinear dissipative circuit* and oscillations will decay as z exceeds the value of the excitation parameter m. In a quasilinear case ($m \le 1$, $g \le 1$), $z(\tau)$ varies by a law nearing to the harmonic law and has small amplitude and period. Near-harmonic oscillations are generated. The parts of transferred and scattered energies are entirely compensated in the mean over one oscillation period. As the nonlinearity is raised, the oscillations become complicated due to the time-synchronization of energy exchange processes at great averaging times. The oscillation period is increased. Then, the situation takes place when an infinitely great time is necessary to synchronize the energy exchange, that means that chaotic generation is born. The results calculated for the quantity $z(\tau) - m$ are shown in Fig.5.14 for two limit regimes which illustrate the above [5.1].

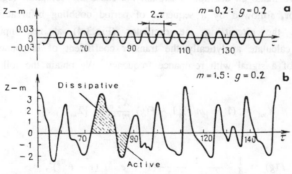

Fig.5.14. The dependence of values $[z(\tau) - m]$ in the regular (*a*) and chaotic (*b*) oscillation regimes.

Such considerations can be accomplished in terms of the stability theory. The regularity of oscillation regime requires the maximum Lyapunov exponent of the LCE spectrum λ_1 to be zero, in the mean, for a period along the trajectory. In the region of

$z < m$, where the system is unstable, the maximum exponent would be positive and for $z > m$, negative. $<\lambda_1>$ must rigorously become zero, in the mean, for an oscillation period. The chaotic regime is realized in the case where the instability of the system in the vicinity of the source, at the equilibrium point will "overpower" the stabilizing action of the dissipative oscillator's elements in the region of $z > m$ removed in the phase space away from the source. The averaging over arbitrarily great times T will give a positive value of λ_1 for the system's attractor.

To verify the above, we perform a simple but interesting computer simulation. Let us introduce into the algorithm of calculating the LCE spectrum an additional condition of independent averaging over the time for two phase space regions separated by the plane $z = m$. The averaged exponent values over the first, and second regions, and the entire phase space are calculated. We obtain the following results. As one would expect, for regular regimes, the exponents in the first and second regions are equal in magnitude and differ in sign. The total exponent is rigorously zero. In the chaos regime, the positive exponent is superior. The total exponent is defined by the appropriate difference and proves to be positive. It is of interest, that the absolute exponent magnitudes of both regions exceed remarkably, as a rule, the value of an exponent averaged over the entire attractor. This phenomenon is common for generating systems having a vector field divergence depending on the phase coordinates. The active flow mixing is performed not over the entire phase space occupied by the attractor, but it is localized in the source region (in this case, in the region of saddle-focus at the origin). The mixing is characterized by the magnitude of the positive exponent that substantially exceeds the mean value over the entire attractor.

Finally, we make an attempt to explain the cause of oscillation regime complication in the oscillator, followed by a sequence of period doubling bifurcations. To do this, we disconnect the circuit in the oscillator (Fig.3.4) at the input of the first amplifier and calculate analytically the transfer coefficient of disconnected system for the amplitude of a signal with resonance frequency. We obtain the following expression [5.1]:

$$X_{out} = (1 + m/b)|1 - F(g)\frac{X_{in}^2}{m + b}|X_{in} , \qquad (5.5)$$

where

$$F(g) = \frac{1}{4} - \frac{1}{\pi g(4 + g^2)} (1 + e^{-\pi g})(1 - e^{-\pi g})^2,$$

b is the constant coefficient depending on the parameters of specified selective oscillator element. It is easy to see that eq. (5.5) describes the one-parametric family of parabola type at the fixed values of parameter g. The appropriate plots for some values of parameters m and $g = 0.2$ are drawn in Fig.5.15. The amplitude characteristic of the inertial amplifier belongs to the class of Feigenbaum maps! As the parameter m is increased, the slope of the amplitude characteristic achieves a critical magnitude and the oscillation regime loses its stability via doubling.

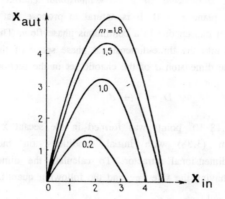

Fig.5.15. Transfer coefficient of inertial amplifier for the harmonic signal with resonance frequency ($g = 0.20$).

The possibility of "unpredictable" evolution is not mysterious from the standpoint of modern insight into the behavior of systems comprising the local regions of exponential instability. The system's response is rigorously determined but is extremely sensitive to small deviations in the initial data in the vicinity of the strong instability region (for example, on the summit of a ball). The entire "economics" of automatic play machine industry is based on this feature. In the oscillator with inertial nonlinearity, the role of a strong instability region is played by the local vicinity of a singular point at the origin. The presence of a saddle-focus loop means, from the physical point of view, that the properties of a nonlinear system have an autonomous power "to throw" a phase trajectory into the vicinity of the singular point. Therefore, the oscillator proposed can be considered as a radiophysical analog of the automatic play machine.

5.4 ON THE DIMENSION OF AN ATTRACTOR

The geometry and probability properties of chaotic attractors are defined by different *dimensions* having a mutual relationship given for the present by the estimates of inequality type and requiring special theoretical investigations in general. The totality of revealed basic properties of the dynamical system (3.38) show a series of computer simulations on calculating various dimensions of chaotic attractors and their detailed quantitative comparison to be possible.

Consider the developed chaotic regime in the system (3.38) with the parameter values $m = 1.5$, $g = 0.2$ for a saddle-focus attractor shown in Figs 5.6 and 5.7. At first, we calculate the *dimensions* of a *two-dimensional chaotic* set in the Poincaré section with the secant plane $x = 0$. It is natural to propose that the dimension would not depend on the secant chosen due to a continuous phase flow. The full dimension D of an attractor, embedded into the three-dimensional phase space of the system, is obtained by adding a unity to the dimension d of the chaotic set in the section

$$D = 1 + d. \quad (5.6)$$

A data file of $3.18 \cdot 10^6$ points was formed in the secant $x = 0$ by numerically integrating the system (3.38) with initial conditions on the attractor. It was considered as a two-dimensional attractor. To calculate the dimensions, the attractor was covered by boxes having the side ε and the following quantities were calculated

$$\begin{aligned} d_\mu &= \lim_{\varepsilon \to \infty} [\ln M(\varepsilon)/\ln(1/\varepsilon)], \\ d_I &= -\lim_{\varepsilon \to \infty} [\sum_{i=1}^{M(\varepsilon)} p_i \ln(p_i)/\ln(1/\varepsilon)], \\ d_c &= \lim_{\varepsilon \to \infty} [\ln [\sum_{i=1}^{M(\varepsilon)} p_i^2]/\ln(1/\varepsilon)], \end{aligned} \quad (5.7)$$

where $M(\varepsilon)$ is the number of boxes covering the attractor, p_i is the probability of the attractor's point to belong to the i-th box, d_μ, d_i and d_c are the *metric*, *information* and *correlation dimensions of the attractor*, respectively.

Since one cannot calculate as $\varepsilon = 0$, the estimate of corresponding dimension, d_ε was calculated depending on the specified magnitude of ε:

$$d_\varepsilon = H_\varepsilon \cdot [\ln(1/\varepsilon)]^{-1}, \quad (5.8)$$

where, depending on definition (5.7)

$$H_\varepsilon = \ln M(\varepsilon), \quad H_\varepsilon = -\sum_{i=1}^{M(\varepsilon)} p_i \ln(p_i), \quad H_\varepsilon = \ln[\sum_{i=1}^{M(\varepsilon)} p_i^2]. \quad (5.9)$$

With ε small enough, it may be considered that

$$\exp H_\varepsilon \approx c \, \varepsilon^{-d_\varepsilon}, \quad (5.10)$$

where c and d_ε are a certain constant and a value corresponding to the dimension, respectively. Then,

$$H_\varepsilon = lnc + d_\varepsilon ln(1/\varepsilon). \quad (5.11)$$

Thus, d_ε depends linearly on $[ln(1/\varepsilon)]^{-1}$ and, in the limit as $\varepsilon \to 0$, tends to an exact value of the attractor dimension. In experiment, one has to confine himself by a small interval of the ε values where the linear dependence (5.11) is approximately valid and to extrapolate the latter into the region of $\varepsilon \to 0$ to find the dimension d.

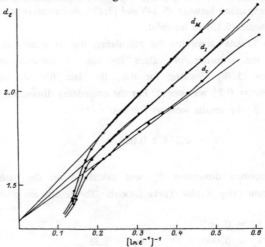

Fig.5.16. Calculated attractor dimensions of the system (3.38) in the Poincaré section as $x = 0$.

The calculated results are presented in Fig.5.16. As seen from the plots, the experimental dependences are close to the linear dependences within the interval of $0.01 < \varepsilon < 0.30$. The quantitative estimation of linear dependence coefficients using the method of the least squares and approximating the appropriate straight lines into the region of $\varepsilon = 0$ have given the following dimensions [5.9]:

$$d_\mu = 1.306 \pm 0.015, \quad d_I = 1.30 \pm 0.013, \quad d_c = 1.277 \pm 0.017, \quad (5.12)$$

where the mean-square deviations of theoretical approximations from experimental data are characterized by an error. The full dimension D is easy to obtain by adding a unity (5.6).

The results derived satisfy the theoretical estimates of $D_\mu > D_I > D_c$. A complete equality is possible, however, with regard to the calculation error:

$$D_\mu = D_I = D_c \simeq 2.29. \quad (5.13)$$

Let us calculate the correlation dimension by means of Grassberger-Takens' method using a discrete sequence of one of two-dimensional attractor's coordinates in the map. To do this, we select a sequence of y coordinates from the data file set up. For the dimension of embedding space, being four, the experimental results are as follows:

$$d_c = 1.292 \pm 0.041; \quad D_c = 1 + d_c. \tag{5.14}$$

As seen from comparison between (5.14) and (5.12), this method provides as well the results enabling to assume (5.13) to be valid.

The Grassberger-Takens' algorithm for calculating the correlating dimension can be applied directly to the sequence of values for one of three-dimensional attractor coordinates of system (3.38). Seeking for this, the data file of coordinate $x(i,\Delta t)$, $i = 0,1,2,\ldots,2\cdot 10^4$, $\Delta t = 0.25$ was stored. For the embedding dimension, being five, and the delay time $k\Delta t = 2$, the results were as follows:

$$D_c = 2.214 \pm 0.026. \tag{5.15}$$

Finally, the Lyapunov dimension D_L was calculated on the base of the entire attractor's LCE spectrum using Kaplan-Yorke formula. The results are given below:

$$\begin{aligned} \lambda_1 &= 0.062 \\ \lambda_2 &= 0.000 \qquad D_L = 2.33 \, . \\ \lambda_3 &= -0.187 \end{aligned} \tag{5.16}$$

The calculation of the correlation (5.15) and Lyapunov (5.16) dimensions for the attractor of system (3.38) provides a somewhat underestimated D_c and a somewhat overestimated D_L values as compared with the data obtained on the base of calculating the attractor dimensions in the map. There are reasons to assume the above differences, which are smaller than 3% in absolute magnitude, to be due to a relatively small length of realization $x(t)$[2].

In spite of certain differences in the quantitative dimension magnitudes, we may uniquely conclude: these differences *are not sufficient* from the view point of physics. Particularly, *any dimension* the calculation of which is more suitable in a specific case for one or another reason, *can be used* to solve a problem on the number of active freedom degrees operated in the specific regime of selfoscillations of the system.

As for the particular example of saddle-focus attractor considered in the three-dimensional system, it is not discarded that, theoretically, all the investigated

---

[2] Data file and averaging time extention demonstrated a trend for D_C growth and D_L decrease.

dimension types have the same quantitative value. This problem is theoretically unsolved so far, as it is known, for the three-dimensional flows of systems having a vector field divergence depending on the phase coordinates.

CHAPTER 6
TWO-FREQUENCY OSCILLATION BREAKDOWN

6.1. GENERAL PROBLEM STATEMENT

Transitions to chaos in non-linear distributed systems and media are preceded, as a rule, by the *selfmodulation* becoming more complicated. The oscillation spectrum is enriched by new and combination frequencies with the growth of control parameter; their number and intensities increase, the number of harmonics is enhanced. As a result, the spectrum becomes continuous showing the oscillations to be irregular. Historically, one of the first theoretical models of *turbulence evolution after Landau* has considerably defined the way of the progression of dynamical chaos theory and, experimentally, it was in agreement with really observed processes of formation of a spectrum of the developed chaos. However, there were no fundamental mathematical concepts at that time to study more deeply the bifurcation phenomena leading to chaos birth under quasiperiodic oscillation breakdown. The direct integration of complicated differential equations was hindered by the absence of suitable computers. It is to note that physical insight into the chaos was based, up to the sixties, mainly on statistical phenomena caused by fluctuations.

The situation was changed after the famous papers of E.Lorenz [6.1], D.Ruelle and F.Takens [6.2,6.3] had been published, as well as other fundamental investigations on the above problem [6.4 - 6.7]. A concept of *strange attractor* and *quasiattractor* was introduced. Novel qualitative methods of analysis were formed allowing one to investigate in detail the processes of appearance and the structure of attracting *hyperbolic subsets* in the phase space of high-dimensional dynamical systems. The theory achievements and the development of the software and fast-action of computers enable the recurrence to the problems of non-linear physics connected with *quasiperiodic oscillation bifurcations* at a principally new level of understanding. Recently, many interesting experimental and theoretical papers have appeared as regards the mechanisms of transition to chaos via two- and three-frequency oscillations. Three directions can be separated in this field. *First*, there is a rigorous theoretical analysis on the base of analytical methods of the dynamical systems theory [6.8,6.9]. *Second*, a model description of the problem of non-linear phenomena leading to the destruction of invariant curves in discrete maps is involved along with the methods of renormalization group transformations and calculation [6.10,6.11]. And *third*, computer simulation and full-scale experiments on investigation of the transitions to chaos via quasiperiodic oscillations in particular finite-dimensional and distributed systems are concerned [6.12,6.13].

The analysis of experimental and theoretical results obtained, points to a series of universal qualitative regularities inherent in transitions to chaos via quasiperiodic regimes independently of specific systems. Moreover, these results demonstrate that, at least with a relatively small exceeding of the criticality threshold, the dynamics of distributed and finite-dimensional systems are similar. This is of extreme interest and of practical importance since this allows one to look forward to make the theoretical bases for an adequate simulation of the turbulence evolution processes of the continuous medium.

As shown in Chapter 5, the presence of a *structurally unstable homoclinic trajectory* of saddle-focus loop type and *robust homoclinic trajectories* for saddle periodic motions is responsible for the development of quasiattractors with appropriate structures.

In case of transition to chaos via quasiperiodic oscillation, such a central object is a *homoclinic structure* initiated by the non-linear processes of *two-dimensional torus breakdown*. In the Poincaré map, the resonant two-dimensional torus is corresponded by an invariant curve formed by the closure of the unstable separatrixes of saddle points to stable nodes. The mechanisms of homoclinic evolution prove to be qualitatively similar to processes in the vicinity of a saddle point of the periodic motion in the Poincaré map as the local vicinity of one of the saddle points on the invariant curve is analyzed. At the same time, the finite number of saddle points and their belonging to the invariant closed curve result in differences in the structure of homoclinic trajectories and, as a consequence, in the special physical properties of the quasiattractor originated.

The two-frequency oscillation regimes are realized and investigated by periodically forcing the stable limit cycle of autonomous systems, in the systems of two interacting active non-linear oscillators, as well as in autonomous non-linear finite-dimensional and distributed systems being able to operate under selfmodulation. To study a number of principal phenomena, it is useful to apply the model representations in the form of discrete systems having an invariant closed curve which is a pre-image of two-dimensional torus in map.

The methods of exciting the two-frequency oscillations suggest how to realize the regimes with three independent frequencies. For example, the three-frequency oscillations are likely to arise when one weakly periodically forces a system with an attractor in the form of two-dimensional torus.

The study of bifurcation phenomena that accompany the transition to *torus-chaos* from the regime of *two-frequency* oscillations, is more easy to begin with analyzing the processes in a self-oscillator driven by the external periodic force. The advantages of such an approach are connected with the low dimension of a dynamical system and the simplicity of changing the *Poincaré winding number* by varying the frequency of a modulation signal.

In Chapters 6 and 7, the basic mechanisms of transition to torus-chaos and its properties will be considered the torus-chaos being born under the breakdown of two- and three-frequency quasiperiodic oscillations. A number of simple autonomous and nonautonomous, differential and discrete, radiophysical systems will be used along with a complex of theoretical, numerical and experimental methods. An autonomous system, the modified oscillator with inertial non-linearity will be a basic model for the goal-directed complications of possible dynamical phenomena in the systems under study. The above properties of the model, as combined with relatively simple computer simulation and full-scale experiments enable one to analyze in detail the complicated dynamical phenomena and to reveal the typical regularities of transition to torus-chaos.

6.2. BIFURCATION DIAGRAM OF NONAUTONOMOUS OSCILLATOR IN THE VICINITY OF BASIC RESONANCE. COMPUTER SIMULATION

We are rather interested in selecting a system with limit cycle, at which to start our problem. For one thing, the dynamics of Van-der-Pole oscillator, forced periodically, represents some features of transitions to chaos via two-frequency oscillation, but far incompletely and depends on the technique of specifying the non-linear characteristics of a model. It is desired to consider as an autonomous system not simply a periodically oscillating system, but a system capable to exhibit both regular and chaotic self-oscillations, i.e., one of the simple systems with complicated dynamics. Therefore, it is useful to begin the study of the mechanisms of the two-dimensional torus breakdown with examining the dynamics of the *nonautonomous* oscillator with inertial non-linearity [6.13-6.15]:

$$\begin{aligned}
\dot{x} &= mx + y - xz + B_0\sin(p\tau), \\
\dot{y} &= -x, \\
\dot{z} &= g[I(x)x^2 - z],
\end{aligned} \quad (6.1)$$

where, as compared to (3.38), the harmonic force with amplitude B_0 and frequency p is introduced in the first equation.

To study possible oscillation regimes of a nonautonomous system, we employ again the totality of methods used when analyzing the self-oscillations in the oscillator, i.e., the construction of bifurcation diagrams in the chosen plane of control parameters and the study of typical oscillation regimes and their bifurcations with the help of computer and physical simulations. An experimentally convenient and more informative two-dimensional surface is to be found in the parameter space for the study of codimension-one and -two bifurcation phenomena in dynamical systems with the number of parameters more than two. Choose the plane of parameters m and p as such a plane by fixing $g = 0.3$ and consider the dynamics of system (6.1) for the selected values of amplitudes of external force B_0 [6.13-6.15].

In Fig.6.1, the results of calculation of the bifurcation lines in the m, p parameter plane are presented [6.15]. The region of existence of stable oscillations with external force frequency (sector 1) is bounded by the bifurcation lines of *neutrality* l_0, *multiplicity* l_1 and *period doubling* l_2^k. The two-dimensional torus is born in a soft manner from the cycle of external synchronization, and two-frequency oscillations arise while crossing the line l_0.

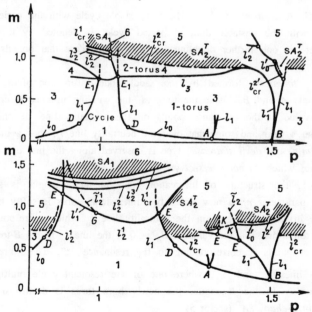

Fig.6.1. Bifurcation diagram of system (6.1). Computer simulation for $B_0 = 0.1$ (*a*) and $B_0 = 0.5$ (*b*); $g = 0.30$.

The winding number Φ changes monotonically along the line l_0. The codimension-two bifurcation points are passed successively that correspond to strong resonances $A(1:4)$

and $B(1:3)$ and to many weak ones $\Phi = \nu : q$, $q \geq 5$. At the bifurcation points D, along with a pair of complex-conjugate multipliers, the third multiplier $\rho = +1$ goes out to a unit circle and line l_0 is interrupted. Above the points D there is a line l_1. The transition from the resonant oscillation zone through the multiplicity line corresponds to the bifurcation of merging and vanishing of *stable and unstable cycles on torus*. When emerging from sector 1 through line $l_{\frac{1}{2}}$, the initial cycle loses its stability with the soft birth of a stable cycle with doubled period. As one moves further upwards in parameter m, the sequence of period doubling bifurcations of a resonant cycle on torus converges, completed by the birth of a quasiattractor of spiral type SA_1 (sector 6).

Doubling lines $l_{\frac{k}{2}}$, ($k = 1,2,3$) and the lines of critical values of parameters l_{cr}^1 on which SA_1 is born are indicated in Fig.6.1. The intersections of lines $l_{\frac{k}{2}}$ and l_1 are marked with points E. As $B_0 = 0.1$, the bifurcation line of the doubled period cycle birth coincides with the period doubling line $l_{\frac{1}{2}}$ in the segment $E_1 - E_1$. The picture is similar for the cycles with higher multiple periods. The increase of B_0 up to the level of 0.5 results in a more complicated diagram. Line $l_{\frac{1}{2}}$ coincides with line l_1' in the segment $E - G$. A segment of $\tilde{l}_{\frac{1}{2}}$, where the stable cycle with the period of external force merges with the unstable doubled period cycle, is denoted by the dashed line. Moreover, there is one further doubling line $l_{\frac{1}{2}}$ in this case that bounds the sector 1 from the left.

Resonant "tongues", formed by the saddle-node bifurcation lines, rest on the bifurcation points A and B. Inside the tongues, the winding number is constant and is equal to the value at the resonance points on the neutrality line. If one follows the winding number Φ near and somewhat over the neutrality line l_0, then a unique step-like dependence $\Phi(m,p)$ of *"devil staircase"* type is observed along the line due to broadening resonant tongues which *do not intersect* yet.

In Fig.6.1, the structure of bifurcation lines in the B-tongue is shown. When increasing the parameter m, a new stable two-dimensional torus is softly born from the resonant cycle inside the tongue on the neutrality line l_0' which corresponds here to a resonant cycle on torus! For the value of $B_0 = 0.5$, the line l_0' in the B-tongue becomes discontinuous at points K corresponding to the resonance 1:2. These are the original points of two lines, l_2 and $\tilde{l}_2$, where one of the resonant cycle multipliers is -1. Above the line l_0' in the B-tongue, the newly born torus is destroyed followed by formation of *torus-chaos* SA_2^T (sector 5).

As seen from Fig.6.1, there are two more types of bifurcation lines l_3 and l_{cr}^2 placed in the region of quasiperiodic oscillations. The transition to torus-chaos via a sequence of many bifurcations is recorded on line l_{cr}^2 which has, in fact, a complicated *fractal structure*. The bifurcations are completed by the torus breakdown followed by the SA_2^T formation. As $B_0 = 0.5$, the *"period-one"* torus (1-torus) is destroyed on line l_{cr}^2. When decreasing the external force amplitude, the torus breakdown is preceded by a finite sequence of *torus doubling bifurcations*. Namely, for $B_0 = 0.1$, one torus doubling takes place as the line l_3 is crossed from below in parameter m. The 2-torus is broken down on line l_{cr}^2. As calculations have shown, the number of torus doubling bifurcations grows with lower intensity of the external force but *remains finite* for the finite amplitudes B of an external signal.

A crucially new *bifurcational phenomenon of torus doubling* will be considered in its own right, and here we take the total for the results of computer simulation. The two-parametric analysis of system (6.1) has enabled determination the typical bifurcations of periodic and quasiperiodic regimes and characteristic regions with topologically different oscillation types. The main oscillation regimes in the nonautonomous oscillator with inertial non-linearity are as follows: 1 is the region of periodic oscillations with the frequency of external force and 2 is that of resonant two-periodic oscillations near $p \approx 2$ (in Fig.6.1 not shown). The nature of resonant two-periodic oscillation bifurcations in the above region, with the winding number $\Phi = 1:2$ is identical at much to the case of the resonance 1:1. Other regimes are: 3 is the region of practically ergodic two-frequency oscillations with various winding numbers, 4 are the regions of 2-torus existence, 5 are the regions of torus-chaos being born under *two-dimensional torus* breakdown, and 6 are the regions of chaos arising as a result of accumulation of *the period doubling bifurcations of resonant cycles on two-dimensional tori* after Feigenbaum. The resonant cycles having suitable rational winding numbers are realized inside the resonance zones resting on the codimension-two bifurcation points on the neutrality line. They lose their stabilities inside the synchronization tongues when varying the parameter m (while going away from their grounds), and the stable two-dimensional tori are born. The breakdown of the tori leads further to torus-chaos again.

6.3. BIFURCATION DIAGRAM OF SYSTEM (6.1). FULL-SCALE EXPERIMENT

The numerical construction of bifurcation diagrams (Fig.6.1) requires considerable machine time consumption. Therefore, further investigations will be performed by using the full-scale experiment with addressing the computer in special cases only. The full-scale experiment can not only help in controlling the adequacy of mathematical and physical models but it may be an independent method for elucidating the complicated bifurcation phenomena. Experimental data will be interpreted in the terms of the bifurcation theory on the context of which brief explanations on the measuring technique are to be given. In the course of an experiment, parameter m was regulated by the resistance $R_m \sim m$ which controls the amplification factor in the circuit of positive feedback of generator. Parameter p was varied by changing the frequency of the external generator $f_1 \sim p$, amplitude B_0 was defined by the voltage of an external signal $V_0 \sim B_0$. Parameter g was fixed at the level of 0.3 as with calculation.

To define the nature of stability loss by the oscillation regimes, we use an oscillograph and a spectrum analyzer to study the oscillation types' rebuildings using the shape of time realizations and the phase trajectory projections onto selected coordinate planes, as well as the corresponding power spectra. The diagnostics of neutrality and doubling lines is not a particular problem in case of supercritical bifurcations. The multiplicity lines are more difficult to define, especially, in the regions exhibiting hysteresis phenomena, however, it is also possible. The torus doubling and breakdown bifurcations are clearly fixed when they are soft while comparing phase projections and the appropriate oscillation spectra under transitions.

In Fig.6.2, a portion of the bifurcation diagram on the R_m, f_1 parameter plane is presented for the external force frequency zone $5.5 \leq f_1 \leq 15$ kHz corresponding to the region where $0.75 \leq p \leq 2.0$ ($p = f_1/f_0$, $f_0 = 7.18$ kHz is a natural oscillator frequency at small R_m values) [6.15]. The diagram involves many characteristic oscillation regimes that qualitatively recur with the extension of the frequency band Δf_1 at various amplitudes V_0. The parameter plane partition in Fig.6.2 demonstrates the bifurcations which were found experimentally in the vicinity of the basic resonance to correspond to the calculation results of Fig.6.1. The experimental diagram shown in Fig.6.2 comprises over 30 sectors formed by the intersection of the bifurcation lines l_i, ($i = 0,1,2.3$) and $l_{cr}^{1,2}$ (designations are the same as in Fig.6.1). As with calculation, bifurcation lines l_3 and l_{cr}^2 were experimentally found which correspond to torus doubling and breakdown. The latter is followed by the formation of SA_2^T representing the non-trivial bifurcations. The line l_{cr}^2 is characterized by an intricate fractal structure. Near l_{cr}^2, the torus breakdown is recorded followed by the transition to torus-chaos.

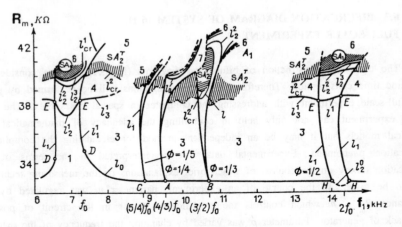

Fig.6.2. Bifurcation diagram of system (6.1). Full-scale experiment for $V_0 = 0.5$ V.

The codimension-two bifurcation points, shown in Fig.6.2 in the number of over forty, are defined by the intersecting bifurcation lines l_i in the experimental diagram. The winding number and, thus, resonance order in the synchronization regions can be determined experimentally for the points belonging to the neutrality line l_0. Many other bifurcation points are not susceptible to diagnostics in experiment in general case. The codimension-two bifurcation points on the neutrality line l_0 are highlighted by the additional conditions that the winding number is rational and provide for the tongues of synchronous oscillations with appropriate Φ. So, $\Phi = 1/4$ and $\Phi = 1/3$ inside the A-tongue, resting on the point A on neutrality line, and the B-tongue, respectively. The zones of weak resonances ($q \geq 5$) are partially observed in experiments but with great errors due to noise. They are not shown in Fig.6.2 to hold visualization.

The analysis of regimes in different diagram sectors allows one to mark out seven topologically different oscillation regimes, which regions of existence are marked with Arabic numerals in calculated and experimental diagrams. In sector 1, a system is in the regime of forced oscillations with external force frequency f_1. Frequency locking and basic tone synchronization are performed near the resonance frequency f_0 as $R_m > 3.3 \cdot 10^4$ Ω. The oscillation period is $1/pf_0$ here. The second harmonic synchronization is performed at the external force frequencies f_1 which are close to the doubled natural frequency $2f_0$ (sector 2). It is of interest that the cycle with external frequency undergoes the period doubling bifurcation when crossing the line l_2^0 from below near $p = 2$ followed by the frequency locking. The cycle period in sector 2 is $2/pf_0$.

In sectors 3, two-frequency oscillations are recorded that have different winding numbers ($q > 5$). The winding numbers can also take rational values corresponding to narrow zones with resonant oscillation. The analysis of these regimes was hindered by resolution of a measuring set restricted by noise. In sectors 4, the phase space trajectories of the system lie on the surface of doubled torus after one of inherent oscillation periods has undergone the period doubling bifurcation. Many narrow resonance zones, realized under parameter variation, make the system's dynamics in sectors 4 similar to that in sectors 3. But the dynamics differs in that the resonance phenomena take place on 2-tori and appear in the narrower zones of parameter plane [6.15].

In sectors 5, SA_2^T is realized which arises when crossing the bifurcation lines l_{cr}^2 due to several different mechanisms leading to the torus breakdown followed by formation of torus-chaos. The movement into regions 5 is consistent with the processes of the SA_2^T structure complication. The processes are accompanied by a smoothed continuous spectrum, autocorrelation decrease, and the growth of integral oscillation power. In sectors 6, lying in the region of main and second harmonic resonances, SA_1 is realized, the quasiattractor appearing as a result of accumulation of the period doubling bifurcations of resonant cycles after Feigenbaum [6.14,6.15].

Sector 7, where the *A*- and *B*-tongues with different winding numbers intersect is characterized by many different regimes exhibiting hysteresis, the *intermittency of "cycle-chaos" and "chaos-chaos"* types that are sensitive to fluctuations. The intermittency regime between the resonant 3-cycle of *A*-tongue and the SA_2^T was more typical in experiments when entering sector 7 from below and was born from a two-dimensional torus on the base of 2-cycle of *B*-tongue (sector 5 in the *B*-tongue).

Fig.6.3 illustrates the oscillation regimes in the bifurcation diagram sectors described above. It shows the phase trajectory projections onto *xy*-plane taken photographically from the oscillograph screen. Note, that the parameter plane points and regime image scales were chosen from visualization considerations.

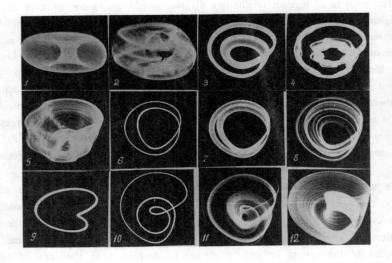

Fig.6.3. Projections of phase trajectories onto *xy*-plane in typical sectors of bifurcation diagram of Fig.6.2.

As seen from the diagrams in Figs 6.1 and 6.2, sectors 5 and 6 correspond to dynamical chaos. However, there are four possible bifurcational transitions to chaos in the system under study. The first proceeds via *resonant cycle doublings* after Feigenbaum *inside the synchronization regions* 1 and 2. The transition is realized by moving in the parameter plane transversely to doubling lines l_2^k. The second transition is performed when crossing l_{cr}^2 from below and is due to the *soft* breakdown *of two-periodic motions* on 1-torus (or on 2-torus if sectors 4 intersect). The third type of transition to chaos takes place under the motion inside any of the synchronization regions toward sectors 5 transversely to the multiplicity lines l_1. The *chaos appears abruptly* with this. Near the line l_1, the *intermittency between* SA_2^T *and synchronization cycles* is observed as well as the hysteresis. The fourth transition type is achieved when moving from sectors 5 to sectors 6 and vise versa. Here, complicated *attractor interaction phenomena* and the *intermittency of "chaos-chaos" type* are registered [6.15]. Such a picture can be observed in the regions of non-linearly interacting oscillation types with different Poincaré winding numbers (in the region of resonance intersection, sector 7).

6.4. TWO-DIMENSIONAL TORUS DOUBLING BIFURCATION. SOFT TRANSITION TO CHAOS

Consider in more detail the transition to SA_2^T under variation of parameter m by recording different values of parameters p and B_0. An example can be the analyzing of oscillation regime rebuildings in the section $f_1 = 13$kHz (Fig.6.2). The 1-torus is softly born on line l_0 from external force cycle. It runs through a series of weak resonances with the growth of m and looks essentially ergodic in experiments. The crossing of line l_3 results in the transition 1-torus → 2-torus; *the two-dimensional torus is doubled*. The doubling of ergodic torus has been found first experimentally in [6.14]. Unlike the resonant cycle doubling on torus, this bifurcation is not understood yet completely from the viewpoint of the stability theory. Nevertheless, it was confirmed by a number of numerical and physical experiments [6.12-6.17].

In Fig.6.4, the Poincaré sections and appropriate oscillation power spectra are indicated that were secured when passing the torus doubling bifurcation point. They were computed for $p = 0.111$, $B_0 = 1.2$, and for the specified parameter m values pointed out [6.15]. There is no doubt in that *the torus doubling proceeds with no resonance* within the calculation accuracy. The map comprises more than 10^3 points, and a long relaxation time is eliminated. Passing the bifurcation point $0.7745 < m^* < 0.7750$ is accompanied by *the soft doubling* of initial invariant closed curve into two independent invariant curves. The curves smoothly move from each other, with the parameter m growth, and remain ergodic. Similar result was obtained by numerically constructing the one-dimensional model map and investigating its evolution with the increase of parameter m. The return map, representing an elliptic closed curve symmetrical with respect to bisectrix, doubles into two closed curves while passing through the bifurcation point m^* which smoothly diverge symmetrically with respect to the initial one.

The numerical and full-scale experiments indicate the *number of torus doubling bifurcations* preceding the transition to SA_2^T to be finite. Their number is defined by the amplitude of external force B_0 and by the distance in frequencies of f_1 from the resonance frequency f_0. It remains yet unclear, whether an infinite sequence of converging torus doubling bifurcations is possible like a cascade of cycle period doublings. Rather, it is not the case. The experiments show that the number of doublings

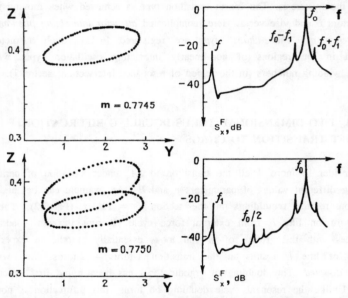

Fig.6.4. Poincaré sections and power spectra when passing through the point of doubling bifurcation of ergodic two-dimensional torus.

grows with decreasing external force intensity and it seems to increase as B_0 tends to zero. But the limit case of torus transition to a cycle is realized here! For the finite amplitudes of external force B_0, the transition to chaos is accompanied by torus doublings, however, they are *not a mechanism of* SA_2^T *birth*.

In Fig.6.5, the experimentally obtained oscillation regime diagrams on the plane of

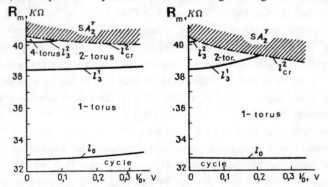

Fig.6.5. Experimental bifurcation diagram of system (6.1) for $f_1 = 8$ kHz (*a*) and $f_1 = 11$ kHz (*b*).

parameters R_m, V_0 are given for the fixed values of the external force frequency f_1. They visually account for the chain of torus doubling bifurcations to be finite and to be interrupted the earlier, the higher the external force amplitude is. What is the bifurcational mechanism of SA_2^T birth actually?

To answer this question, we consider the results of computer simulation performed in the section $p = 0.111$ for $B_0 = 0.3$. In Fig.6.6, the evolution of one of two invariant curves of the 2-torus' Poincaré section with the secant plane $x = 0$ is presented under the parameter m increase. The ergodic 2-torus is doubled once more when increasing m (Fig.6.6b). With this, the onset of invariant curve distortion is seen. Fig.6.6c shows the chain of doublings to be interrupted and the invariant curves *to lose its smoothness*. Above in parameter $m > 1.06$, there exists no longer a two-dimensional torus. In its vicinity (Fig.6.6d), a chaotic set SA_2^T, "*torus-chaos*", appears.

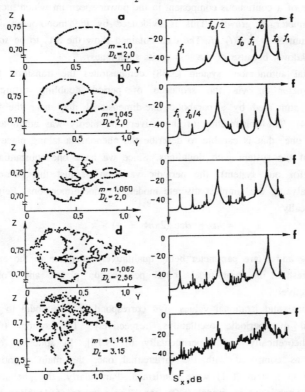

Fig.6.6. SA_2^T development through the loss of smoothness by 4-torus. Computer simulation for $B_0 = 0.3$, $p = 0.111$, $g = 0.3$.

The calculation of the whole Lyapunov exponents' spectrum for transition (Fig.6.6) has shown that Lyapunov dimension D_L on K-tori, $K = 1,2,4$ is 2. Then it grows quickly in the region of $1.06 < m < 1.10$ remaining in the interval of $2 < D_L < 3$, and further, exceeds 3. The analysis of the power spectra $S_x^F(f)$ demonstrates as well *two stages* in the transition to chaos via two-dimensional torus. At the first stage of variation of the parameter m, because of doublings, a half-frequency harmonics $f_0/2$ (2-torus) appear followed by the harmonics $f_0/4$ (4-torus). The second stage starts at gradually developing phenomena of *smoothness loss* by the 4-torus followed by its breakdown. Many spikes of combination frequencies arise softly in the oscillation power spectrum, the latter remaining discrete, however, up to the moment of torus breakdown. The dimension $D_L = 2$ although the Poincaré section looks intricate (Fig.6.6c).

The occurring of a positive Lyapunov exponent in the LCE spectrum is accompanied by the appearance of a continuous component in the power spectrum which increases smoothly with the parameter m growth. This is evidenced by the monotonous increase of the integral spectrum $S_i = \int S(f)\, df$. The data obtained show the SA_2^T to be *softly born* under the torus breakdown because of the loss of its smoothness.

The initial autonomous system (3.38) demonstrates the transition to chaos when varying parameter m via the sequence of period doubling bifurcations and is approximately simulated by a parabolic one-dimensional map near the transition point (Chapters 3, 4). The construction of the universality theory was attempted, similarly to Feigenbaum's one, that is capable to describe the phenomena taking place under periodic perturbation of the system with doublings. Since we have one substantial parameter of the task (m for our system), the periodic variation of just this parameter is to be studied by analyzing the simplest discrete model, the one-dimensional Feigenbaum's map forced periodically:

$$x_{n+1} = a + B\cos(2\pi n\Phi + \Psi) - x_n^2, \qquad (6.2)$$

where a, B, Ψ and Φ are parameter being qualitatively adequate to the exceeding of the generation threshold m, amplitude, initial perturbation phase, and Poincaré winding number, respectively.

The rational and irrational values of Φ correspond in this model to resonant cycles on torus and quasiperiodic oscillations, respectively. The system (6.2) was first investigated theoretically and experimentally in [6.18]. In spite of essential model simplification as compared with the differential one, the main phenomena observed experimentally in system (6.1) were confirmed. Particularly, the phenomenon of torus doubling was established in model (6.2), too. The number of doubling bifurcations are assumed to grow tending to infinity when decreasing the external force amplitude to zero.

Unfortunately, the possibility of analyzing the mechanisms of two-dimensional torus breakdown by using the model maps of a ring or a plane is excluded due to the restrictions for one-dimensional description of the evolution of invariant curves with variation of the parameters.

The bifurcation phenomena, being qualitatively similar to the above ones, take place inside the synchronization regions (in the A- and B-tongues, for example). The resonant cycles with complicated structures (Fig.6.3), lose their stabilities on the bifurcation lines l'_0 under the soft birth of corresponding tori which is broken down when crossing the lines of critical parameter values l^2_{cr}.

A regularity in the sequence of critical phenomena is seen. The resonant cycles lose their stabilities on the neutrality lines l'_0. The synchronization tongues appear at the codimension-two bifurcation points on lines l'_0 where Φ is rational. The more intricate resonant cycles again lose their stabilities inside the tongues with the birth of two-dimensional tori. The resonant cycles' periods increase with every bifurcation of such a type by a factor of q that is an order of resonance. As a result, periodic regimes are born with the period tending to infinity. The cascade of critical phenomena is condensed in parameters and presents a qualitative generalization of the Feigenbaum's similarity regularities for a case of two-parametric families.

From the viewpoint of radiophysics, the task considered is a part of general problem of *synchronizing* the non-linear oscillations with strong interactions in dissipative systems. Here, the phenomena are manifested which are difficult or impossible to be calculated analytically. The model experiments described above assist in understanding these phenomena.

Let us enumerate the major non-linear phenomena found experimentally in the nonautonomous oscillator with inertial non-linearity: *frequency locking* in the region of basic resonance on subharmonics $f_k = pf_0/2^k$, $k = 1,2,...$; *ergodic torus period doubling bifurcation*; *the birth process of tori* and *resonant cycles* that are accumulated in parameters *in the synchronization regions* with the rational Poincaré winding number; chaos *onset* via the cascade of resonant cycle *doublings* and *two-dimensional tori* breakdown preceded by smoothness loss and resonances; as well as the *intermittency* of "*cycle-chaos*" and "*chaos-chaos*" type. As expected, system (6.1) realizes practically all the known transitions to chaos while being the most simple from such class of systems.

6.5. BIFURCATION MECHANISMS OF TORUS-CHAOS BIRTH UNDER TWO-FREQUENCY OSCILLATION BREAKDOWN

System (6.1) seems to be suitable for investigating the problem of two-dimensional torus breakdown in detail. A rigorous conclusion of [6.9] on the mechanisms of the two-dimensional torus breakdown is substantiated by the results of experiments described above. Nevertheless, investigations are desired to interpret the qualitative conclusions in radiophysics language, particularly, by describing the mechanisms of torus breakdown, followed by the SA_2^T formation, in the terms of oscillation power spectra [6.15,6.19].

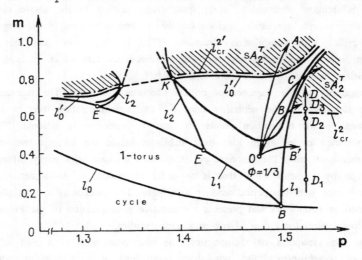

Fig.6.7. Critical phenomena in system (6.1) in the vicinity of resonance 1:3. Computer simulation.

To do this, elaborate a synchronization region having the Poincaré winding number $\Phi = 1:3$ (*B*-tongue in Fig.6.1a) presented in Fig.6.7 on an enlarged scale. We analyze the evolution of oscillation regimes numerically when moving in the directions A, B, B', C and D shown in the figure with arrows. The projections of a *stroboscopic map* in a period of external force onto the plane of variables x and y, as well as the power spectra of phase flow $S_x^F(f)$ and the corresponding map power $S_x^P(f)$ are to be considered. To calculate the spectra, the data file was set by storing a sequence of intersection points with the secant followed by calculating the spectrum $S_x^P(f)$ [6.19].

In Fig.6.8, the evolution of above characteristics is presented under the soft loss of smoothness by 1-torus that corresponds to the motion along the way D in Fig.6.7. The invariant curve L, being smooth near the line of torus birth l_0, is gradually deformed when moving in the direction transversal to l_0. Then it loses its smoothness and is broken down when crossing the line l_{cr}^2. A chaotic attracting set, the "torus-chaos" SA_2^T is born in the vicinity of the torus broken down. Its image is a set of the stroboscopic section points as $m = 0.625$[1]. The destruction of invariant curve L in the map is accompanied by the enrichment of spectrum $S_x^F(f)$ with combination frequencies, their

---

[1]The results obtained by introducing the plane x = 0 (Fig.6.6) and with the help of the phase space mapping onto itself in a period of external force (Fig.6.8) are qualitatively identical.

number increases and their amplitudes grow smoothly. The birth of SA_2^T is adequate to the appearance of a continuous spectrum. The picture of the evolving map spectra $S_x^P(f)$, which represents the variations in a low-frequency part of spectrum more accurately, is especially illustrative and informative. From the physical point of view, the map spectrum corresponds to the spectrum of time realization envelope $x(\tau)$, i.e., that of the detected signal.

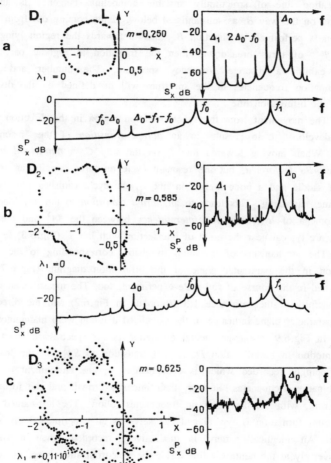

Fig.6.8. Evolution of Poincaré maps and oscillation power spectra during the loss of smoothness by a two-dimensional torus.

The evolution of oscillation regimes that corresponds to the motion along the way A (Fig.6.7), qualitatively follows the pattern shown in Fig.6.8. The differences lie in the arrangement of spectrum lines on the frequency axis which is specified by the winding number Φ at the point of intersection of the neutrality lines l'_0 on the way A.

The coming out from the synchronization region in the direction, indicated in the diagram with arrow B, is attended by transition to SA_2^T under the merging of saddles and nodes followed by their vanishing when the torus is not smooth already at the bifurcation point on line l_1. This transition mechanism is possible provided that the bifurcation line of structurally unstable homoclinic tangency is above the curve of motion on the way B, as substantiated below. If the coming out from the synchronization region is performed in direction B', that is towards the region lying below the critical line l_{cr}^2 of torus breakdown, then the transition to ergodic oscillations is observed that exhibit a discrete frequency spectrum. The number and intensities of the combination frequencies in the spectrum will be defined by the distance in parameter from the bifurcation line l_0.

The movement from the synchronization region in the direction C to the region of the developed chaos results in an *abrupt transition* of the "resonant cycle - chaos" type. When moving towards line l_1 on the way C, a *non-trivial hyperbolic subset of trajectories is formed*, but the resonant cycle on torus remains as an attracting regime. As a saddle and a node merge on line l_1, the cycle vanishes and the hyperbolic subset becomes attracting. This leads to an *abrupt transition* to the developed torus-chaos. The phenomena of *hysteresis and intermittency* between the SA_2^T and the resonant cycle on torus are typical near the point of intersection with line l_1 (Fig.6.3).

The mechanisms of transition to chaos, corresponding to the given directions of motion in the parameter plane of the bifurcation diagram (Fig.6.7), are conclusively recorded in the course of full-scale experiment, too. The measurements were performed in the vicinity of the resonance 1:3 (B-tongue in Fig.6.2) and the directions of motion in the parameter plane indicated in the calculated diagram were maintained.

In Fig.6.9, the experimental oscillation power spectra are given that correspond to the motion in the direction D. A soft transition to SA_2^T via the loss of smoothness is convincingly recorded which is accompanied by the enrichment of the spectrum with combination frequencies. Further, their intensity growth and the formation of continuous spectrum, with δ-spikes at basic frequencies follow. The last-named feature of the SA_2^T spectrum formation is typical and can serve as an experimental criterion for torus-chaos birth. An identical pattern is qualitatively recorded when moving in direction A transversely to the neutrality line inside the synchronization zone.

The oscillation spectrum variations that take place when moving in directions B and B' are illustrated in Fig.6.10. In the first case, entering the SA_2^T region is performed and in the second one, entering the region of ergodic deformed torus proceeds when it is not broken down yet. When we move in the direction C, then, as it was predicted by calculations, an abrupt transition to the developed torus-chaos via intermittency is recorded with the hysteresis phenomena. Detailed comparison of experimental and calculated data enables one to argue that all the mechanisms of transition to chaos SA_2^T

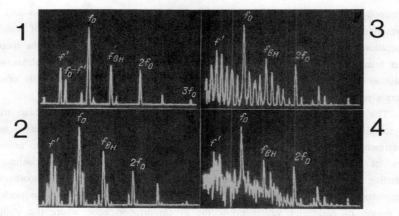

Fig.6.9. Evolution of oscillation power spectra under motion in parameter plane in direction D of the diagram shown in Fig.6.7. 1 - spectrum of smooth torus; 2,3 - development of smoothness loss phenomena; 4 - spectrum of torus-chaos.

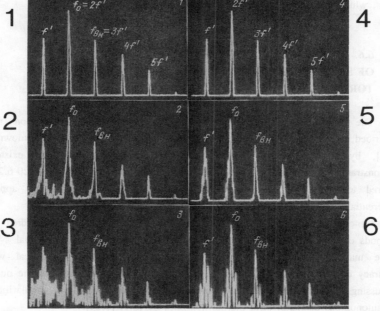

Fig.6.10. Changes in spectra when going out of the region of phase locking in direction B (1,2,3) and B' (4,5,6). Spectra 1 and 4 are identical and correspond to the point O of parameter plane in Fig.6.7.

in the nonautonomous oscillator with inertial non-linearity are realized rigorously according to the *torus breakdown theory* [6.9].

In the course of computer simulation and full-scale experiments described, the loss of smoothness by torus was diagnosed by the invariant curve deformations and by the corresponding spectrum evolution under the motion close to the line of saddle-node bifurcation but outside, and not within, the synchronization zone. As for the theorem of torus breakdown, it was formulated for the resonance region and manipulates with the evolution of the closure of the unstable saddle cycle manifolds to a stable one which forms a robust two-dimensional torus.

The technique of numerical (maps and spectra) and physical (phase trajectory projections and spectra) visualization of the torus smoothness loss, used in experiments, is based on that the resonant invariant manifold behaves, when crossing the lines of saddle-node bifurcations, as if it does not suffer a discontinuity, by inheriting its shape in the region of ergodic oscillations. The smoothness loss phenomenon would turn out to be "invisible" if such investigations are to perform for the region of parameter values inside the synchronization zone where the stable limit cycle exists. Here, the unstable saddle cycle manifolds or unstable saddle point separatrixes in map are to be calculated which would "*develop*" the two-dimensional torus, which is invisible in case of resonance.

6.6. UNIVERSAL QUANTITATIVE REGULARITIES OF SOFT TRANSITION TO CHAOS VIA TWO-DIMENSIONAL TORUS BREAKDOWN

The transition to torus-chaos from the two-frequency oscillation regime is described, along with *qualitative regularities* stated by the torus breakdown theorem [6.9], by a series of *universal quantitative characteristics*. Their existence was demonstrated as applied to one-dimensional circle maps [6.10,6.11,6.20-6.22]. It is desired to reveal to what extent the conclusions on universality are applicable to differential systems having invariant closed curves in the Poincaré map.

General regularities of the soft transition to torus-chaos were studied using the methods of numerical and full-scale experiments on a number of differential systems and plane maps [6.23,6.24]. The results obtained are practically identical within the accuracy of experiments. Therefore, in the present section we shall confine ourselves by discussing the investigations of quantitative torus-chaos transition regularities in the nonautonomous oscillator with inertial non-linearity (6.1) [6.24].

In Fig.6.11, the results of a physical experiment investigating the oscillation regimes and their bifurcations on the plane of parameters V_0 (external signal voltage amplitude) and f_1 (modulation signal frequency) are presented. A qualitatively equivalent picture was obtained as well by numerically calculating the bifurcation diagram of system (6.1) with the values of control parameters corresponding to the experiment.

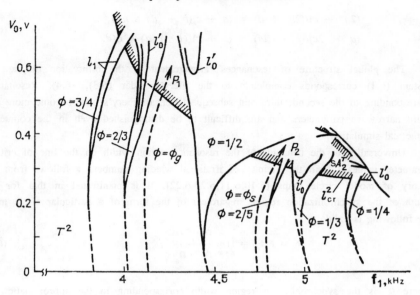

Fig.6.11. Bifurcation diagram for the nonautonomous oscillator (6.1). Full-scale experiment.

The resonance regions of the diagram, shown in Fig.6.11, can be classified according to *Fiery's* rule. The resonances, corresponding to the winding numbers $\Phi = 1:(n + 1)$ belong to the so-called first tier of "*Fiery's tree*". By Fiery's rule of addition, the resonances of the first tier are born by two rational numbers[2]:

$$1/(n + 1) = 1/n \oplus 0/1, \quad n = 0, 1, 2, \ldots \qquad (6.3)$$

The second tier of the Fiery's tree is formed by "adding" two rational fractions $a/b \oplus c/d$. The fraction $(a + c)/(b + d)$ is the only rational number having a minimum denominator on the segment between the numbers a/b and c/d. The value of this rational fraction lies in the middle of the above segment and is called a *mediant* if the condition of unimodularity $|ad - bc| = 1$ is fulfilled.

The mediant $(a + c)/(b + d)$ is unimodular with every of bearing rational fractions a/b and c/d providing the construction of subsequent mediant branches of the Fiery's tree as follows:

---

[2] Symbol $\oplus$ denotes: $1/n \oplus 0/1 = (1 + 0)/(n + 1)$.

$$(2a + c)/(2b + d) = (a + c)(b + d) \oplus a/b,$$
$$(a + 2c)/(b + 2d) = (a + c)/(b + d) \oplus c/d. \tag{6.4}$$

The global structure of resonances' arrangement in the bifurcation diagram of system (6.1) corresponds completely to the Fiery's order (6.3), (6.4). Resonances corresponding to the second, third and subsequent tiers of Fiery's tree, become more and more narrow in parameters and are difficult to be distinguished even in the course of numerical simulation.

Universality in the behavior of the resonance region width on the line of critical parameter values l^2_{cr} in the vicinity of irrational winding numbers Φ follows from the theory of model circle maps [6.10,6.11,6.20-6.22]. It is manifested in that for the width of the synchronization region, regardless of the form of a particular circle map, the following is valid:

$$\delta = \delta(\Phi) = \lim_{n \to \infty} \frac{\Omega_{n-1} - \Omega_n}{\Omega_n - \Omega_{n-1}}, \tag{6.5}$$

where Ω_n is the synchronization region width corresponding to the approximation of irrational number Φ under study by the n-th term of a rational number set which converges to Φ as $n \to \infty$. The universal constant δ depends on the winding number only and equals: $\delta_g = 2.833...$ (*golden section*, $\Phi = 0.5 \ (\sqrt{5} - 1)$), $\delta_s = 6.799...$ (*silver section*, $\Phi = \sqrt{2} - 1$).

One-dimensional curves P_1 and P_2, obtained experimentally, are shown in Fig.6.11, where the winding number is constant and is equal to the golden and the silver sections, respectively. As calculations have demonstrated, $\delta = 6.25 \pm 0.5$ in the vicinity of the silver section being in a satisfactory agreement with theoretical predictions.

A set of blanks between the resonance zones supplementing the set of resonance zones to normalized Lebesgue measure 1 has a *Cantor structure* on the critical parameter value line l^2_{cr}. The measure of a set of these blanks is zero but the fractal dimension D_F is finite. According to the theory, $D_F = 0.867...$ in the vicinity of the golden section. To calculate D_F by the results of computer simulation an full-scale experiments, the following relation can be used:

$$\lim_{n \to \infty} \sum_i [R_i(n)]^{D_F(n)} = 1, \tag{6.6}$$

where $R_i(n) = L_i(n)/\bar{L}(n)$, $\bar{L}(n)$ is the length of a blank between the resonances a/b and c/d on the n-th Fiery's tree tier, and $L_i(n)$, $i = 1,2$ are the blank lengths between every of these resonances and their mediant $(a + c)/(b + d)$.

Based on relation (6.6), the values of fractal dimension D_F were determined using the data of computer simulation and full-scale experiments for a number of systems. In particular, the estimation of D_F using the data of full-scale experiment (Fig.6.11) provides $D_F = 0.78$ by measuring the blank lengths between the resonances 1:2, 3:4 and 2:3 (in the vicinity of the golden section P_1 on line l_{cr}^2). In the vicinity of the silver section (blanks between the resonances 1:2, 1:3, and 2:5), experiments yielded a surprising result, $D_F = 0.86$! A full agreement to the theory is accounted for rather by an accidental coincidence because of experimental errors due to approximation of the curved boundary of l_{cr}^2 with a broken line and with a relatively great error of definition of the resonance region's boundaries. The closeness of experimental dimension D_F values to the theoretical one is evidently not accidental. This is ensured by the results of calculations performed with a higher, as compared to experiment, accuracy. The calculated dimension values D_F as a function of the winding number, considered tier of resonances of the Feiry's tree, and a specific form of the system under study are within the interval $0.85 < D_F < 0.90$ [6.24]. The data obtained do not conflict with the theory at least, if one takes into account that the absolute error of dimension D_F calculations is as low as ± 0.05.

The power spectrum of a circle map at the critical point was shown, by using the renormalization group method, to exhibit a number of universal properties [6.10]. Its nature is defined by the winding number Φ. If Φ can be presented as a periodic continued fraction, then the spectrum would have the property of scale invariance. The golden section $\Phi = \Phi_g = 0.5(\sqrt{5}-1) = <1,1,1,1...>$ is characterized by the simplest expansion into a periodic continued fraction. In this case, the reduced frequencies of spectrum lines $\nu = f/f_1$ satisfy, within the interval [0,1], the following relation

$$\nu = |n_2 \Phi_g - n_1|, \quad n_2 > n_1$$

where n_1 and n_2 are the sequential terms of one of Fibonacci's series. So, for a major spectrum series, comprising the lines with the highest amplitudes, n_2 and n_1 are the sequential terms of the main Fibonacci's series with the base [1.1]. Further, the spectrum series, arranged in decreasing order with respect to their amplitudes, correspond to the Fibonacci series having the following bases: 2nd series - (2,2), 3d one - (1,3), 4th one - (3,3), 5th one - (1,4), 6th one - (2,5) and so on. The spectrum power S_i^j for the lines of a each series is proportional to the square of frequency

$$S_i^j = C_i \nu^2(j), \qquad (6.7)$$

where i and j are the numbers of a series and of the given series' line, respectively. Relation $S_i^j/\nu^2(j)$ for great j is constant for the lines of every series. It becomes practically constant even for $j = 2,3$ for the circle map.

It is interesting to study the quantitative relations in the oscillation power spectrum when the two-frequency oscillations are broken down in a real flow system. The externally harmonically driven oscillator (6.1) seems to be the most convenient model for this.

The spectrum line frequencies were measured in the course of experiment within the interval $[0, f_1]$ near the point of torus breakdown (near the line l^2_{cr}) at different parameter g values. The measurement results are presented in Fig.6.12a,b together with the corresponding spectra photographs. The arrangement of spectrum lines on the frequency axis is in a full agreement with the theory! As for the amplitudes of spectrum lines $a_i^j = 10\lg[S_i^j/S_1^1 \nu^2(j)]$, the regularity predicted is not observed here. The power-to-frequency ratio for the lines of an individual series decreases with the diminishing of frequency. The nature of the spectrum line behavior does not vary qualitatively with variation of parameter g.

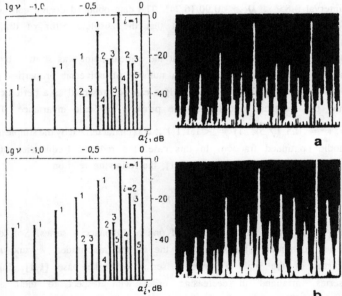

Fig.6.12. Power spectrum of oscillations near the torus breakdown line; a - $g = 0.1$, b - $g = 0.4$.

To compare the results of a physical experiment with the data obtained by examining the mathematical model (6.1), the power spectrum was calculated near the transition to torus-chaos SA_2^T with the winding number being equal to the golden section. The inertia parameter was chosen to be 0.3, the excitation parameter $m = 0.7$ corresponding to a period-one limit cycle in the autonomous oscillator. The external force amplitude was increased up to a value being close to the critical one. Calculation was performed as

$B_0 = 0.24$, $p = 2\pi f_1 = 0.62192$ (f_1 is a normalized frequency).

In Fig.6.13, the computer simulation results are indicated: a - refers to the spectrum of a flow $S_x^F(f)$, b - to the spectrum of a map $S_x^P(f)$ in the section of the flow with the plane $x = 0$, and c - is the invariant closed curve in the section of torus at the instant of its breakdown. As seen from the figure, the spectrum of a flow does not provide a sufficient information to analyze the spectrum lines. The need for considering the spectrum of a map appears.

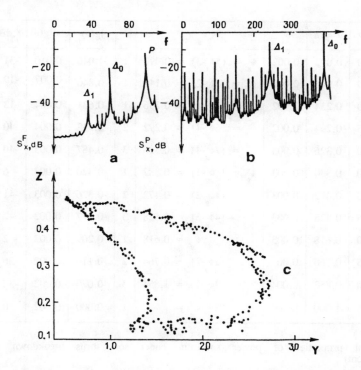

Fig.6.13. Power spectra of flow (*a*) and Poincaré map (*b*). Poincaré section (*c*).

Due to a limited memory of a computer, it turned out to be impossible to construct the spectrum of a map in the band involving the external force frequency f_1. Therefore, one had to confine himself by using the interval $[0, \Delta_0]$ and to take Δ_0 and $\Delta_1 = f_1 - \Delta_0$ as basic frequencies with their ratio being equal to the golden section, too (since these are the frequencies of the adjacent lines of the main spectrum series). For a more detailed comparison of the results of computer simulation and full-scale experiment, the full-scale experiment was repeated for $g = 0.3$ and the corresponding frequency band $[0, \Delta_0]$.

The findings are summed in Table 6.1 (full-scale experiment) and Table 6.2 (computer simulation). Here, the measurement error estimates for frequency $|\pm \delta_1|$ and for amplitude $|\pm \delta_2|$ of spectrum lines are also indicated. To exclude a systematic error due to an inexactly specified winding number Φ_g, the spectrum lines are taken into account in the Tables with the given frequencies differing from theoretical values by not more than the measurement error evaluated.

Table 6.1

f,kHz	$\nu = f/\Delta_0$	δ_1	$\|n_2\Phi_g - n_1\|$	i	$\lg(\nu)$	$\delta\lg(\nu)$	a_i^j,dB	δ_2,dB
0.240	0.092	0.002	$\Phi_g^5 = (5\Phi_g - 3) = 0.090$	1	-1.040	0.010	-31	3
0.380	0.146	0.002	$\Phi_g^4 = \|3\Phi_g - 2\| = 0.146$	1	-0.840	0.007	-19	3
0.610	0.234	0.002	$\Phi_g^3 = (2\Phi_g - 1) = 0.236$	1	-0.631	0.004	-13	3
0.765	0.294	0.002	$= \|6\Phi_g - 4\| = 0.292$	2	-0.532	0.004	-40	3
0.850	0.326	0.003	$= (7\Phi_g - 4) = 0.326$	3	-0.487	0.003	-40	3
0.990	0.380	0.003	$\Phi_g^2 = \|\Phi_g - 1\| = 0.382$	1	-0.420	0.003	-6	3
1.230	0.472	0.003	$(4\Phi_g - 2) = 0.472$	2	-0.326	0.003	-41	3
1.375	0.528	0.003	$\|4\Phi_g - 3\| = 0.528$	3	-0.277	0.002	-42	3
1.610	0.618	0.003	$\Phi_g = 0.618$	1	-0.209	0.002	-2	3
1.995	0.766	0.003	$\|2\Phi_g - 2\| = 0.764$	2	-0.116	0.002	-26	3
2.220	0.852	0.004	$(3\Phi_g - 1) = 1.854$	3	-0.070	0.002	-32	3
2.605	1.000	0.004	$\Phi_g^0 =$	1	0.000	0.002	0	3

Universal properties of power spectrum under soft torus breakdown (full-scale experiment)

More visually the results are represented in Fig.6.14. The nature of the spectrum obtained experimentally (Fig.6.14a) is the same as in Fig.6.12. The spectrum line amplitudes a_i^j decrease when moving towards the low frequency band. This phenomenon can not be accounted for by an error, since the amplitudes were varied monotonously. Furthermore, the difference between the spectrum line amplitudes exceeds the measurement error considerably. The spectrum obtained numerically is different-looking (Fig.6.14b). Amplitudes a_i^j do not diminish with the decrease of frequency. They fluctuate near a certain constant level, the dispersion of their values being small and comparable with the measurement error ($\approx$ 5dB). The latter corresponds to the theory conclusions better than the results of full-scale experiments. The amplitudes of the 2nd and 3d series in the full-scale experiment are lower than that of the main series lines in the average by 30dB while in computer simulation only by 15dB.

Table 6.2

f,kHz	$\nu=f/\Delta_0$	δ_1	$\|n_2\phi_g - n_1\|$	i	$\lg(\nu)$	$\delta\ln(\nu)$	a_i^j,dB	δ_2,dB
23	0.059	0.003	$\phi_g^6 = \|8\phi_g - 5\| = 0.056$	1	-1.230	0.020	-4	5
35	0.090	0.003	$\phi_g^5 = 5\phi_g - 3 = 0.090$	1	-1.040	0.010	-2	5
57	0.147	0.003	$\phi_g^4 = \|8\phi_g - 5\| = 0.146$	1	-0.840	0.010	-5	5
69	0.177	0.003	$10\phi_g - 6 = 0.180$	2	-0.750	0.007	-16	5
80	0.205	0.003	$\|11\phi_g - 7\| = 0.202$	3	-0.690	0.007	-18	5
92	0.235	0.003	$\phi_g^3 = 2\phi_g - 1 = 0.236$	1	-0.630	0.006	-6	5
115	0.294	0.003	$\|6\phi_g - 4\| = 0.292$	2	-0.530	0.005	-12	5
127	0.325	0.003	$7\phi_g - 4 = 0.326$	3	-0.490	0.005	-13	5
150	0.384	0.004	$\phi_g^2 = \|\phi_g - 1\| = 0.382$	1	-0.416	0.004	-4	5
172	0.440	0.004	$\|9\phi_g - 6\| = 0.438$	4	-0.357	0.004	-24	5
184	0.471	0.004	$4\phi_g - 2 = 0.472$	2	-0.327	0.003	-20	5
207	0.529	0.004	$\|4\phi_g - 3\| = 0.528$	3	-0.277	0.003	-29	5
219	0.560	0.004	$\|9\phi_g - 5\| = 0.562$	5	-0.252	0.003	-15	5
242	0.619	0.004	$\phi_g = 0.618$	1	-0.208	0.003	-1	5
276	0.706	0.004	$6\phi_g - 3 = 0.708$	4	-0.151	0.003	-21	5
299	0.765	0.005	$\|2\phi_g - 2\| = 0.764$	2	-0.116	0.003	-6	5
334	0.854	0.005	$3\phi_g - 1 = 0.854$	3	-0.068	0.002	-14	5
357	0.913	0.005	$\|5\phi_g - 4\| = 0.910$	5	-0.040	0.002	-21	5
391	1.000	0.005	$\phi_g = 1.000$	1	0.000	0.002	0	5

Universal properties of power spectrum under soft torus breakdown (computer simulation)

The discrepancy between the results of full-scale experiment and computer simulation can be explained by the formers referring to the spectrum of the flow and the latters to that of the map. Transition from a flow system to the map is connected with the non-linear spectrum transformation being individual for a particular dynamical system. Furthermore, as numerical investigations show, the relation between the spectrum line amplitudes depends on the shape of the mapped curve even for one-dimensional maps. The invariant closed curve in the section of a particular dynamical system has its specific deformation nature on the way to breakdown. To transform this curve into a circle, a certain non-linear (as well as nonsmooth at the critical point) coordinate substitution is required. Accordingly, the amplitudes of spectrum lines suffer the non-linear transformation as well. The invariant curve in the section of the dynamical system (6.1) is far from the circle at the instant of breakdown (Fig.6.13c). Therefore, it may be proposed that the distinction of the spectrum presented in Fig.6.14b from that of a circle map is accounted for not so much by the calculation error for the amplitudes of spectrum lines but rather by the invariant curve deformation in the section.

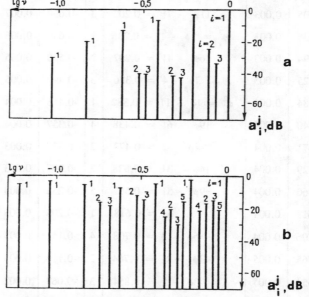

Fig.6.14. Experimental power spectrum of flow (a) and calculated power spectrum of map (b) near the critical point of torus breakdown ($g = 0.3$).

Thus, a distinct non-linear transformation of the spectrum component amplitudes for each particular system is to be performed for going from the spectrum of a specific dynamical system at the instant of torus breakdown when $\Phi = \Phi_g$ to the universal spectrum of a circle map. As for the frequency relations when $\Phi = \Phi_g$, they are observed well on any models since they are completely defined by the value of winding number.

The above experimental results demonstrate a number of universal properties of one-dimensional circle maps to hold for differential systems. A most clear-cut agreement takes place for the spectrum structure of oscillations at the critical point as $\Phi = \Phi_g$.

The theory and experiment results agree within a relatively high accuracy of spectrum measurements. A good agreement with the theory can be stated as well for the fractal dimension of a Cantor set of the irrational winding numbers near the critical line of torus breakdown.

The global structure of resonance regions arrangement on the plane of control parameters is also in keeping with the theoretically described structure, as it follows from the above results, and is defined by the Fiery's rule. The geometric structure of the resonance regions near the critical line l^2_{cr}, defined by relation (6.5), seems to be universal, too. It is to note that the final results cited are of importance and far from being obvious. The matter is that the interrelation between the map in the Poincaré section and the flow is radically not one-to-one. Just due to the above thing considered, the scale invariance of the amplitude relations in the spectrum of quasiperiodic oscillations is not confirmed for differential systems although it undoubtedly takes place for one-dimensional circle maps.

CHAPTER 7

BREAKDOWN OF TWO- AND THREE-FREQUENCY QUASIPERIODIC OSCILLATIONS

7.1 TRANSITIONS TO TORUS-CHAOS IN THE SYSTEM OF TWO COUPLED OSCILLATORS

The evolution of two-frequency oscillations in the nonautonomous oscillator with inertial nonlinearity under variation of parameters provides a rather visual picture of possible mechanisms of two-dimensional torus breakdown and transitions to chaos. But a number of interesting phenomena, being characteristic for autonomous dynamical systems, is not touched upon by it. Moreover, from the physical point of view, substantiated reasons are desired for a qualitative similarity in the behavior of autonomous and nonautonomous systems taking into account the importance of autonomous systems in understanding the mechanisms of turbulence development in the solid medium.

Consider the evolution of quasiperiodic oscillations in an autonomous system. It is natural to address to a model of two oscillators coupled in any manner which are capable to operate individually in complicated self-oscillation regimes.

We refer to a system of two oscillators with inertial nonlinearity coupled by means of inductance having the structure shown in Fig.7.1 as a diagram. Considering the parameters of a partial oscillator as identical, we write the following system of equations [7.1,7.2]:

$$(1 - \gamma^2) \dot{x}_1 = y_1 + x_1(m - z_1) + \gamma[y_2 + x_2(m - z_2)],$$
$$(1 - \gamma^2) \dot{x}_2 = y_2 + x_2(m - z_2) + \gamma[y_1 + x_1(m - z_1)],$$
$$\dot{y}_1 = -x_1, \quad \dot{z}_1 = g[I(x_1) x_1^2 - z_1], \quad (7.1)$$
$$\dot{y}_2 = -x_2, \quad \dot{z}_2 = g[I(x_2) x_2^2 - z_2],$$

where $0 \leq \gamma \leq 1$ is the coupling coefficient.

The case $\gamma = 0$ is trivial. As $\gamma > 0$, two solutions are possible: the *case of general position* when at least one of conditions $x_1 \neq x_2$, $y_1 \neq y_2$, and $z_1 \neq z_2$ is fulfilled for time τ as long as desired; and the degenerate case of motion in the three-dimensional invariant subspace of six-dimensional phase space of system (7.1) where equalities $x_1 = x_2$, $y_1 = y_2$, and $z_1 = z_2$ are complied with for any τ. The general case and the degenerate one will be called *asymmetric* (ASM) and *symmetric* (SM), respectively.

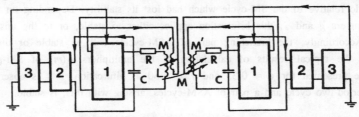

Fig.7.1. Block-diagram of two oscillators coupled by means of inductance. 1 - basic amplifiers, 2 -linear amplifiers of inertial cascades, 3 - one-and-a-half-period square-law detectors with RC-filters.

The critical phenomena for the periodic solutions of the system are to be studied under variation of its parameters. In the SM-case, equations (7.1) degenerate into the three-dimensional system as follows:

$$(1 - \gamma^2)\ \dot{x} = y + x\ (m - z)$$
$$\ddot{y} = -x \qquad (7.2)$$
$$\dot{z} = -gz + I(x)\ x^2,$$

which permits the analysis of the influence of coupling on the partial oscillator's dynamics. Equations (7.2) are reduced, in their form, to the case where coupling is absent, $\gamma = 0$, by a linear substitution of variables and parameters:

$$x = (1 - \gamma)^{1/4}X, \qquad y = (1 - \gamma)^{3/4}Y, \qquad z = (1 - \gamma)^{1/4}Z,$$
$$\tau = (1 - \gamma)^{1/2}t, \qquad m = (1 - \gamma)^{1/2}M, \qquad g = (1 - \gamma)^{-1/2}G. \qquad (7.3)$$

Thus, the dynamics of systems (7.2) and (3.38) are identical. The influence of coupling results in variation of bifurcational values of parameters m and g as a function of γ according to substitution (7.3)[1].

The investigation of SM-regime in the total system of equations (7.1) seems to be of importance. The calculation of SM-cycle multipliers in the six-dimensional system of equations as a function of parameter m, which is varied roughly, and for different g and γ, has shown the change of one of the SM-cycle multipliers via + 1 to take place, with

---

[1] A crucial conclusion follows that chaotic self-oscillation regimes are possible to be completely synchronized in the system of two interacting oscillators. The chaotic synchronization would be realized in experiments provided that SM-chaos is stable with respect to perturbations in the directions being transversal to SM-subspace.

which the cycle under study *does not vanish*! The SM-cycle loses its stability in the directions being transversal to the invariant three-dimensional subspace. With this bifurcation, a pair of ASM-cycles is branched off from the SM-cycle which are mirror-symmetric relative to the SM-cycle which had lost its stability. Depending on the values of parameters g and γ, this bifurcation applies either to stable, or to the unstable SM-cycle. Accordingly, the branching pair of ASM-cycles will be stable or unstable. In Fig.7.2, the typical results of calculation of the multipliers $\rho_i(m)$ of SM-period-two cycle are presented for $g = 0.35$ and $\gamma = 0.30$ as bifurcation + 1 is observed for a stable period-two cycle and a pair of ASM-cycles, arising with this, is stable.

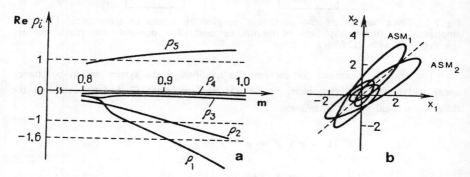

Fig.7.2. Multipliers of the SM period-two cycle of system (7.1) as functions of parameter $m(a)$; projections of the ASM period-two cycles on the plane of variables x_1 and x_2 (b).

The evolution of ASM-cycles under variation of the parameters of system (7.1) turns out to be radically another than for SM-cycles. The calculation of the dependence for the ASM_1-cycle multipliers $\rho_i(m)$ has shown that, as parameter m grows, the loss of stability by a periodic regime is due to the reaching by a *pair of complex-conjugate multipliers* of the unit circle. With this, a stable two-dimensional torus is born in the six-dimensional phase space of the system, *the bifurcation is supercritical*. Similar picture is observed for ASM_2-cycle.

The bifurcation of two-dimensional torus birth in the system under study is typical one-parametric and has *codimension one*, as in general case. Small parameter variations lead this bifurcation not to vanish in the space of parameters m, g and γ but somewhat to shift along the hypersurface corresponding to the given type of stability loss.

In Fig.7.3, the dependences of the modulus of multipliers $\rho_i(m)$ for the period-two ASM_1-cycle for $g = 0.3$ and $\gamma = 0.3$ are cited. At the bifurcation point $m^* = 1.57979... \approx 1.58$, the pair of multipliers reaches the unit circle $\rho_{1,2} = exp(\pm j\varphi)$ where $|\rho| = 1$, $\varphi = 145.0958...°$. Resonance is absent (φ is not a multiple of $360°$), however, the situation is close to the weak resonance 2:5, what is convenient to use. It is known that *torus exists always* in autonomous systems with weak resonances [7.3]. The period of a resonant cycle on the torus depends on winding number and is relatively small for $\Phi = 2:5$. This permits, in practice, to analyze numerically the bifurcations of birth and stability loss for the above cycle on a two-dimensional torus.

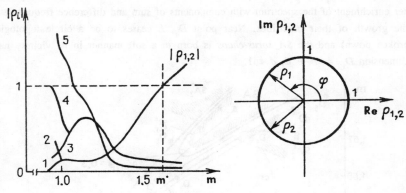

Fig.7.3. Multipliers of ASM_1 cycle of system (7.1) as functions of parameter m (*a*); Multipliers $\rho_{1,2}$ at the critical point $m = m^*$ (*b*).

We analyze the bifurcations of two-dimensional torus in the vicinity of resonance 2:5. Using the point on the neutrality line, found above, as the starting one, construct the line l_0 itself in the plane of parameters m and γ as $g = 0.3$. The *codimension-two bifurcation point* corresponding to the resonance 2:5 on the neutrality line is to be found and a bifurcation diagram in its vicinity is to be constructed. The calculation results are presented in Fig.7.4.

In the resonance region, a resonant cycle P^+ on torus has the spectrum $S_x^F(f)$ with the basic frequencies $f_1 : f_2 = 2:5$ and their harmonics. The presence of the stable cycle P^+ inside the synchronization zone *eliminates the possibility* of observing the phenomenon of the smoothness loss of torus by the form of spectrum $S_x^F(f)$. The spectrum of P^+ is qualitatively identical to that demonstrated in Fig.7.4 everywhere in the resonant tongue. However, if one moves on the way D, where the torus is ergodic, then the smoothness lost by the torus would manifest itself in the characteristic evolution of spectra and Poincaré maps. Near point K and outside the synchronization tongue, the invariant curve L of map projection onto the plane is a smooth curve having the shape of ellipse. The oscillation spectrum is similar to that of resonant cycle but the ratio of frequencies f_1 and f_0 is irrational. Under the motion off from the torus birth line l_0, the curve begins to deform and δ-spikes appear in the spectra at the frequencies

$kf_0 \pm lf_1$. As seen from Fig.7.5, at point D_3, the spectrum lines of the frequency $f' = f_0 - 2f_1$ and of lateral components $f_0 \pm f'$ grow remarkably. New frequencies $f_0 \pm f''$ appear where $f'' = 8f_1 - 3f_0$. Frequency f'' is not seen due to the limited accuracy of calculation at point D_3 but it is clearly displayed at point D_5 (Fig.7.5). On the way from D_3 to D_4, curve L becomes all the more deformed. This is accompanied by further enrichment of the spectrum with components of sum and difference frequencies and by the growth of their intensities. Near point D_4, L ceases to be a circle topologically (is broken down) and the SA_2-*torus-chaos* is born in a soft manner in its vicinity having the dimension $D_L = 2 + d$, $d \ll 1$.

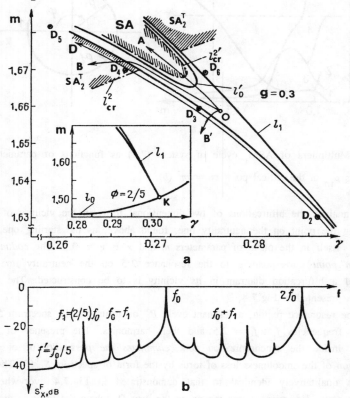

Fig.7.4. Bifurcation diagram of system (7.1) near resonance 2:5 (*a*); the power spectrum of cycle P^+ on torus, calculated for point O (*b*).

A new torus is born when moving in the *direction A* of the diagram (Fig.7.4a) from the resonant cycle on the neutrality line which is further broken down via the loss of stability. A torus-chaos appears, like in case of motion on the way D. *The way B leads* to saddle-node-type periodic motion vanishing and SA_2^T appearing on the multiplicity line l_1, since the torus is broken down already above the point D_4. *The way B' provides the transition to the region of ergodic torus existence.* The movement through the synchronization zone from point D_3 to point D_6 results at first in the vanishing of combination frequencies in the spectrum (resonance) which again appear at point D_6.

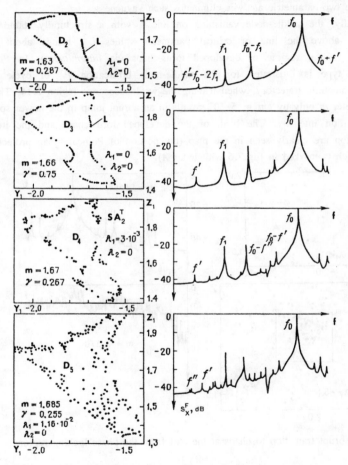

Fig.7.5. The Poincaré map and power spectrum corresponding to the motion in direction D (Fig.7.4).

There is no substantial regime rebuilding. The nonsmooth torus inside the synchronization tongue does not manifest itself practically. The picture of gradual deformation of invariant curve, which is randomly broken by different resonances, is observed usually in the course of computer simulation when using one-parametric approach and the discrete step of calculation in parameters. The high order resonances are overlooked, as a rule, due to their narrow resonance tongues and to the finite step of discretization of control parameter. Depending on the section selected, various bifurcation phenomena sequences can be observed under one-parametric examination which only entangle the general picture. As seen from the discussed bifurcation diagrams in plane, the two-parametric analysis eliminates such vaguenesses.

Finally, the saddle-node vanishing on line l_1 with a structurally stable homoclinic trajectory above the line of critical parameter values l_{cr}^2 brings about an *abrupt transition* to the regime of developed torus-chaos via *intermittency*. In Fig.7.6, this transition type is illustrated for the saddle-node bifurcation P^+ of cycle which has robust homoclinic trajectory, when leaving the weak resonance region 2:14. The variation of parameter m only by $\Delta m = 5 \cdot 10^{-4}$ results in an abrupt jump in the power spectrum and self-oscillation intensity. The birth of the developed torus-chaos and the intermittency phenomenon are clearly seen in the map (the portion of Poincaré map projection on the plane which is marked in Fig.7.6 with letter A).

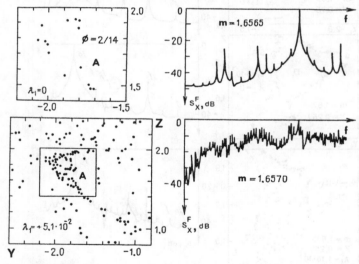

Fig.7.6. Abrupt transition to chaos at the exit from the phase locking region.

7.2 QUALITATIVE DESCRIPTION OF BIFURCATIONS IN THE SYSTEM OF COUPLED OSCILLATORS BY USING A MODEL MAP

As it was confirmed in Chapter 5, the dynamical system having one and a half degrees of freedom ($N = 3$) which realizes a saddle-focus quasiattractor, produces in the secant plane the two-dimensional map of Henon class

$$x_{n+1} = P(x_n, \alpha) + y_n, \quad y_{n+1} = \beta x_n, \tag{7.4}$$

where $P(x_n, \alpha)$ and β are the one-dimensional parabolic map of Feigenbaum class and coefficient defining the equivalent phase flow contraction in the initial differential system, respectively. Calculations show the map of system (3.38) to correspond qualitatively to (7.4) in the secant plane in general case. According to this, it is reasonable to investigate the dynamics of the system of two oscillators by describing it approximately with the help of the discrete system of coupled maps. A discrete system, that models the interaction between two oscillators with inertial nonlinearity, has the form:

$$\begin{aligned} x_{n+1} &= P(x_n, \alpha) + \gamma \varphi_1(x_n, y_n) + \xi_n, \\ y_{n+1} &= P(y_n, \alpha) + \gamma \varphi_1(x_n, y_n) + \eta_n, \\ \xi_{n+1} &= \beta x_n, \quad \eta_{n+1} = \beta y_n, \end{aligned} \tag{7.5}$$

where α is a parameter being faithful to the exceeding of generation threshold, $0 < \gamma < 1$ is the equivalent coupling coefficient, $\varphi(x_n, y_n)$ is a function describing the nature of coupling, and $\beta < 1$ is equivalent dissipation coefficient.

In case of a relatively strong contraction of the initial phase flow in the model system, $\beta \ll 1$ and it is reduced to an irreversible two-dimensional map. By choosing a particular form of parabolic map $P(x_n, \alpha)$, we write the system (7.5) as follows:

$$\begin{aligned} x_{n+1} &= 1 - \alpha x_n^2 + \gamma \varphi_1(x_n, y_n), \\ y_{n+1} &= 1 - \alpha y_n^2 + \gamma \varphi_2(x_n, y_n), \end{aligned} \tag{7.6}$$

The above represents two coupled Feigenbaum's maps.

The dynamics of the discrete system of (7.6) type was studied by many authors to be put to a number of uses. We are interested in illustrating the *mechanisms of invariant curve destruction* when the qualitative similarities in transitions compared with the initial differential system (7.1) are maintained. In [7.4,7.5], the transition to chaos via the destruction of invariant curve is discussed. However, the details of mechanisms of chaos birth are investigated there insufficiently due to the application of one-parametric analysis.

The comparative analysis of the dynamics of systems (7.5) and (7.6) has shown that their bifurcation diagrams in the control parameter plane may be qualitatively similar in the cases of linear, bilinear [7.5], and inertial [7.6] coupling. To achieve the goal stated, the simplest system (7.6) proved to be reasonable for investigation. The question on the influence of coupling types, specified by functions $\varphi(x_n,y_n)$, is far from being trivial, but in our case, one may confine himself to choosing the inertial coupling [7.6] $\varphi_{1,2} = \pm (y_n - x_n)$. The essential condition of the task, the symmetry of the model map (7.6), holds.

Thus, analyze the bifurcation phenomena in a model system [7.7]

$$x_{n+1} = 1 - \alpha x_n^2 + \gamma(y_n - x_n),$$
$$y_{n+1} = 1 - \alpha y_n^2 + \gamma(x_n - y_n), \qquad (7.7)$$

having two parameters being essential for our task: α is a parameter stating for parameter m and γ is a parameter similar to the coupling coefficient in the differential system (7.1).

The map (7.7) is symmetric with respect to substituting the variables $(x,y) \to (y,x)$ and has two SM-fixed points

$$x_{1,2} = y_{1,2} = (2\alpha)^{-1} \cdot (-1 \pm \sqrt{1 + 4\alpha}) \qquad (7.8)$$

and two ASM-fixed points

$$x_{3,4} = \frac{-(1 + 2\gamma) \pm \sqrt{1 - 4\gamma^2 + 4\alpha}}{2\alpha}, \qquad x_{3,4} = \frac{-(1 + 2\gamma) \mp \sqrt{1 - 4\gamma^2 + 4\alpha}}{2\alpha}. \qquad (7.9)$$

Linear analysis of stability provides the expressions for the fixed point multipliers

$$\rho_{1,2} = -\alpha(x + y) - \gamma \pm \sqrt{\alpha^2(x - y)^2 + \gamma^2}. \qquad (7.10)$$

Substituting the coordinates of points (7.9) into (7.10) will give

$$\rho_{1,2} = 1 + \gamma \pm \sqrt{1 - 3\gamma^2 + 4\alpha}, \qquad (7.11)$$

whence it follows that the ASM-points are unstable for any parameter values. For the SM-points, we have

$$\rho_1 = 1 \mp \sqrt{1 + 4\alpha}, \qquad \rho_2 = 1 \mp \sqrt{1 + 4\alpha} - 2\gamma. \qquad (7.12)$$

It is seen that the map (7.7) has the only one SM-point

$$x = y = (2\alpha)^{-1} \cdot (-1 + \sqrt{1 + 4\alpha}\,), \tag{7.13}$$

being stable in a certain region of the parameter plane. We will restrict ourselves by studying this region. Note that the plane of variables x_n and y_n is a discrete analog of the phase space of six-dimensional system (7.1). The one-dimensional space in the form of straight line $x_n = y_n$ in this plane corresponds to the case of degeneracy (7.2).

A fixed point A_c is born on line l_1 as $\alpha = -0.25$ with multipliers $\rho_1 = 1$, $\rho_2 = (1 - 2\gamma)$. With the increase of parameter γ, the point A_c loses its stability via doubling. Line l_2 corresponds to this bifurcation in the control parameter plane as $\alpha = 0.25\,(4\gamma^2 - 8\gamma + 3)$. Both lines are shown in Fig.7.7.

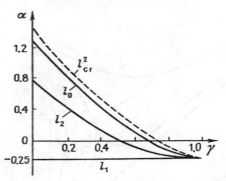

Fig.7.7. The bifurcation lines l_0, l_1 and l_2 of stability loss for the fixed points on the plane of control parameters of map (7.7), l^2_{cr} - the line of destruction of invariant curves L.

As a result of period doubling bifurcation, a period-two point $B_{1,2}$, (2-cycle) of map is softly born that is located symmetrically with respect to bisectrix $x_n = y_n$. With the increase of parameter α, 2-cycle multipliers become complex-conjugate. The multipliers go out to the unit circle on the line l_0 constructed numerically (Fig.7.7). The bifurcation of the invariant curve birth is realized. Stable invariant curves L_1 and L_2 are softly born in the vicinities of fixed points B_1 and B_2 having lost their stabilities. Further parameter α increasing leads to the destruction of invariant curves followed by the chaotic attractor's birth (the dashed line l^2_{cr} in Fig.7.7).

Let us investigate the destruction mechanisms of invariant curves L_1 and L_2 being the model images of two two-dimensional tori which are born from the ASM cycles in the differential system (7.1) [7.7]. To do this, we find the winding numbers Φ of these invariant curves along the bifurcation line l_0 specified by the following expression

$$\Phi = arccos(Re\ \rho)/2\pi, \qquad (7.14)$$

where ρ is the multiplier of the 2-cycle of the map, which is equal to unity in modulus on the neutrality line l_0. The winding number varies continuously on line l_0 taking the values within the interval $0.5 \leq \Phi \leq 1.0$. Having defined the point K of resonance 2:5 on line l_0, we construct a bifurcation diagram for the resonant cycle on the invariant circle. The calculation results are depicted in Fig.7.8 and are in a qualitative agreement with the computational data indicated in Fig.7.4. Now the mechanisms of resonant torus breakdown are to be followed by investigating numerically the behavior of *unstable separatrixes* of saddle fixed points on the resonance curve L_1 illustrated in Fig.7.9.

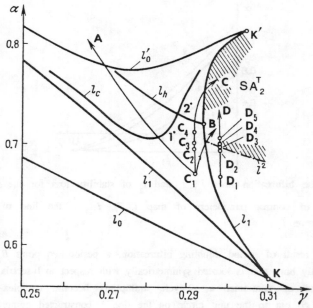

Fig.7.8. Bifurcation diagram of system (7.7) in the vicinity of resonance 2:5.

The invariant manifold for a resonance is formed by the closure of the unstable separatrixes Γ_i^u of saddle points Q_i into stable nodes P_j ($i, j = 1, 2, ..., 5$). Such a closure is smooth near the base of the resonant tongue (point K in Fig.7.8), the derivative in the node does not suffer the discontinuity. The movement in direction C in Fig.7.8 causes the *oscillation of one of unstable separatrix's branches* (points C_2 and C_3 in Fig.7.9). The development of oscillations results in their intersection with nonleading manifolds of nodes (Fig.7.9, point C_4). For the irreversible system (7.7), only the unstable saddle point's separatrixes and vectors being tangential to stable separatrixes can be calculated.

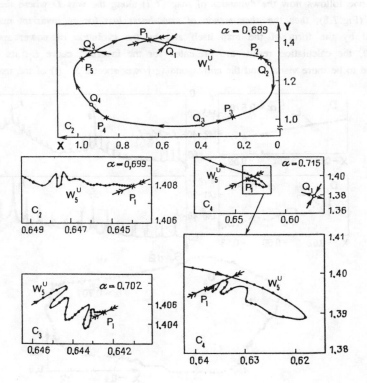

Fig.7.9. Evolution of separated segments of invariant curves L_1 of map (7.7) in the region of resonance 2:5 when moving in direction C (Fig.7.8).

With parameter α growth, a structurally unstable tangency takes place at first. Then, the rough intersection of unstable and stable saddle fixed point's separatrixes proceeds: a *robust homoclinic trajectory appears* for the cycle on the resonant torus. The instance of homoclinic separatrixes' tangency is bifurcational. The invariant curve is destroyed and a *nontrivial hyperbolic set* is born in its vicinity, the forerunner of a chaotic attractor. The bifurcation line in the parameter plane (Fig.7.8) that corresponds to the homoclinic tangency, is marked with l_h. There is no longer the invariant closed curve L_1 over the bifurcation line l_h. In a differential system, this corresponds to the torus breakdown. The stable fixed point (or limit cycle) inside the synchronization zone remains, however. The exit from the resonance region in direction C

will lead to stable fixed point vanishing on the saddle-node bifurcation line l_1 and an *abrupt transition to torus-chaos* regime will take place.

If one follows now the dynamics of map (7.7) along the way D where the torus is ergodic (Fig.7.8), then the *phenomenon of smoothness loss* for an invariant manifold is observed by the form of the map itself and by the evolution of power spectra. In Fig.7.10, the calculation results are presented for the invariant curve L_1, its segments separated to be more visual, and the corresponding power spectra $S_x^P(f)$ of the map under

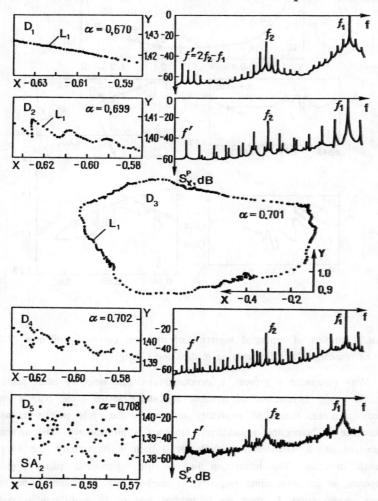

Fig.7.10. Smoothness loss and destruction of invariant curve L_1 under motion in direction D (Fig.7.8).

the motion on the way D of the bifurcation diagram. It is seen that the motion in the parameter plane in direction D leads to the soft destruction of the invariant curve L_1 due to the smoothness loss. The phenomenon is accompanied by the enrichment of the spectrum $S^{P_x}(f)$ with combination frequencies. While crossing the bifurcation line l^2_{cr}, the torus chaos SA^T_2 appears.

One of the main results of the system (7.7) investigation performed is an experimentally substantiated possibility of observing the phenomenon of the torus smoothness loss by the evolution of an ergodic invariant curve *outside the synchronization zone*. The comparison between these calculations for the rational (Fig.7.9, C_2) and irrational (Fig.7.10, D_3) values of winding number Φ testifies visually the *invariant manifold not to suffer a discontinuity* when crossing the bifurcation multiplicity line l_1^2. If the unstable manifolds of saddle cycle in the resonance region are beyond the calculation (this is the typical difficulty in the study of high-dimensional differential systems), one may with confidence diagnose the phenomenon of torus smoothness loss by the Poincaré sections and oscillation power spectra near the multiplicity line l_1 *but outside the synchronization zone*. Such a "visualization" is to be physically explained as follows. A phase trajectory on the torus deformed because of oscillations of unstable manifolds (but which is a resonant one!) *does not visit all* the areas of the torus's surface, in particular, the most "crimped" areas near the node points in the map (Fig.7.9, C_3 and C_4). Therefore, the power spectrum does not reflect the phenomena caused by the nonlinear torus deformations. The map in resonant tongue itself is composed of a finite set of fixed points and eliminates the possibility of invariant manifold to be observed completely (it remains unclear how do the separatrixes go between visible stable fixed points).

The picture changes in principle in the region of ergodic oscillations. Here, the toroidal surface is *densely covered* everywhere by the phase trajectory and, therefore, the temporal realization of any phase coordinate involves the information on invariant manifold deformations. Hence it follows that the diagnostics of torus evolution on the way to the breakdown is feasible with the help of spectra and Poincaré maps.

It is clear, that when the order of resonances increases up to $q \geq 5$ and periods of synchronous cycles become essentially large, the phase trajectories visit the deformed areas of torus surface and their spectra would provide more and more information on the torus smoothness loss phenomenon. The evolution of oscillation spectra in a series of experiments is accounted for just by this circumstance along with the invariant manifolds being continuous. Only rare phase synchronization regions are observed that have relatively small resonant cycle periods (relatively high basic frequencies) and

---

[2]With variation of the parameter, the invariant manifold's shape itself is inherited. With this, the structure of trajectories on it is varied. The resonant cycle of the map makes place everywhere for a dense ergodic curve L_1.

remarkably wide synchronization tongues. Outside these zones, two-frequency oscillations on nonsmooth tori take place with the spectra comprising a great number of combination frequencies. The higher order resonant cycles, defined by large periods, are not recorded because of noise and do not differ from quasiperiodic oscillations in experiments. The lowest of two basic frequencies in such regimes tends to the values of near zero making the measurements more difficult.

Notwithstanding that the problems of two-frequency oscillation breakdown with transitions to chaos were investigated on particular dynamical systems, the results obtained have general character and reflect the phenomena being typical for high-dimensional and distributed systems. This is easy to verify by comparing the results with the qualitative theory conclusions, with numerous computation data as applied to different systems, and, finally, with some experimental results [7.2,7.7].

7.3. TRANSITIONS TO CHAOS VIA THREE-FREQUENCY QUASIPERIODIC OSCILLATIONS

The study of two-frequency oscillation breakdown problem has extended the insight into transitions to chaos in dissipative high-dimensional systems by bringing us nearer to understanding the birth mechanisms of developed turbulence in continuous medium. This was promoted by the rigorous results of qualitative dynamical systems theory. However, the whole variety of critical phenomena, observed experimentally, are not exhausted by the phenomena recorded with two-dimensional torus breakdown. An example is the transition to chaos via the regime of oscillation with three independent frequencies.

The question about the evolution of *three-frequency oscillations* when varying the system parameters, reduced by the mathematical statement to the investigation of the phase trajectory behavior in the vicinity of three-dimensional torus (T^3) is of importance but, unfortunately, not enough investigated. There are objective reasons for this. The approach to the two-dimensional torus breakdown problem is reduced to the qualitative study of critical phenomena as applied to a robust structure on torus in the form of synchronous cycles and, since, it can be successfully based on the theory of periodic oscillation regime stability. The theory of quasiperiodic oscillation regime stability is required to solve the problem of critical phenomena on a three-dimensional torus. The main difficulty consists in that the above theory has not been created yet in the form suitable for applications.

Extremely few data on the dynamics of the systems with three-frequency oscillation are available. They concern different particular systems, are not combined by general theoretical ideas, and have, in general, pure empiric character. The experimental studies of the evolution regularities of three-frequency oscillation, based on certain qualitative similarities or on investigator's intuition, is a more effective way of exploring the above problem. One may hope that clear general regularities, stated in different models and full-scale experiments, would be a base for constructing the appropriate theory.

Consider the critical phenomena under periodic modulation of two-frequency oscillation. With some restrictions, the structural instability of three-frequency oscillation regime (of three-dimensional torus T^3) was demonstrated providing the regime of strange attractor in the vicinity of T^3 [7.8]. But in general case, three-frequency oscillations are possible, without doubt, to be testified by computer simulation [7.9] and full-scale experiments [7.10]. We make use of similarity considerations. A weak periodic perturbation of the autonomous system, having a robust stable limit cycle is needed and is sufficient for realizing the regime of two-frequency oscillations. The regime of stable three-frequency oscillations can be provided if a dynamical system, having the stable regime of autonomous two-frequency oscillations, is influenced by an independent periodic force by similarity. Such a system was considered above, that is an autonomous model of two coupled oscillators with inertial nonlinearities. The case of nonautonomous three-frequency oscillations is to be investigated by introducing perturbation into the right-hand part of the first equation of the autonomous system (7.1):

$$(1-\gamma^2)\dot{x}_1 = y_1 + x_1(m-z_1) + \gamma[y_2 + x_2(m-z_2)] + B_0 \sin(2\pi f_{ex} \cdot \tau),$$
$$(1-\gamma^2)\dot{x}_2 = y_2 + x_2(m-z_2) + \gamma[y_1 + x_1(m-z_1)],$$
$$\dot{y}_1 = -x_1, \quad \dot{z}_1 = -gz_1 + gl(x_1)x_1^2, \qquad (7.15)$$
$$\dot{y}_2 = -x_2, \quad \dot{z}_2 = -gz_2 + gl(x_2)x_2^2,$$

As compared with (7.1), two parameters are added, the external signal amplitude B_0 and frequency f_{ex}. Time τ is explicitly introduced.

For $B_0 = 0$ in (7.15), the stable two-dimensional torus T^2 is born in a soft manner from asymmetric periodic regime. The torus T^2 birth in the plane of parameters m and γ is corresponded by the codimension-one bifurcation line l_0 (Fig.7.4) where a pair of complex-conjugate cycle P^+ multipliers reach the unit circle. Further, we record the values of parameters $m = 1.59$, $g = 0.3$, and $\gamma = 0.3$ corresponding to the regime of T^2 which has two frequencies $2\pi f_1 = 0.684...$ and $2\pi f_2 = 0.283...$, which ratio is irrational as $B_0 = 0$. Then, we study numerically the nonautonomous oscillation regimes in the plane of parameters B_0 and f_{ex} in the region $0 \le B_0 \le 0.7$ and $0.1 \le f_{ex} \le 1.1$ [7.11].

In Fig.7.11, a fragment of bifurcation diagram in the vicinity of one of T^2 natural frequencies f_1 is presented, which displays the most probable oscillation regimes realized. The three-frequency oscillations with basic frequencies $\approx f_1$, $\approx f_2$, f_{ex}' are observed in sector 1. *Resonances on T^3 are possible* when continuously varying f_{ex} in 1. The typical spectrum of Lyapunov characteristic exponents in 1 involves *three* maximum *zero exponents*. Frequency f_1 is locked in sector 2 along with asynchronous suppression of the second of natural frequencies, f_2. The system in sector 2 is synchronized by the external signal (two-fold degeneracy).

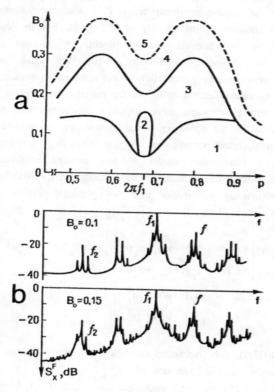

Fig.7.11. Partition of the plane of parameters B_0 and p of system (7.15) into sectors with different oscillation regimes (*a*). The power spectra for the soft destruction of T^3 (transition from 1 to 4 for $p = 0.911$) (*b*).

Sector 3 corresponds to the two-frequency oscillations arising either due to suppression of the frequency f_2 on the base of frequencies f_1 and f_{ex} or because of resonance between f_2 and f_{ex} as the frequency ratio f_2/f_{ex} is rational. When varying B_0 and f_{ex}, the resonant cycles on T^2 seem to appear in sector 3, i.e., the synchronization $T^2 \to T_r^2$ takes place. The two-dimensional torus is broken down at the boundary between sectors 3 and 4 followed by the formation of torus-chaos. The SA_2^T birth mechanism is soft and is due to the smoothness lost by T^2 regime. The LCE spectrum in sector 4 involves one positive exponent, the *Lyapunov attractor dimension* is here $2 < D_L < 4$.

The bifurcation in chaos $SA_2^T \to SA_3^T$ proceeds at the boundary between sectors 4 and 5 resulting in the appearance of the second positive exponent (hyperchaos) in the LCE spectrum, and attractor dimension grows to $3 < D_L < 5$. The exit into sector 4 is performed from sector 1 for the values $2\pi f_{ex} > 0.9$. The transition $T^3 \to SA_3^T$ is recorded. Narrow zones of resonances are not excluded that correspond to one- and two-fold degeneracies. The above transition type is illustrated by the power spectra of realization $x_1(\tau)$ (Fig.7.11b).

Consider the evolution of oscillation regimes as the amplitude of resonant force is increased at the frequency $f_{ex} = f_1$ with the Poincaré map projections and power spectra shown in Fig.7.12. With the growth of $0 < B_0 < 0.05$, the amplitude of frequency f_2 is smoothly decreased in the power spectrum of oscillations and the stable resonant cycle of frequency $f_{ex} = f_1$ is born with the total asynchronous suppression of f_2. The B_0 increase to the value of $\simeq 0.18$ provides an abrupt loss of stability by a resonant cycle followed by the birth of two-dimensional torus T^2 (transition from sector 2 to sector 3 in Fig.7.11a). The surface of two-dimensional torus T^2 is already deformed but the invariant curve in the Poincaré map is not destroyed yet testifying the two-frequency oscillation regime to be present.

The resonance on T^2, $q = 16$ is recorded in a very narrow region of amplitude values close to $B_0 \simeq 0.2$. Then, at the critical point $B = 0.201...$, the transition to SA_2^T proceeds. The two-dimensional torus is broken down because of the saddle-node periodic motion vanishing when a torus at the bifurcation point is nonsmooth. The LCE spectrum is distinguished under this transition by *one positive exponent* appeared. The oscillation power spectrum becomes continuous and has high spikes at frequencies f_1 and f_2, their harmonics, and numerous combination frequencies arising with the torus smoothness loss. Behind the critical point $B_0 > B_0^*$, the dimension of attractor SA_2 grows abruptly from $2 + d$, $d \ll 1$, to $D_L = 3.14$ with $B_0 = 9.2015$ and further, $3 < D_L < 4$.

The *second positive exponent* appears in the LCE spectrum and dimension D_L reaches $4 + d$ as the force amplitude grows $B_0 \leq 0.35$. The transition to hyperchaos $SA_2^T \to SA_3^T$ is performed (transition from sector 4 to sector 5 in Fig.7.11). In sector 5, numerous resonances are observed in the form of the two-dimensional tori or resonant cycles (one- or two-fold degeneracies). The areas of their existence in the parameter plane are relatively small compared with that of chaotic dynamics. The extension into sector 5 with the growing amplitude B_0 of external force is accompanied by the increase of chaotic pulsation intensity. The power spectrum becomes more uniform (Fig.7.12, $B_0 = 0.35$). The dimension of attractor in sector 5 is lower than $D_L \leq 5$ and the number of positive LCE spectrum exponents is ≤ 2. The increase of the amplitude $B_0 > 1.0$ of external force leads, in the end, to the system synchronization in the regime of periodic oscillations.

The results obtained assume the following interpretation. Three-frequency oscillations show a striking tendency to synchronization in the regions of intensive interactions defined by the excitation frequency closeness to one of natural frequencies

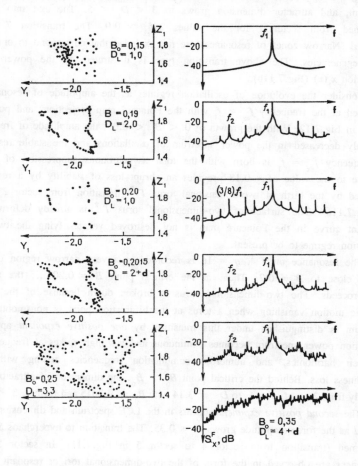

Fig.7.12. The Poincaré map projections on the y_1, z_1 plane and the corresponding power spectra of time series with the increase of the amplitude of external force at frequency $f_{ex} = f_1$.

of the system as well as by the amplitude of external force. The oscillations are synchronized either to two-frequency oscillations (T^2) or to the limit resonant cycle on the three-dimensional torus. The comparison of obtained results to the data on the mechanisms of two-dimensional torus breakdown indicated in Chapter 6 allows one to conclude on the mechanisms of transitions to chaos to be identical to the above ones in a specified parameter value region. The difference is only in that the phenomenon of three-dimensional torus breakdown is preceded, at the beginning, by synchronization $T^3 \to T^2$. Then, the transitions to chaos are possible both via the regime of two-frequency ergodic oscillations and via a resonant cycle because of the saddle-node bifurcation, inducing abrupt transitions with intermittency and hysteresis.

A direct transition to chaos from three-frequency oscillation regime is recorded within the inevitable discretization errors, under weak interactions, when the excitation frequencies are distant in the spectrum from the natural autonomous system's frequencies - the *three-dimensional torus is broken down*. The chaos birth mechanism after Ruel-Takens is realized in experiments. The synchronization phenomena into two-dimensional tori and resonant cycles can take place. However, cycle periods are here, as a rule, very large and the resonances are practically not recorded in the course of experiments. Furthermore, the regions of resonance existence in the parameter space are small. One can assume only, similarly to the case of ergodic two-dimensional torus breakdown, the transition $T^3 \to SA_3^T$ to be rigorously possible but the resolution of numerical calculation is insufficient for such an assertion.

Finally, we discuss the transition to hyperchaos connected with the bifurcation of LCE spectrum signature and the increase of attractor dimension. The phenomenon ascertained is typical for the chaotic high-dimensional system's dynamics. It is quite natural that it has manifested itself just in the six-dimensional case. In dynamical systems having the phase space dimensions $N \gg 3$, a sequence of bifurcations is observed getting complicated in the chaos of type $SA_k^T \to SA_{k+1}^T$, $k = 3, 4, ...$, which is due to the more and more LCE exponents going through zero into the region of positive values. The *developed turbulence* seems to be formed just in such a manner in continuous nonlinear media. An alternative is of principle here: whether the dimension of attractors would grow to 1) infinity, 2) finite, but very high values 3) saturation is conceivable in a chaos characterized by the dimension stabilization at a level of comparatively small values.

Along with the above method for realizing the three-frequency oscillations via a periodical perturbation of the regime of ergodic two-frequency oscillations, it turned out to be more convenient for a radiophysical experiment to study the evolution of three-frequency oscillations arising as the external periodic signals with two independent frequencies drive the oscillator with inertial nonlinearity. The major advantage of such a method is a practical elimination of two-fold degeneracies observed when varying the parameters, i.e., synchronizing into the limit cycle [7.12].

The two-frequency external force is a sum of harmonic signals $B_1 sin(2\pi f_1 t) + B_2 sin(2\pi f_2 t)$ from independent sources driving additionally the oscillator with inertial nonlinearity. The parameters of complex external force B_1, B_2, f_1 and f_2

can be independently varied within the wide limits. As the system's control parameters, $R_m \sim m$, oscillator's excitation parameter, and f_1, the frequency of an external force, were chosen. The other parameters of the system were maintained as constant at different specified levels under the construction the of bifurcation diagrams in experiments. The inertia parameter g was 0.3 everywhere. As in preceding investigations, the diagnostics of oscillation regimes and of their bifurcations was performed by the power spectra, time realizations, and phase trajectory projections onto two-variable planes of interest.

A portion of experimental bifurcation diagram in the plane of parameters R_m, f_1 near the oscillator's resonance frequency f_0 is shown in Fig.7.13. The mutual location of bifurcation lines in the plane qualitatively replicates the picture for the one-frequency perturbation of oscillator (6.1) indicated in Fig.6.2. The meaning of bifurcation lines and regime character is, however, *principally different* in the superficially similar sectors of the parameter plane.

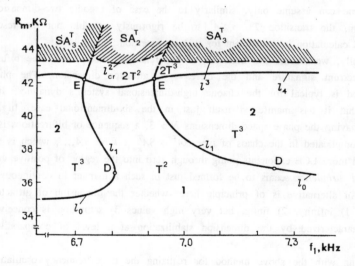

Fig.7.13. Bifurcation diagram of T^2- and T^3- regimes. Physical experiment for $f_1 = 8.97$ kHz, $B_1 = 0.8$ V, $B_2 = 0.15$ V.

Three-frequency oscillations are born on the bifurcation line, marked by l_0 as before, due to excitation of the natural oscillations with the frequency $\simeq f_0$. The two-dimensional torus T^2 exists below the "neutrality line" l_0 (sector 1) corresponding to the oscillations induced by external force. Over the line l_0 (sector 2), the existence region of three-dimensional torus T^3 is located.

At the bifurcation points D, the neutrality line l_0 changes a to "multiplicity line" l_1 that has the following meaning. The natural frequency f_0 is locked by the external signal with frequency f_1 in sector 1 bounded from the left and right by lines l_1. But the second external signal of independent frequency f_2 induces the quasiperiodic regime T^2. Thus, one-fold degeneracy takes place in sector 1, i.e., a partial resonance on the three-dimensional torus when two of three basic oscillation frequencies coincide. In general case, the partial resonance would be observed for the rational relation between two frequencies of T^3. The transition through the line l_1 in sector 2 leads to missynchronized frequencies f_0 and f_1, and *regime T^3 arises abruptly*[3].

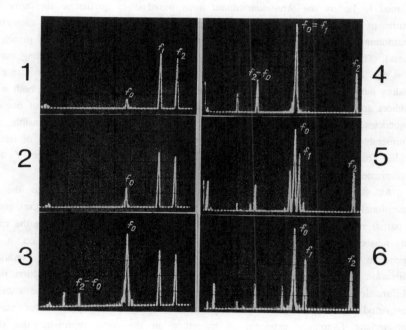

Fig.7.14. Soft (1-3) and abrupt (4-6) bifurcations of three-dimensional torus birth.

The transitions to sector 2 through the bifurcation lines l_0 and l_1 were experimentally detected by variations in the behavior of the power spectra and phase portraits of oscillations. In Fig.7.14, the photographs of oscillation power spectra are presented which correspond to the transition through the neutrality and multiplicity lines l_0 (1 - 3) and l_1 (4 - 6). In the second case, a spectrum line with frequency f_0 is added to frequencies f_1 and f_2 of force. Its intensity grows smoothly from zero level

---

[3] The transition through the line l_0 from sector 1 to sector 2 induces the birth of regime T^3, too, but this bifurcation is supercritical and is due to the soft manner of appearance of oscillations with natural frequency f_0.

on the neutrality line because of the soft character of Andronov-Hopf bifurcation. Combination frequencies based on the frequency f_0 appear sequentially with moving off from the line l_0 deep into sector 2.

The transition through the multiplicity line is accompanied by smoothly missynchronizing frequencies f_0 and f_1 which coincide in the resonance zone (sector 1) *for the finite values* of their intensities. The *above transition is abrupt* in this sense.

The two-dimensional torus inside the region of partial resonance (sector 1) can be assumed to lie *on the three-dimensional torus based* on a qualitative similarity in the picture of bifurcational phenomena in the system considered and in case of the nonautonomous oscillator (6.1) (accounting the above marked differences of principle, of course). The resonance phenomena on T^3 are to be defined by two winding numbers: $\Phi_1 = f_1 : f_0$ and $\Phi_2 = f_2 : f_0$. The partial resonance in sector 1 is due to the rational winding number $\Phi_1 = 1:1$ (basic partial resonance). The full resonance when both winding numbers are rational is eliminated in the case under study. With variation of one of the frequencies of external force, the synchronization $f_1 : f_2 = p : q$ is feasible in the denumerable set of points in the frequency axis. However, the phenomenon of frequency locking (because of the coupling absence between the oscillator frequencies f_1 and f_2) is eliminated and synchronization zones are not realized by experimentation.

As shown, the two-dimensional torus is transformed abruptly into the three-dimensional one at the exit from sector 1 to sector 2 through the bifurcation boundary of partial resonance l_1. Due to the same similarity, this bifurcation can be the merging and subsequent vanishing of stable and saddle two-dimensional tori on a single degenerate three-dimensional torus. In the above sense, the line l_1 is a *"saddle-node bifurcation line for two-dimensional tori"*. There exists no rigorous definition of such a bifurcation for the present. No parameter is available which defines the stability of quasiperiodic oscillation and is adequate to the cycle multiplier. The maximum characteristic Lyapunov exponent of trajectory on the torus represents the stability properties only partially. It is insensitive to the sign but is associated with the varying modulus of small perturbation vector along the quasiperiodic trajectory.

If one moves inside the resonance region T^2 upwards in parameter $R_m \sim m$, then *bifurcations of two-dimensional torus doubling* is recorded. The subharmonics with half-frequency $nf_0/2$ are softly born when crossing the bifurcation line l_3 in the power spectrum of oscillation on T^2. The phenomenon of spectrum enrichment with combination frequencies is recorded here as well and independently testifies to the two-dimensional torus smoothness loss. The sequence of torus T^2 doublings is finite. The number of bifurcations depends on the force amplitudes B_1 and B_2. The two-dimensional 2-tori are broken down on line l_{cr}^2 the mechanism of breaking being the loss of smoothness by tori. The regime of torus-chaos SA_2^T is softly born with the spectrum structure corresponding completely to the case of soft T^2 breakdown analyzed above.

If one follows the evolution of a three-dimensional ergodic torus outside the region of synchronization while moving upwards in parameter R_m, then the *three-dimensional torus doubling effect* is recorded. The oscillation spectrum is discrete at the parameter values below the bifurcation line l_4 (Fig.7.13) and involves the frequencies $nf_0 + kf_1 + pf_2$ (n, k, p are integers). The crossing of line l_4 implies the frequency components $nf_0/2$ to softly appear. *One of three* characteristic *periods* of oscillations on T^3 is *doubled*.

Then, when crossing the bifurcation line l^3_{cr}, a torus-chaos SA^T_3 is born in a soft manner due to the breakdown of ergodic torus T^3 via its smoothness loss. The structure of the power spectra of attractor SA^T_3 near its birth instant replicates the characteristic one for the regime SA^T_2 being distinguished only by δ-spikes in the spectrum of SA^T_3. The spikes are observed on three independent frequencies which correspond to three basic frequencies of the T^3 regime broken down. The spectra of SA^T_3 and SA^T_2 are indiscernible in the case of phase synchronization when two of three frequencies turn out to be rationally coupled (Fig.7.15).

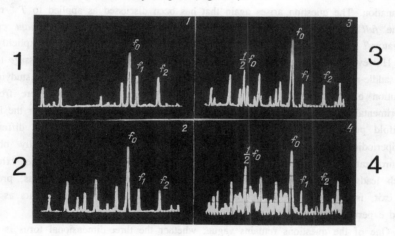

Fig.7.15. Three-dimensional torus period doubling bifurcation in a full-scale experiment.

The motion along the neutrality line l_0 results inevitably in a sequence of resonance phenomena caused by the winding number Φ passing through the rational points. These points are leaned on by the tongues of the resonant three-dimensional tori which have inside the oscillation regimes presented in the form of two-frequency oscillations. Here, a similarity in bifurcational phenomena takes place as well compared with the nonautonomous oscillator if one regards the appropriate bifurcations becoming complicated.

The experimental findings of observation of the transition from quasiperiodic oscillation regime having three, initially independent, frequencies, to chaos, which are obtained numerically on a radiophysical system also, have a common nature. It is seen from the comparison of the observed bifurcational phenomena to the previously reported data on numerical, experimental and theoretical study of this problem. In spite of differences between the systems analyzed, there is a great similarity in the regularities of transition from T^3 regime to chaos. The main mechanism of transition is a torus smoothness loss accompanied by partial or full resonances when varying the parameters and completed by the quasiattractor birth.

If bifurcational phenomena on three-dimensional torus are considered in the Poincaré map that reduces the system dimension by a unity, then dynamical effects on T^3 would be corresponded uniquely by the two-dimensional torus evolution in the map with all following consequences. Though the fact of T^3 regime realization is not typical in autonomous systems from the structural stability considerations (the rough phenomenon of inertial synchronization is more natural), experimental results require examination and explanation. The question arises again that has been discussed as applied to T^2 regime: if the *full resonance* and the following transition to chaos via a *resonant cycle* is obligatory? Experimental data alone are insufficient to answer this question explicitly.

However, the conclusions on continuous invariant manifold under the passage through the saddle-node bifurcation hypersurfaces, which we have drown while studying the evolution of T^2 regime, allow no principal differences to be made here from the experimental point of view. The synchronous cycles with great periods cover the integral manifold practically completely, let it be T^3 or T^2, and they do not differ from quasiperiodic oscillations in experiments. The exceptions are the really observed synchronization regions with relatively small resonant cycle periods, the exit from which leads to abrupt bifurcations. The invariant manifold smoothness loss, preceding this exit, is not seen by the power spectra and Poincaré map but it occurs as it was based experimentally.

One of the questions remains vague: whether the three-dimensional torus is broken down with transitions $T^3 \to SA_2^T$ or the structure of phase trajectories on it goes to chaos when maintaining the integral manifold? The data, obtained in the present paper, are insufficient to conclude with confidence.

CHAPTER 8
SYNCHRONIZATION OF CHAOS

8.1. INTRODUCTION AND DEFINITION OF THE PROBLEM

The phenomenon of *periodic oscillations synchronization* is well known in the classical theory of oscillations. There are two distinguished types of synchronization: the *synchronization caused by external force* and *mutual synchronization* of auto oscillatory systems. Two bifurcational mechanisms of the synchronization of near-harmonic oscillations have been investigated in detail [8.1, 8.2]. They are: 1) *Locking of the frequency* of natural oscillations by an external signal (or mutual locking of the frequencies of two subsystems), associated with the onset of stable and saddle limit cycles on a two-dimensional torus. Simultaneously with this, the phase-locking of oscillations takes place so that the oscillations of the system occur with a constant in time phase shift relative to the external signal. The same process occurs for the phases of oscillations of two coupled subsystems. 2) *Suppression of natural oscillations* by the external signal (or, correspondingly, suppression of oscillations of one of the coupled subsystems by the signal of the other subsystem). This suppression corresponds to the "collapse" of the torus into a limit cycle (i.e., to a bifurcation that is the reverse of a torus birth bifurcation).

The phenomenon of synchronization of more complicated nonlinear oscillations and, particularly, that of chaotic oscillations has been investigated but to a much less extent. From the data obtained in the course of computer simulation and full-scale experiments it is known that two identical autonomous chaotic systems, when coupling between them is introduced, can start oscillating with the same phase with each other; the time waveforms of corresponding coordinates of the subsystems will practically repeat one another. Such a phenomenon is usually called chaos synchronization [8.3 - 8.7]. However, the chaos synchronization is observed only in the limit cases of infinitesimal detuning of the subsystems' parameters, or of an infinitely great coupling; therefore, this phenomenon can be considered as a limit case of synchronization (we shall call it as a *complete synchronization*). The synchronization of chaotic oscillations seems to be much more intricate and is associated with the complexity of attractors in the phase space of dynamical systems.

We propose to generalize the classical concepts about synchronization of oscillations and bifurcational mechanisms of synchronization with regard to a certain class of chaotic systems, those wherein the *basic frequencies* can easily be distinguished in the power spectrum. A spiral type attractor will serve as an example of this type of chaotic systems. A single basic frequency is distinguishable in the power spectrum of oscillations and corresponds to the period of the initial limit cycle (i.e., the cycle that produces a chaotic attractor via a cascade of period doubling bifurcations). Thus, one can postulate about the mechanisms of synchronization by means of locking or suppression of this basic frequency. Things are less obvious with phase-locking. However, in the present work an attempt is made to analyze statistically the phase relationships in the course of synchronization of chaotic oscillations.

The approach to the problem of chaos synchronization mentioned above envisions a finite detuning between the basic frequencies of partial subsystems (or between an autonomous system and the external synchronizing signal). With no detuning between basic frequencies, the chaotic oscillations can be understood as synchronized in the above mentioned sense. Therefore, it is rather difficult to conduct numerical investigations of the synchronization of chaotic oscillations for discrete maps. In this case we are simply at a loss as to how to introduce detuning between basic frequencies.

8.2. EXPERIMENTAL SYSTEM AND ITS MATHEMATICAL MODEL

The modified oscillator with inertial nonlinearity, that has been described in Chapter 3 [8.8, 8.9] was chosen as a partial system. The block diagram of the oscillator with inertial nonlinearity is shown in Fig.8.1.

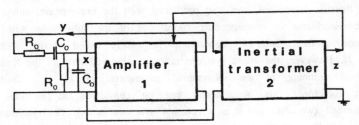

Fig.8.1. Block-diagram of the modified oscillator with inertial non-linearity.

The two-stage amplifier 1 is controlled by an additional feedback circuit incorporating the nonlinear inertial transformer 2. In this case the amplification factor K depends on the properties of the inertial transformer, $K = K(x,z)$. Assuming the amplifier 1 to be linear with respect to x, let us approximate the amplification factor by the following expression

Synchronization of Chaos

$$K(x,z) = K_0 - z \tag{8.1}$$

where K_0 is the amplification factor with no negative feedback, and $z(t)$ is the dimensionless voltage at the output of the inertial transformer.

The inertial transformer is represented by a square-law one-half-period detector with an RC-filter (Fig.8.2), it is described by the following expression

$$\dot{z} = -gz + g\,f(x); \quad f(x) = \begin{cases} x^2, & x \geq 0 \\ 0, & x < 0, \end{cases} \tag{8.2}$$

where $g = \dfrac{RC}{R_\Phi C_\Phi}$ is the ratio of the two characteristic times of the system (parameter of inertia).

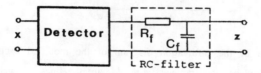

Fig.8.2. Block-diagram of inertial transformer

With regard to all mentioned above it is easy to obtain the following equations for the oscillators with inertial nonlinearity in terms of dimensionless variables (3.38)

$$\begin{aligned} \dot{x} &= mx + y - xz \\ \dot{y} &= -x \\ \dot{z} &= -gz + gf(x) \end{aligned} \tag{8.3}$$

where $m = K_0 - 3$ is the oscillator excitation parameter.

As full-scale experiments and computer simulation have shown (Chapter 3-5), small values of m result in near-harmonic oscillations. With the increase of m the initial periodic regime undergoes an infinite sequence of period doubling bifurcations that is crowned with the appearance of a chaotic attractor of spiral type. A portion of the numerically constructed bifurcation diagram of the system on the plane of parameters g and m is shown in Fig.8.3. The lines $l\,{}^k_2$ represent bifurcation lines of doubling of cycles with periods kT_0, where $k = 2^i$, $i = 0,1,2, \ldots$; T_0 is the period of the initial limit cycle. "kT_0" indicates the periodic regimes with periods kT_0 where $k = 2^i$, $i = 0,1,2,\ldots$. "SA" is the region of the spiral type attractor which evolves from the period T_0 limit cycle.

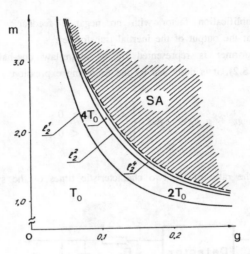

Fig.8.3. A portion of the bifurcation diagram of the modified oscillator with inertial nonlinearity on the plane of parameters g, m. l_2^k are the lines of cycle period doubling, SA is the region of a chaotic attractor (computer simulation).

In the present work we study a system of two interacting modified oscillators with inertial nonlinearity with both unidirectional and mutually symmetrical types of coupling. By controlling the capacitances of the resonant RC filter in one of the oscillators, we provide a way to detune the partial subsystems' basic frequencies. The block-diagram of the investigated experimental installation is shown in Fig.8.4.

The mathematical model of interacting oscillators in a general form is represented by the dynamical system in $\mathbb{R}^6$ [8.10] as follows:

$$\begin{aligned}
\dot{x}_1 &= (m_1 - z_1)x_1 + y_1 + G_1(x_2 - x_1 + y_1 - y_2/p) \\
\dot{y}_1 &= -x_1 \\
\dot{z}_1 &= g_1(f(x_1) - z_1) \\
\dot{x}_2/p &= (m_2 - z_2)x_2 + y_2 + G_2(Bx_1 - x_2 + y_2 - Bpy_1) \\
\dot{y}_2/p &= -x_2 \\
\dot{z}_2/p &= g_2(f(x_2) - z_2)
\end{aligned} \quad (8.4)$$

where $p = \dfrac{\omega_2^0}{\omega_1^0} = \dfrac{C_1}{C_2}$ is the ratio of resonance frequencies of the Wien bridges that determine detuning between the partial subsystems' basic frequencies; $G_1 = \dfrac{R_1}{R_c}$ and $G_2 = \dfrac{R_2}{R_c}$ characterize the degree of coupling; and B is the coefficient of buffer transmission.

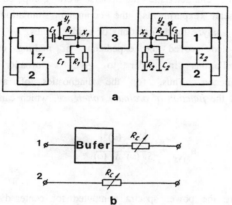

Fig.8.4. (a) Block-diagram of the system of two coupled oscillators. (3 is the coupling block); (b) The structure of the coupling block for unidirectional (1) and symmetrical (2) coupling.

The experimental installation and its mathematical model (8.4) allow us to analyze the interactions of completely identical partial systems ($m_1 = m_2 = m$, $g_1 = g_1 = g$, $p = 1$), systems with different basic frequencies ($p \neq 1$), systems with unidirectional coupling ($G_1 = 0, B = 3$), and systems with mutually symmetrical coupling ($G_1 = G_2 = G$, $B = 1$).

8.3. METHODS OF INVESTIGATION

The dynamics of interacting auto-oscillating systems was studied through both computer simulation and full-scale experiments. The following control parameters were selected: coupling G_1, G_2, detuning p and the partial systems' excitation parameters m_1, m_2. The inertial parameters g_1, g_2 of the partial systems were fixed at 0.3 or 0.2. In the framework of the full-scale experiment, the bifurcation diagrams were constructed on the plane of two control parameters, the remainder of the parameters having constant values. Bifurcational transitions were diagnosed with the help of power spectra, phase trajectories projected onto the selected planes of two variables, and Poincaré maps projected onto these planes.

Bifurcational analysis of the complexity of the attractors was led numerically on a computer. The above oscillation characteristics were also calculated, along with several traditional quantitative characteristics of attractors: the Lyapunov spectrum of characteristic exponents, Lyapunov dimension, and the auto-correlation function. For periodic oscillations, the characteristic multipliers were computed as well. On the contrary to the full-scale experiments, the computer simulation, with the stable regimes, permitted us to analyze the bifurcations of saddle cycles.

The investigations led proved the expediency of making use of some other characteristics of oscillations in order to give a broader understanding of synchronization phenomenon. Thus, for the diagnostics of a complete chaotic synchronization we used the *function of ordinary coherence*, which can be written as

$$\theta^2_{x_1 x_2} = \frac{|S_{x_1 x_2}(\omega)^2|}{S_{x_1}(\omega) S_{x_2}(\omega)}, \tag{8.5}$$

where $S_{x_1}(\omega)$, $S_{x_2}(\omega)$ are the power spectra computed for centered processes $x_1(t)$ and $x_2(t)$, respectively; and $S_{x_1 x_2}(\omega)$ is the complex mutual spectrum determined as the Fourier-image of the mutual correlation function of processes $x_1(t)$ and $x_2(t)$. As follows from (8.5), the function of coherence is determined within the interval [0,1]. For statistically independent processes, θ^2 is equal to 0 and is 1 in the case of a linear functional relationship between the time waveforms of x_1 and x_2 components. In general case, when $0 < \theta^2 < 1$, x_1 and x_2 are related to each other in a more complicated way and the results of measurements include the action of processes unaccounted for.

In a range of the problems considered here *the density of phase spectra difference distribution* (DPSDD) of $x_1(t)$ and $x_2(t)$ is also an important characteristic [8.11]. For nonperiodic oscillations the spectral density of amplitudes can be determined only for a process $x_T(t)$ of finite duration and is as follows:

$$F(\omega) = \int_{-T/2}^{T/2} x_T(t)\, e^{-j\omega t} dt, \quad \varphi(\omega) = \arg F(\omega) \tag{8.6}$$

In the course of computer simulation the integral is replaced by the sum. The phase spectrum would also be discrete

$$\varphi(\omega_i) = \arg F(\omega_i), \tag{8.7}$$

where $\omega_i = \omega_0 i$, $i = 0, \mp 1, \mp 2, \ldots$; $\omega_0 = \dfrac{2\pi}{T}$ determines the step along the frequencies.

For appropriate processes $x_{T_1}(t)$ and $x_{T_2}(t)$ a set of periodograms of T-duration was computed: $\left\{ x_{T_1}^{(k)}(t) \right\}$, $\left\{ x_{T_2}^{(k)}(t) \right\}$, ..., k = 1,2, ...,L.

Let us determine the phase spectra and their difference for each of the periodograms:

$$\begin{aligned}\varphi_{1,2}^{(k)}(\omega_i) &= \arg F_{1,2}^{(k)}(\omega_i), \\ \Delta\varphi_k(\omega_i) &= \varphi_1^{(k)}(\omega_i) - \varphi_2^{(k)}(\omega_i).\end{aligned} \tag{8.8}$$

The totality of $\Delta\varphi_n(\omega_i)$ over all L periodograms defines the distribution $p(\Delta\varphi,\omega_i)$ at the frequency ω_i (ω_i is a parameter). The results of the DPSDD computations are displayed on the computer in the form of a two-dimensional surface $p(\Delta\varphi,\omega)$, or portrayed on the plane of variables $\Delta\varphi$, ω. In the latter case one may examine the law of distribution of $p(\Delta\varphi,\omega)$ by the brightness of corresponding cells on the display screen [8.11].

8.4. FORCED SYNCHRONIZATION OF CHAOS

To study the *forced synchronization*, unidirectional coupling was provided ($G_1 = 0$, $G_2 = G$, $B = 3$). In this case the first oscillator signal is an external independent exciting force on the second oscillator.

The full-scale experiments were led to investigate the influence of the signal from the first oscillator on the second oscillator's dynamics. By changing the excitation parameter of the first oscillator we varied the nature of the forcing signal. We also varied the detuning of the two oscillators' basic frequencies (parameter p) and the degree of coupling (parameter $G_2 = G$). For forcing signals of different nature (from near-harmonic to chaotic) we have constructed the bifurcation diagrams of oscillation regimes in the second oscillator under variation of parameters p and G.

Figure 8.5 shows a portion of bifurcation diagram in the vicinity of the major region of synchronization ($p \approx 1$). The diagram illustrates the simplest case of forcing by near-harmonic signal (the limit cycle with period T_1) on the oscillator in the regime of developed chaos of spiral type.

The diagram shows a typical "*tongue of synchronization*" but it has a more complicated structure than in the case of harmonic oscillations. In this tongue of synchronization there are regions of regular synchronized regimes (the limit cycles with kT_1 periods) and regions of synchronized chaos of the spiral type (SA_1) with the basic

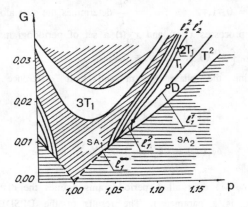

Fig.8.5. A portion of the bifurcation diagram for unidirectionally coupled oscillators as $m_1 \simeq 0.7$, $m_2 \simeq 1.20$, $g_1 = g_2 = g = 0.3$ (full-scale experiment); l_0 is the line of bifurcation of torus T^2 birth from the cycle with period T_1; $l_{1,2}^k$, $k = 2^i$, $i = 0,1,2,...$ are the lines of tangency bifurcations and bifurcations of doubling, respectively, for cycles with periods kT_1; "SA_1" is the region of synchronized chaos, "SA_2" is the region of non-synchronized chaos; l_1^∞ is the boundary between the regions of synchronized and non-synchronized chaos.

frequency f_1, determined by the forcing frequency ($f_1 = \dfrac{1}{T_1}$). Lines l_2^k, $k = 2^i$, $i = 0,1,2, ...$, correspond to period-doubling bifurcations of synchronized cycles with kT_1 periods. Outside the tongue of synchronization one can observe the oscillations with two basic frequencies f_1 and f_2 that correspond to the basic frequencies of the uncoupled oscillators. They can be both chaotic (region SA_2) and quasiperiodic (region T^2 with a higher degree of coupling).

The boundary of the tongue of synchronization corresponds to the two classical mechanisms of synchronization mentioned in the introduction to this chapter. Below point D, the boundary of the tongue is determined by the locking of the basic frequency of natural oscillations of the second oscillator. This boundary consists of segments l_1^k representing tangency bifurcations of kT_1 cycles to kT_2 tori and segment l_1^∞ at the base of the tongue. (Torus kT^2 is k-times doubled with respect to one of the basic periods). When crossing the lines l_1^k inside the tongue there appear resonant cycles on the kT^2 tori. The stable cycles kT_1 thus formed are in fact the observed synchronized regimes. Line l_1^∞ separates the region of non-synchronized chaotic oscillations SA_2 with two basic frequencies f_1 and f_2 from the region of synchronous chaos SA_1 with one basic frequency f_1. In Fig.8.5 line $l\,_1^\infty$ is marked by the dashed line because the bifurcational mechanism of transition $SA_2 \Rightarrow SA_1$ cannot be defined in the framework of a full-scale experiment. The investigation of the boundary $l\,_1^\infty$ of the tongue requires more detailed computer simulation. The results of such experiments will be discussed below.

Above point D, the boundary of the tongue of synchronization is determined by the suppression of the second oscillator's basic frequency. Prior to synchronization, there is a gradual degradation of the chaotic attractor SA_2 into a regular two dimensional torus T^2.

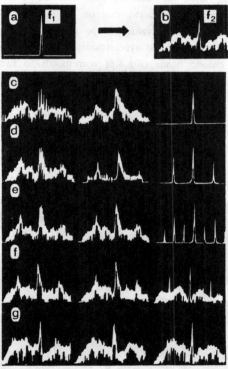

Fig.8.6. Spectra of $x_2(t)$ oscillations, illustrating the forced synchronization through locking of the basic frequency of chaotic oscillations when $m_1 \simeq 0.7$, $m_2 \simeq 1.2$, $g_1 = g_2 = 0.3$ (full-scale experiment). (a) Forcing signal spectrum; (b) Spectrum of autonomous oscillations of the second oscillator; (c) Spectra of oscillations of the second oscillator as $p = 1.08$; (d) $p = 1.058$; (e) $p = 1.043$; (f) $p = 1.039$; (g) $p = 1.034$. The parameter of coupling G is increased from left to right.

The line $l\,_0^1$ corresponding to the birth of torus T^2 from cycle T_1 is the boundary of the tongue in the case of synchronization via suppression. With the crossing of the line l_0^1 inside the tongue the torus T^2 collapses and a stable synchronized cycle T_1 appears.

In Fig.8.6 the photographs of the oscillators' spectra for variables $x_1(t)$ and $x_2(t)$ in the case of near-harmonic excitation are shown. Fig.8.6a shows the external input signal $x_1(t)$ with basic frequency f_1, while the spectrum in Fig.8.6b corresponds to the autonomous oscillations of $x_2(t)$ in the chaotic regime. Fig.8.6c-f illustrates the mechanism of chaos synchronization by means of locking the basic frequency of the second oscillator for various values of detuning between the basic frequencies of the oscillators. According to the bifurcation diagram in Fig.8.5, there can be different synchronized regimes for different values of p; these include cycles of period kT_1, $k = 2^i$, $i = 0,1,2,...$ (Fig.8.6c-d) and synchronous chaotic attractors of spiral type with different numbers of chaotic bands (Fig.8.6f corresponds to two bands of a chaotic attractor, and Fig.8.6g to one band). The change in the chaotic spectrum due to approaching of the two basic frequencies f_1 and f_2 is clearly seen in Fig.8.6c-d. Simultaneous with the two basic frequencies drawing together is the observed suppression of other spectrum components that results in the degradation of chaos into quasi-periodic oscillations; the synchronized regime appears to be a regular limit cycle.

Figure 8.7 shows a series of spectra photographs and $x_2 y_2$ - projections of phase trajectories. The system is composed of unidirectionally coupled oscillators with the

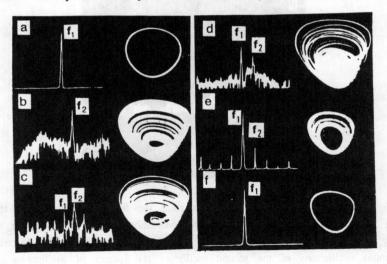

Fig.8.7. The spectra and phase trajectories of oscillations, illustrating the forced synchronization via suppression of the basic frequency of chaotic oscillations as $m_1 \simeq 0.7$, $m_2 \simeq 1.2$, $p = 1.13$ (full-scale experiment). (a) Forcing signal; (b) Oscillations of the second oscillator for $G = 0.01$ (SA_2); (c) $G = 0.019$ (SA_2); (d) $G = 0.028$ (T^2); (e) $G = 0.03$ (T_1).

first oscillator in the near-harmonic regime (see Fig.8.7a). Parameters m_1, m_2, and $g_1 = g_2$ correspond to those in Fig.8.5, with $p = 1.13$. The photographs in Fig.8.7b-f illustrate the synchronization of chaotic oscillations via the bascc frequency suppression. Figure 8.7 gives an idea about the transformation of the chaotic spectrum of the second oscillator under the influence of the first oscillator's signal. When the coupling is weak (Fig.8.7b), the frequency f_1 in the spectrum of the second oscillator is practically negligible. With the strengthening of the coupling, the amplitude of the spectrum line at the basic frequency f_2 of the second oscillator is decreased, while the amplitude of the spectrum line at f_1 is increased (Fig.8.7c,d). Strengthening of the coupling further results in chaotic oscillations with two basic frequencies f_1 and f_2 degenerating into the two-dimensional torus T^2 (Fig.8.7e). Finally, the frequency f_2 suppression takes place and in the region of synchronization a limit cycle T_1 is observed (Fig.8.7f).

The increase of complexity of the forcing signal (i.e., the dynamics of the first oscillator) results in a more complex bifurcation diagram on the plane of parameters p,G. If the forcing signal has period $2T_1$ (i.e., the initial limit cycle in the first oscillator underwent a period doubling bifurcation), there appear two families of synchronized attractors (regular and chaotic ones). This phenomenon is associated with the presence of subharmonics $f_1/2$ in the spectrum of the forcing signal. In the full-scale experiment abrupt changes in the dynamics of the second oscillator are observed, due to transitions from one family of attractors to another, as well as and the phenomenon of hysteresis; this makes the construction of bifurcation diagrams extremely difficult. However, despite of the spectrum enrichment for the forcing chaotic signal the mechanisms of chaos synchronization remain the same, namely, the locking or suppression of the second oscillator's basic frequency f_2 by the forcing signal. It should be noted that when the second oscillator is forced by a complicated signal, the synchronization of oscillations via the frequency f_2 suppression results in a complicated synchronized regime similar to the forcing signal, and no simpler regimes in the zone of synchronization are observed. Thus, if the forcing signal is represented by a double period cycle ($2T_1$), then the frequency f_2 suppression results in the appearance of the synchronized regime $2T_1$ and in the region of synchronization the limit cycle T_1 would no longer exist. For a chaotically driven oscillator, the frequency f_2 suppression results in a synchronized chaotic regime. In this case regular synchronized regimes are not observed (with the exception of windows in the zone of synchronized chaos).

The influence of the forcing signal nature upon the phenomenon of synchronization is shown in Fig.8.8. The figure shows a fragment of bifurcation diagram on the plane of control parameters G and m_1 (parameter m_1 controls the degree of the forcing signal complexity). The symbols in the figure correspond to those used earlier in Fig.8.5. As shown in Fig.8.8, the boundary of the region of synchronization on the bifurcation diagram consists of segments of lines l_0^k, $k = 2^i$, $i = 0,1,2...$, representing the bifurcations of cycles kT_1 into tori kT^2. The segment of the boundary marked by dashed line (line l_0^∞) corresponds to the transition from a non-synchronized chaos outside the region of synchronization to the synchronized one inside the region of synchronization. The bifurcational mechanism of such a transition was established by numerical investigation of the mathematical model (8.4), as with the bifurcational mechanism of the $SA_2 \Rightarrow SA_1$ transition on line l_1^∞ (see Fig.8.5).

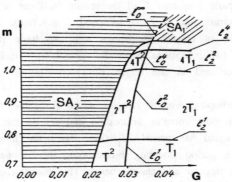

Fig.8.8. A fragment of the bifurcational diagram for unidirectionally coupled oscillators on the plane of parameters G, m_1 as $m_2 \simeq 1.2$, $g_1 = g_2 = 0.3$, $p = 1.2$ (full-scale experiment); l_0^k, $k = 1,2,4$ are the segments of the boundary of the region of synchronization, corresponding to the birth of tori kT^2 from cycles kT^1; l_0^∞ is the boundary between the regions of synchronized and non-synchronized chaos.

8.5. BIFURCATIONAL MECHANISMS OF SYNCHRONIZATION IN THE REGION OF CHAOS

Let us consider the nature of lines l_1^∞ in Fig.8.5 and l_0^∞ in Fig.8.8. These lines correspond to transitions from *non-synchronized chaos* to the *synchronized chaos* in the case of the locking and suppression of frequency f_2. To analyze the bifurcational mechanisms of these transitions in the region of chaos, we led computer simulation with the mathematical model of the system.

Figure 8.9 shows a small fragment of the bifurcation diagram on the plane of parameters p and G in the vicinity of a point of the synchronization tongue calculated at $m_1 = 0.6, m_2 = 1.19, B = 3, g_1 = g_2 = 0.3$. The symbols in Fig.8.9 correspond to those used earlier, while the symbol " $\sim$ " marks the bifurcation lines of saddle cycles. Points Φ_k divide the synchronization region boundary into segments that correspond to the tangency bifurcations of various periodic stable cycles. From Fig.8.9 it is seen that the structure of bifurcation lines features the property of scaling. The accumulation of the lines $\tilde{l}_i^k$ (the tangency bifurcation lines of saddle cycles of all possible periods kT_1, $k = 2^i$, $i = 0,1,2, ...$) corresponds to the "non-synchronous chaos ⇒ synchronous chaos" transition through the locking of the basic frequency.

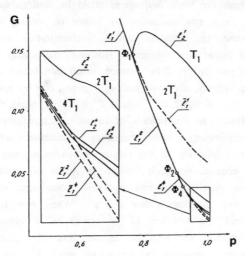

Fig.8.9. A fragment of the bifurcational diagram for unidirectionally coupled oscillators on the plane of parameters p,G as $m_1 = 0.6$, $m_2 = 1.19$, $g_1 = g_2 = 0.3$ (computer simulation).

Numerical investigations have demonstrated that the bifurcation diagram near the boundary of the synchronization region in the case of the "non-synchronous chaos ⇒ synchronous chaos" transition via the suppression of the basic frequency has similar structure. In this case the boundary of chaotic synchronization (l_0^∞) is determined by the accumulation of lines $\tilde{l}_0^k$ of saddle cycles with periods kT_1. On these lines the pairs of complex-conjugated multipliers of saddle cycles lie on the unit circle. A self-similar structure of bifurcation lines is observed.

8.6. MUTUAL SYNCHRONIZATION OF SYMMETRICALLY COUPLED OSCILLATORS

For investigations of *mutual synchronization* we have provided a symmetrical coupling between partial systems ($G_1 = G_2 = G$, $B = 1$). In a full-scale experiment, the elements of the partial systems were selected to maximize the equality of the partial systems' control parameters: $m_1 = m_2 = m$, $g_1 = g_2 = g$. Thus, the auto-oscillating regimes of the partial systems were distinguished only in their basic frequencies (or, more precisely, in their time scales).

The experiments have revealed that the mechanisms of mutual synchronization of two chaotic systems, in which the basic frequencies could be distinguished, correspond to the classical concepts about synchronization by means of locking or suppression of frequencies. In this sense, the study of mutual synchronization confirms the results obtained in the case of forced synchronization. However, the mutual synchronization of chaotic systems has its peculiarities. For instance, when introducing a mutual coupling between partial chaotic systems, the synchronized regimes can be regular (limit cycles with periods kT_0, $k = 2^i$, $i = 0,1,2, \ldots$, where $T_0 = 1/f_0$, f_0 being the basic frequency of synchronous oscillations). In the process of mutual synchronization via suppression of one of the basic frequencies, a higher frequency is commonly suppressed (this was confirmed by computer simulation). In the tongue of synchronization many families of synchronized attractors coexist. For each family, the region of synchronization has its own boundaries. These multistability and hysteresis phenomena make the detailed construction of bifurcation diagrams rather difficult. When $p = 1$ (in the case of identical partial systems) the increase of G results in a bifurcation; the limit manifolds of trajectories that are attractive in the invariant subspace S: $x_1 = x_2$, $y_1 = y_2$, $z_1 = z_2$, become attractive in the whole space $\mathbb{R}^6$. As a result of this bifurcation, the oscillations of partial systems are not only synchronized (in the sense of the basic frequencies being equal) but also have the same phase (i.e. they completely reproduce each other). If $p \neq 1$, the invariant subspace S does not exist. However, at any value of detuning p an increase of coupling G induces an asymptotic convergence $x_2 \rightarrow x_1$, $y_2 \rightarrow py_1$. In a limit case, the variables x_1 and x_2, y_1 and y_2 appear to be linearly dependent, which can also be qualified as complete synchronization (for variables z_1 and z_2, the complete synchronization in the sense mentioned above is not observed).

Figure 8.10 shows a fragment of the bifurcation diagram for the system of symmetrically coupled oscillators when the partial systems represent the developed chaos of spiral type evolving from a cycle with a tripled period. The nomenclature in Fig.8.10 is the same as in Fig.8.5. Lines l_{1}^{k}, $k = 2^i$, $i = 0,1, 2, \ldots$ correspond to mutual locking of the basic frequencies of the partial systems (f_1 and f_2), while lines l_{0}^{k}, $k = 2^i$, $i = 0,1,2,\ldots$ correspond to the suppression of one of the basic frequencies (frequency f_1 at $p < 1$ and f_2 at $p > 1$). Transition into the region of synchronization through l_{0}^{k} is preceded by the degeneration of non-synchronized chaos to tori. The synchronization tongues of higher orders evolve from the line l_{0}^{1}. Two tongues with rotation numbers 5/4 and 4/3 are shown in Fig.8.10. In the major tongue there is a region marked in Fig.8.10 by double hatching. In this region we observe both multistability and the regime of hyperchaos SA_3, arising through the merging of several families of attractors. Figure 8.11 shows the phase projections of chaotic attractors of the two families (a,b) and that of a merged attractor SA_3.

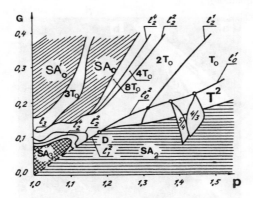

Fig.8.10. The bifurcation diagram of the system of two symmetrically coupled oscillators. The initial chaotic regimes of the partial systems correspond to the developed chaos evolved from a period three cycle. The parameter values: $m_1 \approx m_2 \approx 1.4$, $g_1 = g_2 = 0.3$. (Full-scale experiment).

Fig.8.11. The $x_2 y_2$ - projections of phase portraits of chaotic attractors (full-scale experiment), (a,b) Attractors of two different families; (c) United attractor SA_3.

Region SA_0 in Fig.8.10 corresponds to the synchronized chaos of spiral type with the basic frequency $f_0 \in [f_1, f_2]$, and region SA_0' corresponds to a synchronized chaos evolving from a cycle with a tripled period, that is similar to chaos in the partial systems in the absence of coupling. Regions SA_0, SA_0' are separated by the window of stability of cycle $3T_0$. Line l_3, a corresponds to an abrupt transition from one family of attractors to another. When $p = 1$ (identical partial systems) on line l_3 complete mutual synchronization is observed and above the line l_3 the synchronized chaotic attractor appears in the invariant subspace S. The nature of bifurcation on the line l_3 at $p \simeq 1$ has not been clarified so far.

Asymptotic convergence towards the complete synchronization of chaos when $p \neq 1$ is shown in Fig.8.12. The figure represents the evolution of the $x_1 x_2$ projection of the system's phase trajectory with the increase of the coupling parameter. This process is also characterized by the evolution of the coherence function θ(*f*) displayed in Fig.8.13 (from computer simulation).

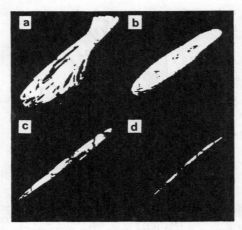

Fig.8.12. The $x_1 x_2$ - projections of phase trajectories in the system of symmetrically coupled oscillators for $p = 1.1$ and various G (full-scale experiment), (a) $G = 0.02$; (b) $G = 0.076$; (c) $G = 0.12$; (d) $G = 0.15$.

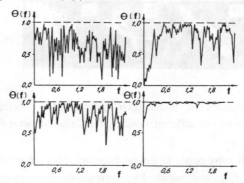

Fig.8.13. Dependence of coherence function on the degree of coupling G (computer simulation at $m_1 = m_2 = 1.19$, $g_1 = g_2 = 0.3$, $p = 1.1$).

Computer simulation confirmed the general character of the bifurcational diagram fragment constructed in the course of full-scale experiment. Fig.8.14 shows a fragment of the bifurcation diagram for the system as $m = 1.19$ that corresponds to a periodic window. This window is in the region of developed chaos that has evolved from the initial limit cycle of the partial systems. The nomenclature is the same as in Fig.8.10. The lines of saddle cycle bifurcations are marked with the symbol " ~ " The cross-hatched small square corresponds to a region with a complex structure of bifurcation lines that has not been fully investigated yet. This square represents the vicinity of point D (see Fig.8.10). As computation have shown, point D can be nonexistent and the boundary of the synchronization region cannot be represented as a continuous line. In the region where

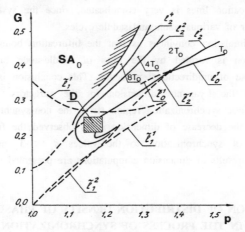

Fig.8.14. A fragment of the bifurcational diagram for symmetrically coupled oscillators at $m_1 = m_2 = 1.19, g_1 = g_2 = 0.3$ (computer simulation). Region D corresponds to a complex structure of bifurcation lines.

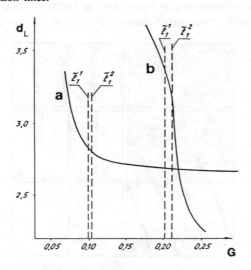

Fig.8.15. Evolution of the attractor's Lyapunov dimension in the system of symmetrically coupled oscillators at $m_1 = m_2 = 1.19, g_1 = g_2 = 0.3$ (computer simulation). (a) $p = 1.1$; (b) $p = 1.22$. Dotted vertical lines correspond to intersections of bifurcational lines l_1^1 and l_2^2 (see Fig.8.1).

216 Dynamical Chaos — Models and Experiments

the change of synchronization mechanisms takes place (from locking to suppression) the boundary of the tongue is very intricate and could not be observed in full-scale experiments. Figure 8.14 shows only a small number of existing bifurcation lines. The full picture of bifurcation lines is very complicated, since the system is characterized by an infinite number of various stable and saddle cycles.

The computer simulation convinces us that the bifurcation boundary of the chaotic synchronization region is formed by the lines of saddle-node bifurcation of saddle cycles, as in the case of unidirectional influence. This conclusion is also confirmed by the computations of the Lyapunov dimension for the attractors. When crossing the boundary of the chaotic synchronization region (from the non-synchronized chaos to the synchronized chaos) the decrease of dimension d_L is observed the from values $d_L > 3$ outside the region of synchronization to the values $d_L < 3$ inside the region of synchronization. The results of dimension computations are presented in Fig.8.15.

8.7. EVOLUTION OF DISTRIBUTION DENSITY OF PHASE SPECTRA DIFFERENCE IN THE PROCESS OF SYNCHRONIZATION

Fig.8.16a,b,c illustrate the evolution of the distribution density of phase spectra difference (DDPSD), corresponding to mutual chaotic synchronization by means of basic frequency f_2 suppression. The partial oscillators were set in the spiral type chaotic

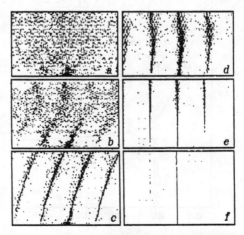

Fig.8.16. DDPSD diagram illustrating the process of mutual chaos synchronization through one of the basic frequencies at parameter values $m_1 = m_2 = 1.65$, $g_1 = g_2 = 0.2$, $p = 1.22$, (a) $G = 0.15$; (b) $G = 0.30$; (c) $G = 0.35$; and during complete chaotic synchronization at the following parameter values: $m_1 = m_2 = 1.5$, $g_1 = g_2 = 0.2$, $p = 1.5$, (d) $G = 0.18$; (e) $G = 0.9$; (f) $G = 2.0$.

region SA_0 ($m_1 = m_2 = 1.65, g_1 = g_2 = 0.2$). The detuning between basic frequencies was fixed at $p = 1.2$. With a weak coupling (prior to the onset of synchronization) the disordered chaotic structure of DDPSD corresponds to a complex non-synchronized attractor (see Fig.8.16a). With the increase of coupling we observe the "self-organization" of the DDPSD, accompanied by concentration of the distribution density in the vicinity of linear structures (see Fig.8.16b). Crossing the line l_0^∞ results in a phase difference distribution in the form of distinct lines, separated by the intervals $\Delta\varphi = \pi$ (see Fig.8.16c).

The rebuilding of the DDPSD structure in the process of complete synchronization represented in Fig.8.16d,e,f is of great interest. With the increase of coupling, the linear structure smoothly disappears and, thus, the DDPSD takes the form of a practically vertical single line, corresponding to $\Delta\varphi = 0$ for all frequencies. Phase spectra $\varphi_1(\omega)$ and $\varphi_2(\omega)$ coincide in a wide frequency range, as a consequence of the linear functional relations between $x_1(t)$ and $x_2(t)$ in the state of complete chaotic synchronization.

A qualitatively similar picture of DDPSD evolution is registered in the case of transition to synchronization through the locking of basic frequencies. However, in the region of parameters where the mechanism of mutual frequency locking is observed, the above mentioned phenomenon of multistability of oscillations takes place. The presence of several attractors in the phase space can complicate the observed DDPSD evolution.

Thus, the analysis of chaos synchronization using the DDPSD method shows that, on the contrary to the classical case of synchronization, the linear structure of the distribution $p(\Delta\varphi,\omega)$ is formed first. The stabilization of the mutual phase of oscillations is formed only in the case of complete chaotic synchronization.

Lastly, let us note the following. The DDPSD used as a characteristic of the nonlinear subsystems' interaction offers a quantitative criterion for the degree of self-organization. In fact, the entropy of non-synchronized chaotic oscillations is fully correlated with the chaotic attractor dimension decrease and the emergence of a regular DDPSD structure in the process of synchronization noted above.

The series of full-scale experiments and cmputer simulation led allows the following conclusions to be drawn which are valid for chaotic systems with distinguishable basic frequencies in the oscillators power spectrum:

1. The interaction of subsystems results in chaos synchronization via locking or suppression of basic frequencies in *full accordance with the classical theory* of oscillations.

2. In the region of synchronization of chaotic systems both chaotic and regular synchronized regimes can exist. For regular regimes the boundary of synchronization is determined by the bifurcation lines of synchronized oscillations (i.e., by lines of tangency bifurcations of limit cycles or the lines of tori birth). In the case of

chaotic synchronized regimes, the *accumulation of bifurcation lines* of various periodic *saddle cycles* provides the *boundary of the synchronization region*. These can be both the lines of tangency bifurcations and the lines of tori birth.

3. The *synchronized attractors* in the system of interacting oscillators are *topologically equivalent* to the attractors of the initial partial subsystems. During a transition into a synchronization region the structure of non-synchronized oscillations becomes simpler.

4. With the increase of coupling in the region of chaotic synchronization, a *complete chaotic synchronization* of the oscillations is observed which leads to a linear functional dependence of corresponding variables in the partial subsystems. In the case of complete chaotic synchronization, the coherence function (8.5) of the subsystems is identically equal to a unity, while the distribution of the phase spectra difference is stabilized at zero.

The results presented in this chapter are applicable to the interaction of chaotic attractors appearing in the vicinity of a saddle-focus separatrix loop as in Shilnikov's theorem [8.12]. Because of the presence of a horseshoe-like map, the cascade of period doubling bifurcations is always realized in such systems. Thus, we can easily distinguish the basic frequencies in the chaos spectrum. A qualitatively equivalent behavior can be observed in a system of two coupled Rössler's oscillators [8.13] or under interaction of two systems from the Chua's circuit family [8.14]. The investigations of Chua's systems is of special interest; due to the nonlinear characteristic symmetry, the partial subsystem has two saddle-foci and, thus, it has two chaotic attractors of spiral type. With coupling between two Chua's systems, one can expect phenomena like those considered in this chapter, but more complicated phenomena are caused by the possible interaction of the four spiral type attractors.

The matter of applicability of the presented ideas to the analysis of chaotic oscillations with practically uniform spectra (e.g., Lorenz' attractor) is left open. How the type of coupling influences chaotic synchronization also demands further investigation [8.15].

CHAPTER 9

NONLINEAR PHENOMENA AND CHAOS IN CHUA'S CIRCUIT

9.1. DEFINITION OF THE PROBLEM

The system (3.38) examined in detail in Chapters 3-7 which simulates the dynamics of oscillator with inertial nonlinearity is an example of this simplest class of three-dimensional systems exhibiting chaotic behavior. The topology, properties and bifurcations of regular and chaotic attractors are defined in such systems by only one equilibrium state which has a structurally unstable homoclinic trajectory of the saddle-focus separatrix loop type. The qualitative regularities of the dynamics of this class of systems are classified almost entirely by Shilnikov's theorem [9.1,9.2] and the contents of Chapters 3-5 have substantiated this conclusion.

Novel interesting properties can be observed from dynamical systems exhibiting *three or more equilibrium states*. Here, there exist *homoclinic or heteroclinic trajectories* which define the structure of attractors, their bifurcations and properties. Additional specific features are observed in *dynamical systems which possess some symmetry*.

In this chapter we considered and investigated the well-known *Chua's circuit* as an example of the class of systems mentioned above [9.3-9.6]. The surprising simplicity of the circuit diagram used to construct Chua's circuit results in a relatively simple mathematical model. The *"simple" Chua's circuit* models exhibits, however, a very rich set of *complicated* regular and chaotic *regimes* and of their bifurcations which are far from being simple! The studies of Chua's circuit dynamics, in spite of a very large number of publications (see, for instance, references to [9.4,9.6]), are not to be regarded as completed. Some new results obtained when investigating the nonlinear dynamics of Chua's circuit are discussed in this chapter.

9.2. CHUA'S CIRCUIT

Consider the electronic circuit shown in Fig.9.1a. It contaitns four linear elements (capacitors C_1, C_2, inductor L and resistor $R = \frac{1}{G}$) and one nonlinear element N_R. The current-voltage characteristic of the nonlinear element (called Chua's diode in recent literature) is shown in Fig.9.1b.

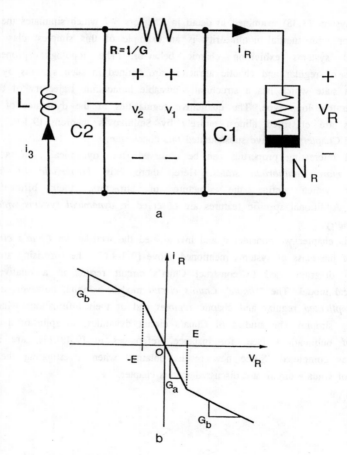

Fig.9.1. Chua's circuit, (a), v-i characteristic of the nonlinear N_R (drawn with $G_a < G_b \leq 0$), (b).

As seen from Fig.9.1b, this Chua's diode has two negative conductances $-G_a(|V_R| < E)$ and $-G_b(|V_R| > E)$. These conditions made it possible to maintain undamped oscillations in the circuit of Fig.9.1a.

The differential equations describing Chua's circuit have the form [9.4,9.5]:

$$\begin{aligned}
\dot{V}_1 &= \frac{1}{G_1} [G(V_2 - V_1) - f(V_1)], \\
\dot{V}_2 &= \frac{1}{G_2} [G(V_1 - V_2) + i_3], \\
\dot{i}_3 &= -\frac{1}{L} V_2,
\end{aligned} \qquad (9.1)$$

where the function $f(V_1)$ is defined by the nonlinear characteristic of Chua's diode (Fig.9.1b).

By substituting the following variables and parameters

$$\begin{aligned}
x &= V_1/E, \quad y = V_2/E, \quad z = i_3/(EG), \\
\tau &= tG/C_2, \quad a = 1 + G_a/G, \quad b = 1 + G_b/G, \\
\alpha &= C_2/C_1, \quad \beta = C_2/(LG^2),
\end{aligned} \qquad (9.2)$$

to equation (9.1) we obtain the following normalized (dimensionless) form, which is more convenient for analytical and numerical studies:

$$\begin{aligned}
\dot{x} &= \alpha[y - h(x)], \\
\dot{y} &= x - y + z, \\
\dot{z} &= -\beta y,
\end{aligned} \qquad (9.3)$$

where

$$h(x) = \begin{cases} bx + a - b, & x \geq 1, \\ ax, & |x| \leq 1, \\ bx - a + b, & x \leq -1. \end{cases} \qquad (9.4)$$

Let us transform equation (9.3) into the form of (3.2). For our main variable, we choose $z(\tau)$ which is proportional to the current i_3 through the inductor L. Removing the variable y from equation (9.3) yields

$$\begin{aligned}
\ddot{z} + \dot{z} + \beta z &= -\beta x, \\
\dot{x} &= -\frac{\alpha}{\beta} \dot{z} - \alpha h(x).
\end{aligned} \qquad (9.5)$$

Equation (9.5) describes the oscillation process $z(\tau)$ in the dissipative LC-circuit which is driven by the force $x(\tau)$. The action of $x(\tau)$ on the circuit is *performed inertially since* the variables x and z are coupled by a differential operator (see the second equation in (9.5)). The inertial excitation mechanism of the oscillations in (9.5) is *qualitatively equivalent to the mechanism of the oscillation excitation in the Lorenz model*. This is easy to see if one compares equation (9.5) to equation (3.9) or (3.11). The oscillations in system (9.5) occur as a result of the *inertial parametric source action on the elements of the dissipative circuit* rather than by compensating for the losses via positive feedback as in the oscillator with inertial nonlinearity of (3.38) and (5.4).

The dynamical system (9.3) specifies a vector field in the three-dimensional space $\mathbb{R}^3$ which is invariant to the symmetry transformation

$$(x,y,z) \to (-x, -y, -z). \tag{9.6}$$

The symmetry of the system is defined by the form of the characteristic $h(x)$ of the nonlinear element which separates the phase space into three regions by the planes $x = \pm 1$:

$$\begin{aligned} D_1 &= \{(x,y,z) : x \geq 1\}, \\ D_0 &= \{(x,y,z) : |x| \leq 1\}, \\ D_{-1} &= \{(x,y,z) : x \leq -1\}, \end{aligned} \tag{9.7}$$

provided that the parameters a and b are non-zero. In each of the regions D_0, $D_{\pm 1}$, the system is linear. This important feature of the Chua's circuit enables one to obtain a number of results analytically [9.4, 9.6].

Consider the divergence of the vector field F of system (9.3):

$$\text{div} F = -\alpha h'_x - 1, \quad h'_x = \frac{dh(x)}{dx}, \tag{9.8}$$

where $h'_x = b$ for $x \in D_{\pm 1}$ and $h'_x = a$ for $x \in D_0$.

Let us choose the following parameter values for the characteristic $h(x)$:

$$a = -0.143, \quad b = +0.286.$$

The condition $a < 0$ signifies, as seen from (9.2), that $|G_a| > G$. In this case, the negative conductance of the source is greater than the active conductance of the circuit! The condition $b > 0$ means that $|G_b| < G$ and that the active losses in the RLC-circuit are greater than that introduced by the source. To maintain an undamped oscillations, a time-averaged balance between the positive and the negative losses is needed. Hence, an important conclusion can be made: *any phase trajectory of system (9.3) may not be localized only in the region* D_0 *or* $D_{\pm 1}$. The phase trajectories must visit either all the three regions D_0 and $D_{\pm 1}$ or two of them: D_0 and D_{+1} or D_0 and D_{-1}. In the latter case, pairs of mutually symmetric attractors will be realized. The above considerations turn out to be useful for understanding some bifurcational properties of system (9.3).

Equilibrium states of system (9.3). Setting the right-hand parts of equation (9.3) to zero, we obtain the following coordinates of the three equilibrium states:

$$P^+ = (K, \ 0, \ -K) \in D_1,$$
$$P^0 = (0, \ 0, \ 0) \in D_0, \qquad K = \frac{b-a}{a} = 1.5. \qquad (9.9)$$
$$P^- = (-K, \ 0, \ K) \in D_{-1},$$

The stability of the equilibrium states is specified by the eigenvalues s_i, $i = 1,2,3$ which satisfy equation (1.20); namely,

$$\det \begin{bmatrix} -\alpha C - s & \alpha & 0 \\ 1 & -1-s & 1 \\ 0 & \beta & -s \end{bmatrix} = 0, \qquad (9.10)$$

where $C = a$ (for P^0), $C = b$ (for $P^{\pm}$).

From (9.10) we obtain the *characteristic equation* for the eigenvalues:

$$s^3 + (1 + \alpha C) s^2 + (\beta - \alpha + \alpha C) s + \alpha C \beta = 0. \qquad (9.11)$$

Let us fix the parameter value $\beta = 14.286$ and examine the stability of the equilibrium states as a function of the magnitude of parameter $0 < \alpha < 15$ by solving equation (9.11).

The equilibrium state at zero with coordinates P^0 is a *saddle-focus* with a one-dimensional unstable and a two-dimensional stable manifolds. They correspond to one positive eigenvalue $s_3 > 0$, and a pair of complex-conjugate eigenvalues (see Fig.9.2a):

$$s_{1,2} = \text{Re } s_{1,2} \pm j \text{ Im } s_{1,2}, \qquad \text{Re } s_{1,2} < 0. \qquad (9.12)$$

Note that the saddle "coefficient" $\sigma(\alpha) = 2\text{Re } s_{1,2} + s_3$ of the equilibrium state P^0 is negative for $\alpha < 7.2$. With $\alpha_0 \cong 7.2$ $\sigma(\alpha_0) \cong 0$ and, for all $\alpha > 7.2$, the saddle coefficient $\sigma(\alpha) > 0$.

The dependence of the eigenvalues on the parameter α is illustrated in Fig.9.2b for the equilibrium states $P^{\pm}$. As seen from Fig.9.2b, the equilibrium states $P^{\pm}$ are also of *saddle-focus type*. For $\alpha < 6.8$, the saddle-focus is stable, Re $s_{1,2} < 0$, $s_3 < 0$. For $\alpha > 6.8$, the saddle-focus $P^{\pm}$ exhibits a one-dimensional stable and a two-dimensional unstable manifolds. At the point $\alpha = 6.8$, a pair of eigenvalues $s_{1,2}$ *crosses the imaginary axis at a non-zero rate*. The third eigenvalue is separated from the imaginary axis ($s_3 < 0$). For the smooth dynamical systems, this situation corresponds to the *Andronov-Hopf* bifurcational and leads to the birth of a limit cycle. The stability of the limit cycle will depend on the sign of the first Lyapunov quantity L_1.

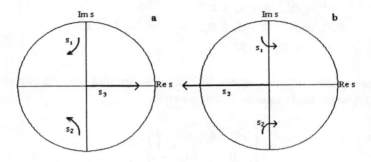

Fig.9.2. Eigenvalues for equilibrium states P^0(a) and $P^{\pm}$(b) as a function of the parameter α.

System (9.3) displays the following features: 1) the equations of system (9.3) correspond to a continuous flow but with a discontinuos first derivative at the break points, and 2) system (9.3) *is linear* in each of the regions D_0, D_{-1} and D_{+1}. Due to the linearity, *all Lyapunov quantities* which are defined by nonlinear terms are *equal to zero*. In this case, nothing can be said concerning the stability (instability) of the limit cycle. Moreover, due to the lack of continuous differentiability, one can not exploit the Andronov-Hopf bifurcation in system (9.3) in a rigorous sense.

Nevertheless, when passing the bifurcation value $\alpha \approx 6.8$ as we vary the parameter α, *stable periodic oscillations are excited* in system (9.3). The limit cycle appears *abruptly* with a finite amplitude, as predicted by an analysis of the divergence (9.8). In Fig.9.3, the phase portraits are presented for two symmetrical periodic attractors in the vicinity of the bifurcation point $\alpha \geq \alpha_0 = 6.8$.

<u>Homoclinic trajectories and attractors of system (9.3)</u>. A remarkable feature of system (9.3) is the existence of a *homoclinic trajectory* which has the form of a *saddle-focus separatrix loop* at the equilibrium state P^0, along with the existance of a

Nonlinear Phenomena and Chaos in Chua's Circuit

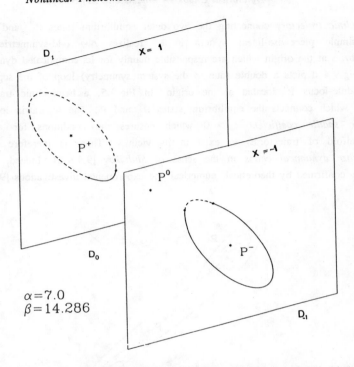

Fig.9.3. Two symmetrical limit cycles near bifurcation point of their birth.

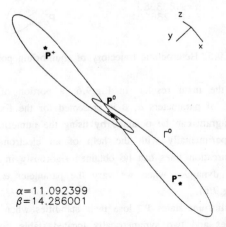

Fig.9.4. Saddle-focus separatrix loop for equilibrium point P^0.

heteroclinic trajectory connecting the two outer equilibrium states P^+ and P^-. In fact, this simple piecewise-linear system in $\mathbb{R}^3$ realizes *two* odd-symmetric *homoclinic trajectories* at the origin which are responsible mainly for its complicated dynamics!

Fig.9.4 depicts a double (due to the system symmetry) loop of the separatrix Γ_0 at the saddle-focus P^0 located at the origin. In Fig.9.5, a heteroclinic trajectory Γ is shown which connects the equilibrium states P^+ and P^-. The separatrix loop Γ_0 has a *positive saddle coefficient* $\sigma > 0$ which ensures the conditions for a hyperbolic submanifold of trajectories to exist in its vicinity. There is therefore a *possibility to realize dynamical chaos* in the sense of *Shilnikov* [9.1,9.2]. Indeed, it has been entirely confirmed by theoretical, numerical and experimental investigations [9.6].

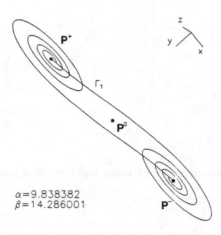

Fig.9.5. Heteroclinic trajectory of equilibrium points $P^{\pm}$.

Illustration of the main results: In Fig.9.6, a portion of the system bifurcation diagram in the plane of parameters α, β is presented for the fixed values a = - 0.143, b = 0.286. This diagram can be obtained by using the numerical methods described in Chapter 2, and experimentally with the help of an electronic circuit realization of Fig.9.1. Certain bifurcation lines can be obtained rigorously in *analytical* form. Let us examine the system dynamics when we vary the parameter α along the straight line $\beta = 14.286$ (see Fig.9.6).

The stable equilibrium states $P^{\pm}$ lose their stabilities when crossing the Andronov-Hopf bifurcation lines and two symmetrically located stable limit cycles with a finite amplitude are born abruptly in the system. Then, as the parameter α increases, these limit cycles undergo a cascade of period-doubling bifurcations in full accordance to the

Feigenbaum's universality law [1]. As a result, a pair of symmetrical *chaotic Rössler*-like

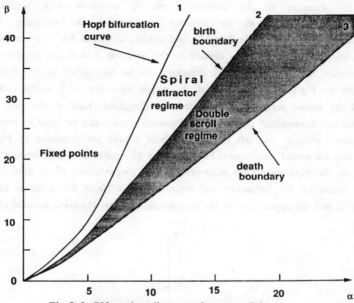

Fig.9.6. Bifurcation diagram of system (9.3).

$\alpha = 8.5000000$
$\beta = 14.286001$

Fig.9.7. Spiral Chua's attractors ($SA_1^{\pm}$) for the system (9.3).

---

[1]The sequence of the points of the period-doubling bifurcations converge to the critical line, which gives birth to a Rössler-like spiral Chua's attractor, in a geometric progression with Feigenbaum's constant $\delta = 4.6692...$[9.6].

spiral Chua's *attractors* are born upon crossing the line of critical parameter values l_{cr}. The structure of the attractors and its evolution with increasing α are qualitatively equivalent to the attractor structure in the oscillator with inertial nonlinearity in all details. In Fig.9.7, two chaotic attractors $SA_1^{\pm}$ are shown for $\alpha=8.5$. Let us introduce a secant plane defined by the condition $y = 0$ and construct numerically the Poincaré section for the $SA_1^{\pm}$ attractors. To be illustrative, in the Poincaré section indicated in Fig.9.8a, all points of the phase trajectory $SA_1^{\pm}$ which cross the plane $y = 0$ are shown regardless of the crossing directions. Each of the four submanifolds near the one-dimensional curves in the Poincaré section can be used to construct a *one-dimensional map* $x_{n+1} = \phi(x_n)$. The calculated results are presented in Fig.9.8b. They verify that the model return maps have the form of a parabola with a smooth peak. Hence, they meet the requirements for *maps* belonging to *Feigenbaum's class*. Both the mechanism in the formation $SA_1^{\pm}$ attractors and their evolutions (from the spiral to the screw-type attractors, and the appearance of the appropriate stability windows) become clear.

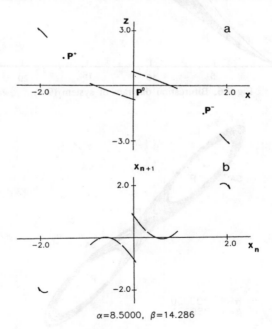

$\alpha=8.5000, \ \beta=14.286$

Fig.9.8. Poincaré sections (a) and one-dimensional return maps for attractors $SA_1^{\pm}$ (b).

With further increase of the parameter α, the phase trajectories of the Rössler-like attractors begin to come closer to the equilibrium states P^0 in the region D_0. As a result, they merge and form a new attractor SA_0 in the form of a "*double scroll*" (DS). The phase portrait of the double scroll Chua's attractor SA_0 is presented in Fig.9.9 for the value $\alpha = 9.0$. This attractor type is realized within the interval $8.8 < \alpha < 10.00$ and its evolution is *typical* for that observed in *quasiattractors*. The SA_0 topology is defined by the existence of both a separatrix loop Γ_0 and a heteroclinic trajectory Γ. This is visualized by comparing the form of the SA_0 attractor phase portraits in Figs. 9.9 and 9.10 and the type of the appropriate homoclinic orbits Γ_0 and Γ in Figs. 9.4 and 9.5.

Fig.9.9. Attractor SA_0 of the "double-scroll" type.

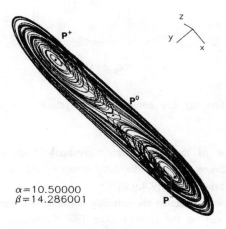

Fig.9.10. Attractor SA_0 of a more complicated type.

The gradual increase in the complexity of the structure of the SA_0 attractor for increasing values of α is mirrored by the corresponding evolution in the Poincaré section, and in the character of the associated one-dimensional map. The typical form of the attractor section cut by the plane $y = 0$ is shown in Fig.9.11a. The characteristic form of the SA_0 section in the vicinity of the equilibrium P^0 accounts for the notion of "double scroll". Fig.9.11b depicts the increased complexity as compared to Fig.9.8, of the form of the 1-D map, which has many discontinuities.

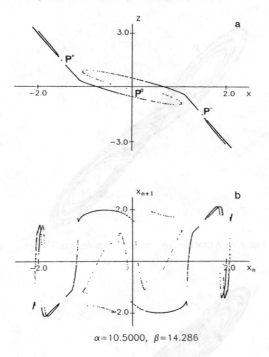

$\alpha = 10.5000, \beta = 14.286$

Fig.9.11. Poincaré section (a) and one-dimensional return map (b) for attractor SA_0 presented in Fig.9.10.

Continued evolution of the SA_0 attractor eventually leads to an *attractor crisis* forced upon by a collision between the SA_0 attractor and the saddle cycle which encircles all three equilibrium states [9.4,9.6].

The phenomena described above are naturally not able to cover the entire variety of bifurcation phenomena and of the corresponding fine distinctions in the topologies of the resulting attractors especially since the number of attractors *is infinite*! In particular, the basic homoclinic orbits shown in Fig.9.4 and 9.5 give rise to a system of more complicated and multiple loop homoclinic orbits in its vicinity. These orbits will give rise to a corresponding set of attractors, both regular and chaotic. Here, we are again faced with the fundamental property of a quasihyperbolic system exhibiting homoclinic trajectories of the saddle-focus separatrix-loop type which were discussed in detail in Chapter 5.

Thus, the existance of three equilibrium states and the symmetry property (9.6) of system (9.3) provide us with a broader set of oscillation regimes in comparison with system (3.38). In particular, in system (9.3) the interaction between two symmetrical attractors $SA_1^{\pm}$ leads to a *"chaos-chaos" type of intermittency* which we discuss next.

9.3. CHAOS-CHAOS INTERMITTENCY AND 1/F NOISE IN CHUA'S CIRCUIT

Regular and chaotic limit set bifurcations of different types are realized in nonlinear quasihyperbolic systems when their parameters are varied [9.7]. Among them, various phenomena arising from the interaction between attractors have been observed, including *intermittency*. Three typical and relatively simple intermittency mechanisms have been distinguished and classified in the theory of chaos [9.8]. However, more complicated intermittency mechanisms can take place in the general case, such as the "chaos-chaos" type of intermittency [9.9-9.11]. The intermittency phenomenon is characterized by a *strong sensitivity to perturbations* [9.12-9.13]. In addition to a purely dynamic origin, the intermittency phenomenon can also occur via a *mechanism similar* to that of a phase transition *induced by noise* [9.14-9.16].

One of the remarkable statistical properties of the chaos-chaos intermittency is the slope of its power spectrum in the low-frequency region. As shown in a number of papers [9.17,9.19], the power spectrum in the region near the onset of intermittency is described by a *universal law* $1/\omega^{\delta}$ ($0.8 < \delta < 1.4$). Such a spectrum shape is due to a finite probability in the existence of arbitrarily long *"laminar"* flow near the onset of intermittency. The 1/f spectrum is observed in many processes having different origins: the fluctuations of the current in electron devices, the fluctuations of the Earth's rotation frequency, the fluctuations of the intensity in sound and speech sources, the fluctuations of the muscle rhythms in the human heart [9.8,9.18], etc.

The *chaos-chaos intermittency* has been discussed in a number of papers [9.17-9.20] along with its characteristic 1/f noise. Most of these papers dealt with discrete maps or nonautonomous systems of the Duffing type as models. However, the influence of external noise on this intermittency type remains insufficiently investigated.

This section deals with the study of chaos-chaos intermittency in the Chua's circuit modeled by the autonomous system of ordinary differential equations (9.3).

Equation (9.3) has a number of important properties which make it an *ideal model for studying the chaos-chaos intermittency*: it is autonomous, it has an $N = 3$ and it is symmetrical relative to a change in the variables (9.6). On the α-β parameter plane, there is a bifurcation line corresponding to the birth of the double-scroll Chua's attractor from two symmetrical Rössler-like spiral Chua's attractors (Fig.9.6). The chaos-chaos intermittency phenomenon is observed at this point of bifurcational transition: phase trajectories evolve for a long time around each symmetrical spiral Chua's attractor, followed by relatively rare transitions between them. During a full-scale experiment, this effect is recorded as flickers in the two symmetrical components of the double scroll Chua's attractor on the screen of the oscillgraph.

In this section, we will study thoroughly the above intermittency phenomenon in Chua's circuit and investigate the *influence of external noise* and numerical accuracy on its *statistical properties*.

To analyze this intermittency phenomenon, we will make use of not only such standard characterizations as Lyapunov exponents, power spectrum, and stationary probability density, but also other characteristics associated with the statistics of the "laminar" phases. Here, we shall mean by "laminar phase" the time period where a phase trajectory remains on one of the two symmetrical double scroll components. Although there exists a number of other characteristics for some specific properties of this intermittency phenomenon [9.20], we shall dwell in this section only in the calculation of the power spectrum and the statistics of the "laminar" phase.

In a recent paper [9.21], a technique for determining the main intermittency characteristics has been proposed which is based on the mean-first-passage-time theory of Markovian processes. We will see that such stochastic process appears naturally in the statistics of the laminar phase associated with the chaos-chaos intermittency in Chua's circuit. A separatrix surface in the three-dimensional phase space of system (9.3) forms a boundary which separates the basins of attraction of the symmetric components of the double scroll Chua's attractor. Unfortunately, it is impossible in our case to obtain analytical results due to the difficulty of explicitly specifying the boundary separating the two basins of attraction. Therefore, all process characteristics in our study were determined by numerical simulation.

Equation (9.3) is integrated by a 4-th order Runge-Kutt method. As the numerical integration is being calculated the probability density $P(T)$ of the residence times on the two symmetrical double scroll components and the moments of this distribution $<T^q>$ ($q = 1,2...$) were also calculated under the assumption that the calculated motion is ergodic. Then, the power spectrum is calculated and its shape examined. In addition, we also investigated the associated two-dimensional Poincaré map.

All calculations were carried out with the following fixed parameter values:

$$(\beta, a, b) = (14.3, -1/7, 2/7).$$

Only the parameter α was varied. The following bifurcational value of the parameter $\alpha = \alpha^*$ was calculated numerically to a 8-digit precision by detecting the point in the α-parameter interval when the two spiral Chua's attractors suddenly merged into the double scroll Chua's attractor:

$$\alpha^* = 8.813232... \pm 10^{-7}.$$

While varying the parameter α, the averaged characteristics were determined as a function of the *overcriticality* defined by

$$\varepsilon = \alpha - \alpha^*. \tag{9.13}$$

To investigate the effect of external noise, a δ-correlated noise source of intensity D was added to the right-hand part of the third equation of (9.3). The inclusion of this external noise will also be particularly convenient in our full-scale experiments. The resulting system of stochastic differential equations has the form:

$$\begin{aligned} dx/dt &= \alpha[y - h(x)], \\ dy/dt &= x - y + z, \\ dz/dt &= -\beta y + \xi(t), \\ <\xi(t)\xi(t+s)> &= 2D\delta(s). \end{aligned} \tag{9.14}$$

The extent of the noise influence on the different process characteristics was determined by varying the parameter D with fixed system parameters.

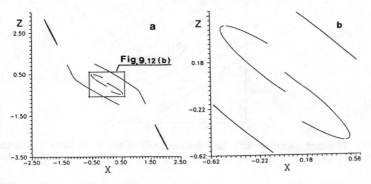

Fig.9.12. Poincaré section of the spiral Chua's attractors calculated with $\alpha = 8.8133$ ($\alpha < \alpha^*$).

Poincaré map. Consider first the Poincaré section[2] of the system (9.3) with the plane $y = 0$ corresponding to system parameters chosen just before the onset of intermittency. The results are given in Fig.9.12(a-b).

Observe from this figure that the map consists of two symmetrical parts, obtained with two different initial points. Each part corresponds to the intersection of one spiral Chua's attractor with the plane $y = 0$.

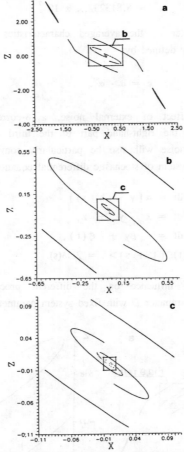

Fig.9.13. Poincaré section of the double-scroll Chua's attractor calculated with $\alpha = \alpha^* = 8.813232$.

---

[2] All intersection points of the phase trajectory with the plane $y = 0$ in Fig.9.12 and Fig.9.13 are for trajectories coming from the top region $(y > 0)$ and from the bottom region $(y < 0)$.

Consider next the system parameters chosen at the onset of double scroll Chua's attractor. The corresponding results are presented in Fig.9.13 (a-c). Figure 9.13(a) corresponds to the complete Poincaré map. Observe from this figure that the map is practically one-dimensional. Let us magnify a small area near the origin $x = 0$, $y = 0$ (enclosed by a square in Fig.9.13(a)). As shown in Fig.9.13(b), the map repeats its structure in the vicinity of the origin, i.e., it *demonstrates a scaling property*; namely, *self-similarity*, which is confirmed by several increasing levels of magnification in the vicinity of the origin where the same structure keeps repeating itself.

It is of interest to note that, in contrast to systems having an attractor for which the two-dimensional Henon map is typical [9.9], the Poincaré map in Figs.9.12-9.13 is approximately *one-dimensional* [9.22] where the structure of the unstable manifold of the saddle equilibrium state is repeated at all scales. If the scaling in the Henon map manifests itself in the repeated lamination of its fine structure, then the map structure in the vicinity of the origin in Figs.9.12-9.13 is also repeated here.

Power spectrum. Let us now investigate the power spectrum and other statistical intermittency characteristics. When integrating Eq.(9.3) and Eq.(9.14), the integration step was chosen automatically according to the accuracy specified. The number N of the calculated solution points in the time series $x(t)$ was fixed and can be as high as $N = 4 \times 10^4$. The maximum integration time T_{max} and the discretization time ΔT were determined from the maximum frequency component ω_g in calculating the power spectrum; namely,

$$T_{max} = \frac{\pi N}{\omega_g}, \quad \Delta T = T_{max}/N. \tag{9.15}$$

Since the maximum frequency component ω_g in Chua's Circuit is relatively small, it is possible to analyze the *low-frequency region* of the power spectrum $S_x(\omega)$.

The power spectrum was calculated under the assumption that the motion is ergodic by using the method of periodograms.

Consider the power spectrum $S_x(\omega)$ at the critical parameter point $\alpha^* = 8.813232...$. The low-frequency spectrum region is shown in Fig.9.14 (points), plotted on a double logarithmic scale.

Observe that on the left, closer to the zero frequency, there is an interval where $\Delta\omega \approx 10^{-3}$ is almost constant, hence forth referred to as the "*spectral plateau*". We will show later that the plateau is an artifact of numerical errors and external noise and is *not* a part of the actual spectrum. Beyond the plateau, however, lies the correct spectrum which obeys the following law:

$$S_x(\omega) \propto \omega^{-\delta}, \quad \delta = 1.1 \pm 0.1 . \tag{9.16}$$

Observe that the points cluster and cling to each side of the *ideal* 1/f line, which corresponds to points with δ = 1.

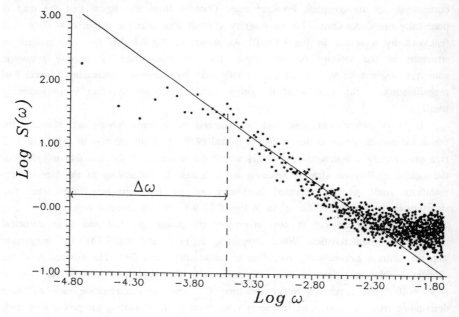

Fig.9.14. Power spectrum of the x(t) with $\alpha = \alpha^* = 8.813232$. The solid line gives the approximating function $\omega^{-\delta}$, $\delta = 1.1$.

<u>Analysis of the spectral plateau</u>. Let us examine first the dependence of those quantities which characterize the power spectrum in the low-frequency region as a function of the overcriticality measure defined by $\varepsilon = \alpha - \alpha^*$, $\varepsilon \ll 1$. The plateau width $\Delta\omega$ is shown in Fig.9.15 as a function of ε on the double logarithmic scale, where the data points are shown in asterisks.

Observe that the $\Delta\omega(\varepsilon)$ dependence can be approximated by the following step function (solid line in Fig.9.15):

$$\Delta\omega(\varepsilon) \propto \varepsilon^\gamma, \quad \gamma = 0.62 \pm 0.02 \qquad (9.17)$$

As we increase the parameter α from α^* to higher values, the plateau is seen to extend closer to the zero frequency. This observation can be explained by the fact that the mean length of the laminar phase is finite, and it decreases as ε increases [9.8]. As for the index δ in (9.16), it maintains its value within our limits of calculation accuracy for all small overcriticality variations $\varepsilon \leq 10^{-4}$.

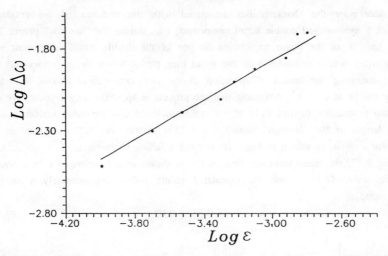

Fig.9.15. Dependence of the spectral plateau width $\Delta\omega$ on the overcriticality measure $\varepsilon = \alpha - \alpha^*$. The solid line gives the approximating function ε^γ, $\gamma = 0.62$.

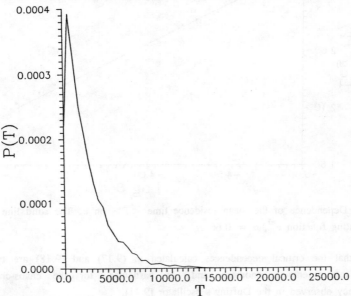

Fig.9.16. Probability density of the residence times P(T) for $\alpha = \alpha^* = 8.813232$.

Consider now the characteristics associated with the statistics of the residence times on each symmetrical double scroll component, i.e., during the "laminar" phase. The residence time T of the phase trajectories on one of the double scroll component is a random quantity which coincides with the mean time period when the trajectory hits the boundary separating the basins of attraction of the two symmetrical chaotic attractors when they merge at $\alpha = \alpha^*$. Assuming that the process is approximately ergodic, we can estimate the probability density $P(T)$ of the residence times and the mean residence time (or mean length of the "laminar" phase) $<T>$. The distribution $P(T)$ calculated at the critical point $\alpha = \alpha^*$ is shown in Fig.9.16 and has a falling character as $T \to \infty$.

In Fig.9.17, the mean residence time $<T>$ is shown as a function of ε in a double logarithmic scale. Observe that the calculated points follow approximately a straight line with slope ϑ:

$$<T> \propto \varepsilon^{-\vartheta}, \quad \vartheta = 0.66 \pm 0.02 \tag{9.18}$$

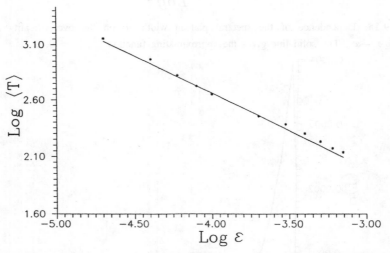

Fig.9.17. Dependence of the mean residence time $<T>$ on ε. The solid line gives the approximating function $\varepsilon^{-\vartheta}$, $\vartheta = 0.66$.

Observe that the critical dependences calculated in (9.17) and (9.18) are typical for all the intermittency phenomena and, in particular, for the chaos-chaos type intermittency observed in the Duffing's oscillator [9.11].

As we increase the parameter α further beyond the critical value α^* corresponding to the birth of the double scroll Chua's attractor, the shape of the power spectrum in the low-frequency region follows approximately a Lorenzian law:

$$S_x(\omega) \propto <T>/[(\omega<T>)^2 + 1]. \tag{9.19}$$

This relationship is shown in Fig.9.18 for $\alpha = 8.84$.

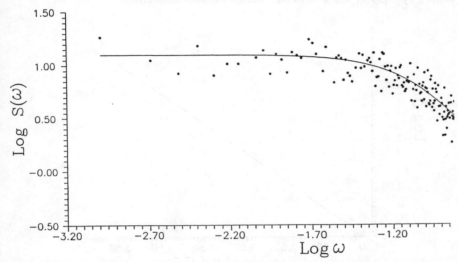

Fig.9.18. Power spectrum of x(t) for $\alpha = 8.84$. The solid curve gives the Lorenzian function.

Our computer simulation have shown that the power spectrum near the zero frequency always has a plateau whose width $\Delta\omega$ increases with ε. In other words, the experimental 1/f noise characteristic is obtained only to the right of the plateau. This low-frequency experimental limitation is due to unavoidable perturbations in all physical systems. These perturbations are caused by numerical integration and round-off errors. To illustrate this problem, our calculations of the statistical characteristics as a function of the integration accuracy σ are presented below. In Fig.9.19, the width of plateau (flat portion) $\Delta\omega$ is shown as a function of calculation accuracy σ in double logarithmic scale. Observe that this relationship follows approximately a *degree law* (a straight line of slope in double-logarithmic scale):

$$\Delta\omega \propto \sigma^\mu, \quad \mu = 0.32 \pm 0.02 \tag{9.20}$$

Again, the index δ in the spectral dependence Eq.(9.16) is found to be independent of the integration accuracy, relative to the accuracy of our numerical methods.

Observe that as we increase our calculation accuracy (decreasing σ), the width $\Delta\omega$ of the plateau decreases in accordance with that given by Eq.(9.20). Since round-off errors are inevitable, so too is the plateau which limits the range of our measurable 1/f power spectrum in the low-frequency region. In fact, the formula given by Eq.(9.20) can be considered as a fluctuation-dissipation relation [9.23] for the dynamical system defined by Eq.(9.3) which takes into account the calculation errors.

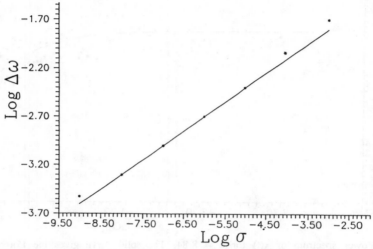

Fig.9.19. Dependence of the spectral plateau width $\Delta\omega$ on the integration accuracy σ. The solid line gives the approximating function σ^μ, $\mu = 0.3$.

The results of our investigation on the influence of the external noise (stochastic equation Eq.(9.14)) shows approximately *the same* dependence as Eq.(9.20), namely, as we increase the noise intensity D, the plateau width in the power spectrum also increases.

Concluding remarks. As a result of our numerical investigations, we have confirmed that near the bifurcation point $\alpha = \alpha^*$ which gives birth to the double scroll Chua's attractor, a *chaos-chaos type intermittency is realized* in Chua's circuit. This intermittency phenomenon is caused by two dynamically interacting symmetrical spiral Chua's attractors. The *power spectrum* associated with this chaos-chaos type intermittency *is characterized by a 1/f divergence* in the low-frequency region. Nearer to the zero frequency, however, there is a *flat plateau* in the power spectrum whose width depends on the overcriticality measure ε via Eq.(9.17), on the calculation accuracy via Eq.(9.20), and on the external noise intensity which also follows Eq.(9.20)[9.24].

9.4. DYNAMICS OF NON-AUTONOMOUS CHUA'S CIRCUIT

In this paragraph we consider oscillations in the *non-autonomous Chua's circuit* when excited by a harmonic signal externally. The non-autonomous oscillations in the Chua's circuit are of interest from several points of view. The introduction of a harmonic force extends the phase space to the dimension N = 4. New oscillation regimes and observation of their bifurcations when varying parameters become possible. For Chua's circuit, this means a chance to excite and examine two-frequency oscillations (two-dimensional tori) and its bifurcations. Furthermore, the non-autonomous Chua's circuit enables us to study the general problems of *forced synchronization*, including the problem of *chaos synchronization* considered in Chapter 8. Finally, the comparison of the dynamics of the non-autonomous Chua's circuit and of that of the non-autonomous oscillator with inertial nonlinearity will allow us to evaluate the level of generality of dynamic regularities in systems exhibiting a saddle-focus separatrix loop.

In Fig.9.20, the electrical scheme of the *non-autonomous Chua's circuit* is presented where the voltage source $V_{ex}(t) = V \sin(2\pi f_1 t)$ is connected with the inductance L via a coupling resistance R_c.

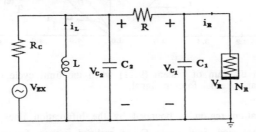

Fig.9.20. Nonautonomous Chua's circuit.

The dynamical equations of the system in dimensionless variables have the following form:

$$\dot{x} = \alpha[y - h(x)],$$
$$\dot{y} = x - (1 + \gamma)y + z + \gamma E_0 \sin(p\tau), \qquad (9.21)$$
$$\dot{z} = -\beta y.$$

We have used the normalization relation (9.2), where $\gamma = R_c/R$ is coupling coefficient; $E_0 = V/V_0$; $V_0 = 1$ V, the normalized forcing amplitude; and $p = 2\pi R C_2 f_1$, the normalized circular frequency of external force.

Consider the *initial regime* where the autonomous Chua's circuit has a *periodic oscillation* with frequency f_0. This regime is realized with the following parameters in (9.21): $\alpha = 10.0$, $\beta = 22.0$, $\gamma = 0.1$; and $E_0 = 0$. Let us examine the dynamics of the nonautonomous system (9.21) in the plane of the external forcing parameters E_0 and p *in the vicinity of the basic resonance* $f_0 : f_1 = 1 : 1$. The calculation results are shown in Fig.9.21.

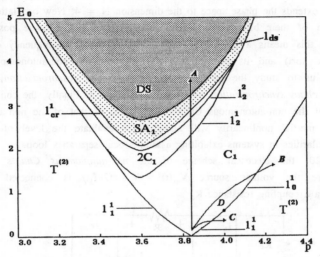

Fig.9.21. Bifurcation diagram of system (9.21) when the limit cycle of the autonomous system is driven by an external forcing signal.

The synchronization region is bounded by the bifurcation lines l_1^1 and l_0^1. Inside this region, a synchronized limit cycle C_1 is realized with a frequency which agrees rigorously with that of external forcing. The cycle C undergoes a cascade of period-doubling bifurcations along the direction A, followed by the birth of the spiral Chua's chaotic attractor SA_1 along the bifurcation line l_{cr}^1. Then the attractor SA_1, similar to the autonomous case, turns into a double scroll Chua's attractor when crossing the line l_{ds}.

Exit from the synchronization region through the bifurcation lines of the saddle-node bifurcations l_1^1 with a low forcing amplitude . *Synchronization is destroyed* and the two-frequency regime of oscillations appears (a two-dimensional torus $T^{(2)}$ is abruptly born). The boundary of the synchronization region on the left involves two bifurcation lines l_1^1 and l_0^1 separated by a *bifurcation point D having codimension* 2.

On the line l_0^1, a pair of complex-conjugate multipliers of the synchronized limit cycle C lie on the unit circle:

$$\rho_{1,2} = |\rho_{1,2}|e^{\pm j\phi}, \quad |\rho_{1,2}| = 1, \; 0 \le \phi \le 2\pi. \tag{9.22}$$

A two-dimensional torus is *softly* born from the limit cycle C_1 when crossing l_0^1 in the direction B and synchronization is destroyed. If one moves along the path B, but in the reverse direction, the *first* classic synchronization *mechanism* is realized because of the *suppression of the second independent frequency*.

The bifurcation point D is characterized by the following properties. When approaching the point D from above along the line l_0^1, the multipliers (9.22) of the limit cycle C_1 which differ only by the sigh of ϕ, come nearer to each other since ϕ is tending to zero. At point D, the phase ϕ becomes zero. Two multipliers merge into one, namely, $\rho_1 = +1$. The bifurcation line l_1^1 corresponds to this condition beyond point D. The crossing of the line l_1^1 in the direction C leads to a loss of synchronization and the abrupt burth of quasiperiodic oscillations. If one moves along the path C but in the reverse direction, *the second classic mechanism of synchronization is* realized due to *frequency locking* by the forcing signal.

The synchronization phenomena caused by frequency suppression and locking, which correspond respectively to crossing the lines l_0^1 and l_1^1 (motion in the directions B and C, respectively), illustrate the oscillation power spectra $S_x(f)$ presented in Fig.9.22.

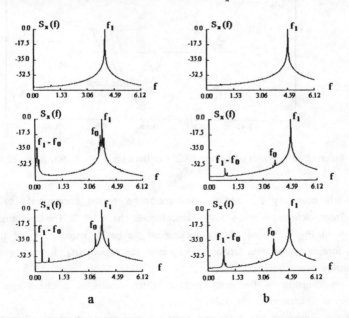

Fig.9.22. Power spectrum evolution along the route B (a) and C (b) of the diagram from Fig.9.21.

Any periodic regime kC_1 can be chosen as the initial regular oscillation in the autonomous Chua's circuit. The structure of bifurcation diagram suffers no qualitative variation.

Let us examine the phenomenon of weak *chaos in the form of spiral Chua's attractor* SA_1 realized with the following parameter values of the system (9.21): $\alpha = 11.98$, $\beta = 22.0$, $\gamma = 0.1$, $E_0 = 0.0$ (i.e., autonomous regime). We will study the synchronization events in the nonautonomous system ($E_0 > 0$) in the vicinity of the basic resonance. To do this, let us construct an appropriate bifurcation diagram by using numerical methods. The result are shown in Fig.9.23.

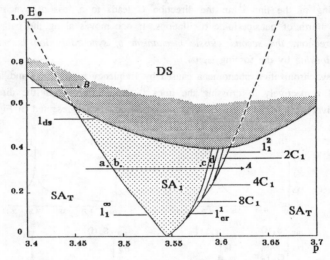

Fig.9.23. Bifurcation diagram of system (9.21) in the case $\alpha = 11.98$, $\beta = 22.0$, $\gamma = 0.1$ and $E_0 = 0.0$

The left boundary l_1^∞ of the synchronization region corresponds to the phase transition "nonsynchronous chaos SA_T - synchronous chaos SA_1". The evolution along the direction A in Fig.9.23 leads to the locking of the basic frequency f_0 of the attractor SA_T by a forcing signal with frequency f_1. Inside the region SA_1, $f_0 = f_1$ and a *chaotic synchronization* event is observed.

The phenomenon of frequency locking which results in a chaos synchronization phenomenon is illustrated in Fig.9.24.

Here the power spectra $S_x(f)$ and projections of the stroboscopic Poincaré sections are presented as we move in the direction A of the diagram in Fig.9.23 from point a to point b (crossing the line l_1^∞).

In Fig.9.24a, the SA_T *torus-chaos regime* is clearly diagnosed which is based on the broken down doubled torus $2T^{(2)}$. The basic frequencies of the torus are close to each

Nonlinear Phenomena and Chaos in Chua's Circuit

other but it is seen that $f_0 \neq f_1$. The chaotic synchronization via frequency locking (bifurcational transition in chaos $SA_T \to SA_1$) is demonstrated in Fig.9.24c. At the intermediate point (Fig.9.24b), the *intermittency phenomenon* is observed between the regimes SA_T and SA_1 which occurs directly near the bifurcation line l_1^∞.

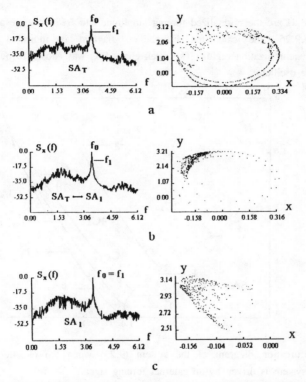

Fig.9.24. Evolution of power spectra and projections of Poincaré section along the route A of the diagram Fig.9.23.

The right boundary of the synchronization region (Fig.9.23) is formed by the line of the saddle-node bifurcation l_1^2 of the limit cycle $2C_1$. Then the synchronized oscillations $2C_1$ undergo a cascade of period-doubling bifurcations, followed by the synchronized chaos SA_1. The line of critical values l_{cr}^1 lies between points c and d (see Fig.9.23).

Let us compare the results obtained from the Chua's circuit with the data on external synchronization for the oscillator with inertial nonlinearity (see Chapter 8). With identical switching of the modulation source through the coupling resistance, the nonautonomous oscillator is described by the following system of equations:

$$\dot{x} = mx + y - xz + \gamma E_0[\sin(U\tau) - \frac{1}{U}\cos(u\tau)],$$
$$\dot{y} = -x, \qquad (9.23)$$
$$\dot{z} = -gz + gF(x),$$

where E_0 and U are the normalized forcing amplitude and frequency, respectively.

In Fig.9.25, the calculated bifurcation diagram is indicated for the basic oscillator's synchronization region. The weak chaos SA_1 was chosen to be the autonomous regime of the oscillator.

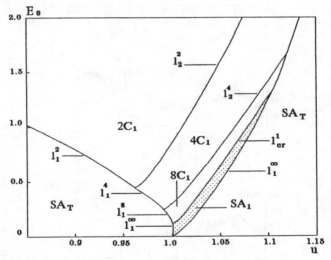

Fig.9.25. Bifurcation diagram of the system (9.23) when the chaotic regime SA_1 of autonomous system is driven by an external forcing signal.

The *synchronization processes are qualitatively equivalent* in the Chua's circuit and in the oscillator for low forcing amplitudes (at the bottom of the synchronization zone). The distinction in the dynamics of Chua's circuit and that of the oscillator (9.23) is as follows. In the former, the synchronous chaos SA_1 evolves into the regime DS with the increase of amplitude of the external modulation. Even for high forcing amplitudes (see Fig.9.23), we observe no chaos suppression phenomenon. *The regime DS is not destroyed* by the external signal.

In the latter, as seen from the diagram in Fig.9.25, the region of chaos SA_1 is bounded from above by a sequence of "*period halving*" bifurcations: $SA_1 \to \ldots \to 8C_1 \to 4C_1 \to 2C_1$. When influenced by an external signal, *the chaos SA_1 regime collapses* and becomes *regular oscillations* kC_1. Investigations have demonstrated that these distinctions are due to the features of the nonlinear characteristics of the

Chua's circuit and the oscillator.

The calculation results presented in Fig.9.21 and Fig.9.23 are qualitatively reproduced in the course of a *full-scale experiment*. The data of physical measurements are presented in Figs 9.26 and 9.27.

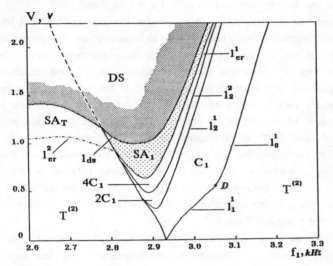

Fig.9.26. Results of full-scale experiments illustrating the conditions of Fig.9.21

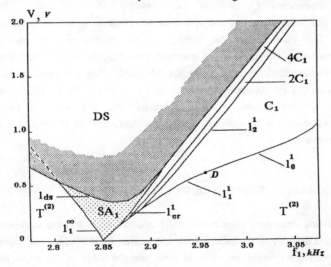

Fig.9.27. Results of full-scale experiment illustrating the conditions of Fig.9.23.

Note that in the course of full-scale experiments on Chua's circuit, the phenomenon of the destruction of the DS regime is realized for a high external forcing amplitude, and followed by the birth of kC_1 limit cycles (*synchronous chaos suppression*). The reason is the difference of the real nonlinear element's characteristic from the piecewise-linear one for high oscillation amplitudes in a physical circuit.

A comparison of the data cited with the results described in Chapter 8 for the case of the external synchronization, enables one to conclude *on the generality of the bifurcation phenomena* in systems with saddle-focus separatrix loops. Inside the synchronization zone we observe a picture of bifurcations of synchronization regimes which is topologically equivalent to the bifurcations in the autonomous system. In the autonomous Chua's circuit, unlike the oscillator, the "chaos-chaos" transition $SA_1 \to DS$ is realized because of the circuit's symmetry. This transition also takes place in the non-autonomous circuit in the synchronization regions.

Outside the synchronization regions we deal with the evolution of quasiperiodic oscillations. Here a qualitative agreement is also observed between the oscillator and the Chua's circuit. The regularities of the bifurcation and the breakdown of the two-dimensional tori are qualitatively equivalent. We will return to this problem at the end of this chapter (see paragraph 9.6).

It should be noted as well that the results presented for the regions of basic resonance 1:1 are qualitatively reproduced for the regions of strong resonances 1:2, 1:3 and 2:1 which is confirmed both by calculations and by full-scale experiment.

At the boundaries l_{ds} defining the transition $SA_1 \to DS$, the "chaos-chaos" intermittency phenomenon is realized which is described in detail in paragraph 9.3 of this chapter. The external periodic forcing in the vicinity of the above transition induces an interesting phenomenon of resonant amplification of the modulation signal. Accordingly, higher signal-to-noise ratios are observed. This phenomenon is associated with the *phenomenon of stochastic resonance* which is typical for bistable systems under the conditions of regular and noise modulations. We will see that the stochastic resonance phenomenon can be realized in Chua's circuit, both *in the presence and in the absence of external noise.*

9.5. STOCHASTIC RESONANCE IN CHUA'S CIRCUIT

The *stochastic resonance phenomenon* (SR) is observed in bistable dynamical systems driven simultaneously by a periodic signal and noise. In this case, the *signal-to-noise ratio* (SNR) reaches its maximum value at some specific value of the noise intensity D. This intensity value D_0 depends on the parameters of the modulation signal, and on the properties of the nonlinear system [9.25-9.29].

For dynamical systems which bifurcates to chaos via *different types of intermittency*[3] it is possible to obtain a bistable behavior [9.8-9.11].

The intermittency results from the dynamical interaction between a regular (fixed point, limit cycle, torus) and a chaotic attractor in quasi-hyperbolic systems. Such interaction gives rise to a *random switching process* between two attractors. Hence, it can be considered as a generalized form of bistable behavior of nonlinear dynamical systems [9.30].

If the parameters of the dynamical system are chosen slightly less than the bifurcational value μ^* which leads to an intermittency phenomenon, then the process of attractor interaction can be initiated by an external noise. In this case, it is reasonable to refer to this form of chaos as an intermittency induced by external noise [9.30-9.32]. Note that this form of chaos is no longer deterministic.

Therefore, chaotic systems having this attractor interaction capability are, in a generalized sense, bistable systems. Hence, they can be used as a vehicle for the investigation of the SR-phenomenon.

Consider Chua's circuit and choose the (α, β) parameters so that two co-existing symmetrical spiral Chua's attractors are almost colliding with each other, thereby giving birth to the well-known double scroll Chua's attractor. Let us induce an interaction between these two attractors by connecting a small sinusoidal signal voltage source $v_s(t) = \varepsilon_0 \sin\omega_0 t$ and a noise source $\xi(t)$ of intensity D in series with the inductor, thereby obtaining the following dimensionless three-dimensional system [9.33]:

$$\frac{dx}{dt} = \alpha(y - h(x)),$$

$$\frac{dy}{dt} = x - y + z, \quad\quad\quad (9.24)$$

$$\frac{dz}{dt} = -\beta y + \varepsilon_0 \sin\omega_0 t + \xi(t),$$

$$<\xi(t)> = 0, \quad <\xi(t)\xi(t+\tau)> = D\delta(\tau),$$

where $<\cdot>$ and $\delta(\tau)$ denotes the averaging operator and the delta function, respectively.

The parameter region of interest is shown here in Fig.9.6, bounded by the bifurcation curves 1 and 2. The autonomous (i.e., $\varepsilon_0 = 0$, $D = 0$) *interaction* between the

---

[3]Intermittency in this paragraph includes the phenomenon when two nearby attractors of possibly different types (e.g., limit cycle, torus, strange attractor) collided with each other at some critical parameter μ^* in the absence of noise. Such bifurcational transitions include the "cycle-chaos" type intermittency, the "torus-chaos" type intermittency, and the "$chaos_1$-$chaos_2$" type intermittency.

two spiral Chua's attractors is *impossible* in this region. *The SR-phenomenon can take place* in this region *only by applying an external noise excitation* to the circuit. As we move further to the right, the two spiral Chua's attractors eventually merge at the bifurcation curve 2 to form the double scroll, thereby initiating the onset of a "chaos-chaos" type intermittency in the autonomous Chua's circuit. Therefore, *the SR-phenomenon can be observed* in the region of the α-β parameter plane where we cross the bifurcation curve 2 from the spiral Chua's attractor regime to the double scroll Chua's attractor regime (shown shaded in Fig.9.6).

Our numerical computation is based on the *"two-state"* dynamics method [9.29]. In this case, the time-dependent $x(t)$ is replaced by a random sequence of square pulses with amplitude + 1 or - 1. The duration of each pulse is equal to the amount of time a trajectory resides in the vicinity of either the first (+ 1), or the second (- 1), interacting spiral Chua's attractor. A special filter is needed to implement this procedure in full-scale experiments [9.29].

Figure 9.28 shows the probability density $p(\tau)$ of the residence time of the phase trajectory in one of the spiral Chua's attractors, interacting under the influence of a noise with intensity $D = 0.03$. Here τ is the normalized time, i.e., $\tau = t/T_0$, $T_0 = 2\pi/\omega_0$. The probability density $p(\tau)$ consists of a series of sharp peaks having a finite width of $\Delta\tau \approx 0.22$ on the τ-axis. In the absence of noise, the imaginary envelope which goes approximately through the extremum points (peaks) of the function $p(\tau)$ decreases approximately as an exponential function of τ (Fig.9.28). In the presence of a sinusoidal modulation, the probability density $p(\tau)$ can change dramatically. Details of the structure of the distribution $p(\tau)$ will be discussed in the end of this paragraph.

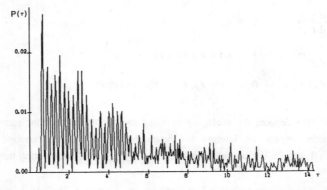

Fig.9.28. The residence time probability density P(τ) in the absence of a modulation signal for $\alpha = 8.6$, $\beta = 14.286$, $D = 0.03$. Here $\tau = t/T_0$, $T_0 = 10.46$.

From the distribution $p(\tau)$ we can estimate the mean residence time of a trajectory in one of the attractors and its corresponding switching frequency:

$$T_s = <t> = \int_0^\infty t\, p(t)\, dt, \quad \omega_s = 2\pi/T_s \qquad (9.25)$$

Plots of the mean switching frequency ω_s versus both the noise intensity D (the noise induced intermittency, $D > 0$) and the parameters α and β in the double scroll region (the dynamical intermittency, $D = 0$) are shown in Fig.9.29. Observe from Fig.9.29 that the mean switching frequency ω_s in system (9.24) can be controlled within a range of $0.1 \leq \omega_s \leq 0.5$ by varying either the noise intensity D, or the values of the parameters α and β.

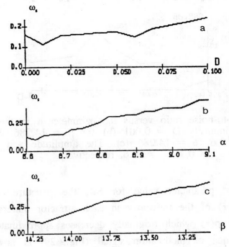

Fig.9.29. The dependence of the mean switching frequency ω_s on the noise intensity D for $\alpha = 8.6, \beta = 14.286$ (a), on parameter α for $D = 0, \beta = 14.286$ (b) and on parameter β for $D = 0, \alpha = 8.6$ (c) in the presence of a modulation signal $\varepsilon_0 = 0.01, \omega_0 = 0.6$.

Figure 9.30 demonstrates the possibility of producing the SR-phenomenon in Chua's circuit. Results of our computation of the signal-to-noise ratio (SNR) as a function of the parameter α in the double scroll regime (the dynamical intermittency) are shown both in the absence of noise (a), and in the presence of noise with a fixed intensity $D = 0.001$. These curves are calculated for fixed parameters $\alpha = 8.6, \beta = 14.286$ in the double scroll regime (noise-induced intermittency). These results confirm the existence of optimal parameter values for α, β and D which result in a maximum SNR for the fixed values of the amplitude and frequency of the periodic excitation $\varepsilon_0 = 0.1$, $\omega_0 = 0.6$.

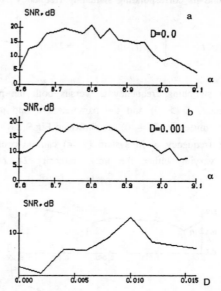

Fig.9.30. The signal-to-noise ratio versus the parameter α in the absence of noise (a) and with the noise intensity $D = 0.001$ (b) for $\beta = 14.286$. SNR versus the noise intensity D at $\alpha = 8.6$, $\beta = 14.286$ (c). The amplitude and the frequency of the modulation signal are $\varepsilon_0 = 0.1$, $\omega_0 = 0.6$, respectively.

At this optimum parameter value for SR, the structure of the residence time probability density $p(\tau)$ of the trajectory in each attractor undergoes a transformation; namely, the envelope $p(\tau)$ which originally decreases approximately as an exponential function in the absence of modulation signal (Fig.9.29) is now transformed into a multimodal distribution function, as shown in Fig.9.31.

Observe that under this optimum SR situation, the extremum points (maxima) of $p(\tau)$ are located near the discrete set

$$\tau_n = (2n+1)/2, \quad n = 0,1,2,\ldots \tag{9.26}$$

in view of the symmetry of the system (9.24). The maximum $p(\tau_n)$ decreases approximately along an exponential function of n. In this case, the most probable switching frequencies are connected deterministically, i.e., synchronized, with the external modulation frequency ω_0:

$$\omega_s^n = 2\omega_0/(2n+1), \quad n = 0,1,2,\ldots \tag{9.27}$$

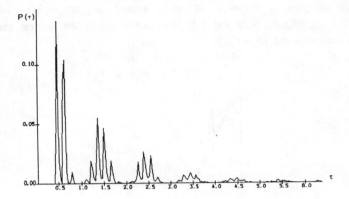

Fig.9.31. The residence time probability density $p(\tau)$ under the conditions of SR for $\alpha = 8.6$, $\beta = 14.286$, $D = 0$, $\varepsilon_0 = 0.1$ and $\omega_0 = 0.4$.

At $n = 0$, the mean switching frequency ω_s^0 is equal to the double modulation signal frequency ω_0. The optimal condition for SR takes place under this situation and the SNR-value reaches its maximum value.

The power spectrum of the random switching process during a chaos-chaos type intermittency, in the absence of the external modulation, has the typical $1/f^\alpha$ shape shown in Fig.9.32 (dashed curve). Applying a small sinusoidal modulation signal (under some optimal SR condition to be discussed below), causes a greatly amplified peak to appear at the modulation frequency ω_0 while a major portion of the characteristic is depressed below the original curve, as shown in Fig.9.32 (solid curve). Our calculation shows that the integral over the entire spectrum in Fig.9.32 is approximately a constant in both cases. This means that the amplification of the modulation signal intensity takes place at the expense of a transfer of a part of the noise energy to the energy of the input signal.

There is an important characteristic feature of the phenomenon of the "chaos-chaos" type intermittency which is absent from the classical transitions in bistable systems with two equilibrium points. This new feature is the presence of an additional characteristic time constant related to the dynamical properties of the interacting attractors. For Chua's circuit, this time constant is equal to the mean duration of one rotation of the phase trajectory around the spiral Chua's attractor and has the value $T_R \approx 2.31$. This corresponds to a normalized time constant $\tau_R = T_R/T_0 \approx 0.22 \approx \Delta\tau$, where $T_0 = 2\pi/\omega_0$. Since the transition in Chua's circuit is most probable after some integer numbers of trajectory rotations around each interacting attractor, this characteristic time constant results in a discretization of the probability density $p(\tau)$, the structure of which contains a sequence of peaks located in the vicinity of $\tau = k\tau_R$, ($k = 1,2,3,...$). Figures 9.28 and 9.31 demonstrate clearly the discrete nature of $p(\tau)$ with the same step size $\Delta\tau \approx 0.22$.

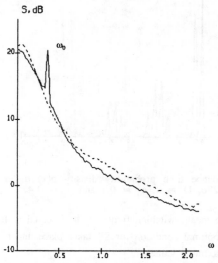

Fig.9.32. The power spectra of switching processes in the absence of a modulation signal for $\alpha = 8.875$, $\beta = 14.286$, $D = 0.005$ (dashed curve) and under the conditions of SR in the presence of the modulation signal $\varepsilon_0 = 0.15$, $\omega_0 = 0.35$ (solid curve).

Hence, *bistable chaotic systems* of the chaos-chaos intermittency type have *three characteristic time constants* (or three characteristic frequencies) in the nonautonomous regime: the *modulation frequency* ω_0, the *mean switching frequency* ω_s, and the *mean frequency of oscillations* $\omega_R = 2\pi/T_R$ inside each interacting attractor. A coherent interaction between the modulation frequency ω_0 and the mean switching frequency ω_s is responsible for the SR-phenomenon. However, even more complicated phenomena related to the coherent interaction among all the three frequencies can take place.

Finally, we remark that the SR-phenomenon has been observed numerically in the system (9.24) by varying either the parameter α, or the parameter β, in the vicinity of the bifurcation curve 2 of the diagram in Fig.9.6, Moreover, the familiar *classical resonance phenomenon* can also take place in the system (9.24) due to the coherent interaction between the other two frequencies ω_0 and ω_R. In this case, a smooth transition to the SR-phenomenon can be realized by increasing the noise intensity.

9.6. Confirmation of the Afraimovich-Shilnikov torus-breakdown theorem via Chua's torus circuit

The appearance of chaos following the *breakdown of a two-dimensional torus* is a typical and interesting scenario of the transitions to chaos. Ruelle and Takens [9.34], and later, Newhouse, Ruelle and Takens (NRT) [9.35] had provided the seminal results on this subject. Compared with the Landau-Hopf scenario [9.36], the NRT-scenario specifies a *finite bifurcation sequence* in the route to chaos: a stable equilibrium point (FP) ⇒ a stable limit cycle (LC) ⇒ a stable two-dimensional torus (T^2) ⇒ chaos:

$$FP \Rightarrow LC \Rightarrow T^2 \Rightarrow Chaos \qquad (9.28)$$

The sequence of attractors (9.28) has been confirmed both by computer simulations of many simple models and by full-scale experiments on Couette flow, hydrodynamical instabilities, etc. (look at chapter 6 and its references). However, the NRT-scenario *does not give a full description of all possible bifurcation scenarios* leading to the destruction of the torus T^2. In particular, it does not address the following questions. Does the T^2-breakdown always lead to chaos? If not, what additional conditions are required? What is the structure of the resulting chaotic attractor?

Many authors have investigated the phenomena of T^2-breakdown theoretically. One of such results which will be applied in this section is a *theorem by Afraimovich and Shilnikov* on two-dimensional tori breakdown [9.37].

In this section we will use a member of the Chua's circuit family, namely, the *Chua's torus circuit* [9.38,9.3-9.5] as a vehicle to illustrate the main bifurcational phenomena, which cause transitions to chaos, as predicted in the Afraimovich-Shilnikov theorem.

Chua's torus circuit was chosen for the following reasons. The phase space of the associated state equations has the dimension necessary for the transition $T^2 \Rightarrow$ Chaos. Moreover, the *circuit is autonomous* and the T^2-*attractor arises without any external periodic excitations*. Since the state equation is piecewise linear, some important results can be obtained analytically. Besides, it would be interesting to investigate whether the bifurcation scenarios predicted by the Afraimovich-Shilnikov theorem, which was proved only for smooth dynamical systems, also occur in piecewise-linear systems.

The Afraimovich-Shilnikov torus breakdown theorem. Consider the dynamical system:

$$\dot{x} = F(x,\mu) \qquad (9.29)$$

in $\mathbb{R}^N$, where $N \geq 3$. Here $x \in \mathbb{R}^N$, $\mu \in \mathbb{R}^K$ and $F(x,\mu)$ is assumed to be a sufficiently smooth function of x and of the parameter μ. Let us suppose that a smooth attracting torus $T^2(\mu_0)$ exists in some region G of the state space of system (9.29) at $\mu = \mu_0$ and assume that the torus $T^2(\mu_0)$ is formed by the closure of the flow of (9.29) originating from the unique stable limit cycle Γ^+, the saddle-type limit cycle Γ^- (which can be proved to always exist), and the unstable manifold W^u of Γ^-. Assume further that the multiplier of the stable limit cycle Γ^+ on the torus $T^2(\mu_0)$ which has the smallest absolute value is simple and real.

Consider a continuum set of continuous curves $H = \{\mu(s): \mu \Rightarrow \mathbb{R} \Rightarrow \mathbb{R}^K\}$ in the parameter space $\mathbb{R}^K$ of system (9.29), $0 \leq s \leq 1$. Assume that for $\mu(0)$ a smooth stable torus corresponding to $\mu(0) \in H$ exists in the region G for system (2), but for $\mu(1)$ the torus does not exist. Then the Afraimovich-Shilnikov theorem asserts the following three distinct breakdown scenarios for the torus:

1. For some intermediate parameter value s_1, where $0 < s_1 < 1$, the torus exists, however, either $T^2[\mu(s)]$ *loses its smoothness* due to the nonsmooth behavior of the unstable manifold W^u in the vicinity of the stable limit cycle Γ^+ for $s > s_1$, or the pair of multipliers $\rho_{1,2}$ of the stable limit cycle Γ^+ becomes complex-conjugate inside the unit circle at $s = s_1$.

2. There exists a value s^*, $s^* > s_1$, such that the attracting torus $T^2(\mu)$ no longer exists for $s > s^*$. In this case, there are three possible routes leading to the destruction of the torus:

A. At $s = s^*$, the *stable limit cycle* $\Gamma^+(\mu)$ *loses its stability* via some typical scenarios of bifurcation of periodic solutions.

B. *A structurally unstable homoclinic trajectory* of the saddle-type limit cycle Γ^- *occurs* due to the presence of a tangency between the stable manifold W^s and the unstable manifold W^u of the saddle cycle.

C. At $s = s^*$ the *stable* and the *unstable* (saddle type) *limit cycles* on the torus *merge* into a saddle-node periodic solution *and then disappear*. The *torus is nonsmooth* at the bifurcation parameter $s = s^*$.

Fig.9.33 shows a sketch of the qualitative bifurcation diagram of phase-locking on the torus T^2 on the two-parameter $\mu_1 - \mu_2$ plane. The direction of the paths on the parameter plane corresponding to routes A, B, and C of the Afraimovich-Shilnikov theorem is shown in Fig.9.33.

The phase-locked region is formed by the two bifurcation curves l_1, corresponding to the merging and annihilation of the saddle-type limit cycle Γ^- and the stable limit cycle Γ^+ on the torus $T^2(\mu)$. The phase-locked region originates from the codimension-2 bifurcation point K on the bifurcation curve l_0. The curve l_0 corresponds to the bifurcation of a torus T^2 spawned from a limit cycle Γ_0 having a pair of complex-conjugate multipliers on the unit circle.

Consider *first the route* PA of the diagram in Fig.9.33. The limit cycle Γ^+ becomes unstable when crossing the curve l_2, via one of the several possible bifurcations. The torus $T^2(\mu)$ does not exist above the curve l_2. A loss of smoothness of the torus precedes the torus breakdown as we approach l_2 from the lower side. For example, the transition to chaos along the route PA can come from a period doubling bifurcation. If a new two-dimensional torus appears from the cycle Γ^+ on the curve l_2, which is possible in some systems [9.38,9.40], then the above scenario of torus breakdown will occur again.

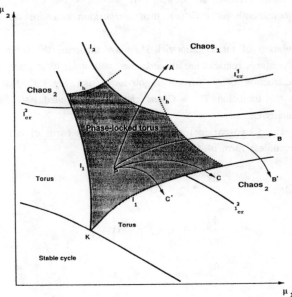

Fig.9.33. The qualitative illustration of bifurcation of the two-dimensional phase-locked torus breakdown. The region of the torus regime is bounded by curves l_1, l_h and l_2 (the shaded region). Routes related to transitions to chaos on the parameter plane are labelled by A, B and C.

Next, let us *move along the route* PB. The phenomenon of homoclinic tangency of the stable and the unstable manifolds of the saddle cycle Γ^- on $T^2(\mu)$ takes place on the bifurcation curve l_h. In this case, the *torus $T^2(\mu)$ is destroyed, but the stable cycle Γ^+ remains as an attractor*. A structurally stable homoclinic structure emerges above the curve l_h and between the curves l_1 and l_2, but it is not an attractor. In this region the system (9.29) exhibits a *metastable chaos* having a finite life time [9 38-9.40]. A transition to a true dynamical chaos can be realized by moving along the route PB'. In this case a *chaotic attractor emerges abruptly* at the moment we cross the saddle-node bifurcation curve l_1. In this case, a subset of hyperbolic trajectories is transformed into a chaotic attractor, and the stable limit cycle Γ^+ disappears.

Finally, *moving along the route PC* results in a saddle-node bifurcation when crossing the curve l_1. Two cases can be realized here. Along the route PC' a transition from the phase-locked torus to an ergodic one (i.e., quasi-periodic solution) takes place. Although the structure of the trajectories on T^2 is changed when crossing l_1, in this case, the torus remains an attractor. If we cross the curve l_1 into the region Chaos$_2$ of the nonsmooth torus $T^2(\mu)$, then a *transition to chaos via intermittency* will take place.

The formulation of the Afraimovich-Shilnikov theorem, the bifurcation diagram of Fig.9.33, and the above remarks are all concerned with these bifurcation phenomena which lead to the destruction of $T^2(\mu)$. It is obvious from Fig.9.33 that the phenomena associated with the transition $T^2 \Rightarrow$ Chaos can be investigated *only by carrying out a two-parameter analysis*.

<u>Computer-aided two-parameter analysis of Chua's torus circuit</u>. Consider the following dimensionless form of *Chua's torus circuit*:

$$\frac{dx}{dt} = -\alpha f(y - x)$$

$$\frac{dy}{dt} = -z - f(y - x) \qquad (9.30)$$

$$\frac{dz}{dt} = \beta y$$

The piecewise-linear function describing the nonlinear resistor (Chua's diode [9.3-9.5]) in the circuit is described by:

$$f(\xi) = -a\xi + 0.5(a + b)(|\xi + 1| - |\xi - 1|) \qquad (9.31)$$

where as in reference [9.38], we choose the parameters

$$a = 0.07, \quad b = 0.10 \qquad (9.32)$$

for comparison purposes.

We will use a computer to analyze the bifurcation phenomena of (9.30) over some strategically selected region in the $\alpha - \beta$ parameter plane.

First, a cycle Γ_0 of system (9.30) which gives birth to a torus $T^2(\alpha,\beta)$ will be found. Then the torus bifurcation curve l_0 will be obtained via a continuation method. On this curve the pair of complex-conjugate multipliers of Γ_0 lie on a unit circle;

namely, $\rho_{1,2} = \exp(\mp j\varphi)$, $|\rho_{1,2}| = 1$. There is a point K on the curve l_0 corresponding to the phase-locked region having a rotation number 1:6. By finding a stable limit cycle Γ^+ on $T^2(\alpha,\beta)$ and by calculating the multipliers $\rho(\alpha,\beta)$ under various bifurcation conditions, we can construct the bifurcation diagram of system (9.30) in the region of interest.

To identify the type of the attractors, we take a Poincaré section on the plane $z = 0$ and calculate the corresponding power spectra $S(\omega)$ associated with the variable $x(t)$. To analyze the phenomenon occurring during the loss of torus smoothness, we *introduce additive sources of white noise* into equation (9.30). The full spectrum of Lyapunov exponents and the dimension D_L of the attractors were computed for diagnosing the phenomenon of homoclinic tangency (curve l_h).

Bifurcation diagram in the region with 1:6 rotation number.

The bifurcation diagram for system (9.30) in the region corresponding to a rotation number of 1:6 is shown in Fig.9.34. The curve l_0 ($\alpha = 1$) corresponds to the bifurcation which gives birth to the torus $T^2(\alpha,\beta)$ associated with the limit cycle Γ_0. This result was derived theoretically in reference [9.38]. The two curves labelled as l_1 are boundaries of the 1:6 phase-locked region and correspond to a saddle-node bifurcation of the stable limit cycle, henceforth denoted by Γ^+, on $T^2(\alpha,\beta)$. They were calculated by imposing the condition $\rho_1 = +1$. On the curve l_2 one of the multipliers ρ_1 of the cycle Γ^+ is equal to -1. A soft period-doubling bifurcation of the limit cycle Γ^+ takes place on this curve. The curve l^1_{cr} corresponds to a transition to chaos due to the period-doubling bifurcation of the limit cycle Γ^+. On the curve l^2_{cr} a smooth transition to chaos resulting from the loss of smoothness of the torus $T^2(\alpha,\beta)$ is initiated. The bifurcation curve l_h in Fig.9.34 corresponds to a homoclinic trajectory spawned by an intersection of the stable and the unstable manifolds of the saddle-type limit cycle Γ^-. To the right of the curve l_h, the torus $T^2(\alpha,\beta)$ does not exist.

There is some peculiarity in the birth of the torus $T^2(\alpha,\beta)$ in the vicinity of the bifurcation curve l_0; namely, when crossing the curve l_0 ($\alpha = 1$) from the left to the right in Fig.9.34 an ergodic torus is born abruptly, i.e., the associated waveform $x(t)$ suddenly changes from a periodic to a quasi-periodic function defined by two independent and incommensurate frequencies ω_1 and ω_2 with finite amplitudes. The phase portrait and the power spectrum of a typical torus $T^2(\alpha,\beta)$ near the curve l_0 are shown in Fig.9.35. The oscillation dynamics is nonlinear in principle even in the vicinity of the torus birth curve l_0, as verified by the power spectrum of $x(t)$. Observe that besides the two basic frequencies $\omega_1 = 0.05$ and $\omega_2 = 1.00$, various harmonics $m\omega_1$ and $n\omega_2$ and combination frequencies $\omega_{m,n} = m\omega_1 \pm n\omega_2$ are clearly seen in the spectrum at least for $n \leq 3$, $m \leq 6$ (in Fig.9.35 only the part of the spectrum up to the second harmonic of ω_1 is shown).

Observe that the phenomenon of an abrupt appearance of both quasi-periodic and periodic oscillations is a typical property of piecewise-linear systems.

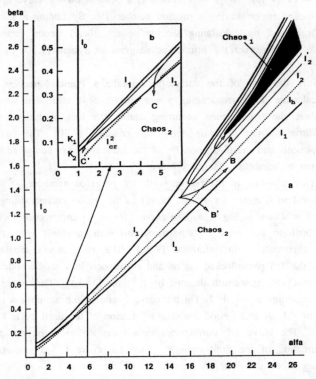

Fig.9.34. The experimental bifurcation diagram of the system (9.30) illustrating the Afraimovich-Shilnikov theorem: (a) the complete diagram, (b) the fragment of the diagram for $\alpha \leq 6$, $\beta \leq 0.6$. The curve l'_2 is related to the period-doubling bifurcation of the phase-locked doubled cycle, $[K_1, K_2]$ is the segment of the curve l_0 related to the resonance 1:6 on the torus. All other notations are the same as in Fig.1.

There is another peculiarity in the bifurcation diagram of system (9.30). As shown in the diagram in Fig.9.34, the 1:6 phase-locked region does not originate from a single point K, but rather over an interval $[K_1, K_2]$. Such a situation can not take place in smooth dynamical systems. Hence, this phenomenon is also a consequence of the piecewise-linear nature of $f(\xi)$ in (9.31). Figure 9.36 shows a typical picture associated with a periodic limit cycle Γ^+ having a Poincaré rotation number 1:6. Observe that there are 6 points in the Poincaré map corresponding to the secant $z = 0$. Observe that only harmonics $n\omega_2$ of the minimum frequency $\omega_2 = 0.22$ exist in the power spectrum. The phase-locked condition is $\omega_1 = 6\omega_2$.

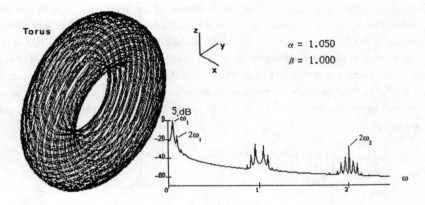

Fig.9.35. The torus $T^2(\mu)$ in the system (9.30) near its birth bifurcation and the power spectrum calculated for the variable x(t).

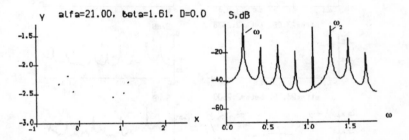

Fig.9.36. The Poincaré section and the power spectrum of the phase-locked cycle Γ^+ on the torus $T^2(\mu)$ in the phase-locked 1:6 region.

To investigate the main points of the Afraimovich-Shilnikov torus breakdown theorem it is sufficient to investigate the dynamics of system (9.30) in the vicinity of the 1:6 phase-locked region.

Torus breakdown and routes of transition to chaos. Let us investigate in detail the evolution of the oscillatory regimes in system (9.30) along the routes A,B,C predicted by the Afraimovich-Shilnikov theorem and determine the conditions where chaos originates from a torus breakdown.

A. *Torus breakdown due to the period-doubling bifurcation of phase-locked limit cycle Γ^+ on the torus $T^2(\alpha,\beta)$*. Our computer simulation in Fig.9.34 shows that if we move along the direction PA, we would cross the period-doubling bifurcation curve l_2 of the limit cycle Γ^+. The torus is destroyed on the curve l_2. This destruction is preceded by a loss of smoothness as evidenced from the distortion of an invariant curve in the Poincaré map. The loss of smoothness is caused by oscillations of the unstable separatrix of the saddle point as it approaches the stable node point. We should note that a curve l exists in the phase-locked region at a very short distance from l_1 inside of the Arnold tongue, where the multipliers of cycle Γ^+ change from real to complex-conjugate on this curve[4]. A rotation leads to the loss of smoothness, thereby resulting in the destruction of the torus on the curve l_2. We shall come back to this phenomenon in the next paragraph.

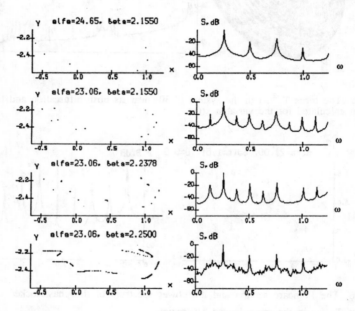

Fig.9.37. Transition to Chaos$_1$ due to period-doubling bifurcations along the direction A of the bifurcation diagram of the system (9.30) (Fig. 9.34, a)

The cascade of period-doubling bifurcations of the stable limit cycle Γ^+ and the transition to chaos occur above the curve l_2. But the route of period-doubling sequence is finite in system (9.30), where the transition $T_0 \Rightarrow 2T_0 \Rightarrow 4T_0 \Rightarrow$ Chaos$_1$ takes place (here T_0 is the period of the limit cycle Γ^+). Some results are presented in Fig.9.37 to illustrate this transition. The Poincaré sections and the corresponding power spectra clearly show the typical scenario as we move into the chaos$_1$ region and cross the curve

---

[4]This curve is not plotted in Fig.9.34 because it practically coincides with the curve l_1 inside the phase-locked region.

l_2 transversally. Observe that the finite number of period-doubling bifurcations is not due to any computation difficulties, but is a genuine phenomenon caused by the piecewise-linear character of the function $f(\xi)$ in system (9.30). The Lyapunov exponents in the Chaos$_1$ region in the vicinity of curve l^1_{cr} in Fig.9.34 for $\alpha = 23.06$, $\beta = 2.25$ are:

$$L_1 = + 0.021, \quad L_2 = - 0.00009, \quad L_3 = - 0.200 . \quad (9.33)$$

This results in a corresponding value of the Lyapunov dimension $D_L = 2.104$.

B. *Torus breakdown due to the appearance of a homoclinic trajectory of the saddle-type limit cycle Γ^-: the abrupt transition to chaos.* Our goal here is to construct the bifurcation curve corresponding to a tangency of the stable and the unstable manifolds of the saddle limit cycle Γ^- on the torus. This problem can be solved directly by calculating the unstable and the stable separatrixes of the corresponding saddle points in the Poincaré section, as it was done, for example, in [9.39]. However, this method is rather time consuming and we therefore devise a different method. Our method consists of adding an additive source of white noise to Eq.(9.30); namely,

$$\frac{dx}{dt} = - \alpha \; f(y - x) + \xi_1(t)$$

$$\frac{dy}{dt} = - z - f(y - x) + \xi_2(t) \quad (9.34)$$

$$\frac{dz}{dt} = \beta y + \xi_3(t)$$

where $<\xi_i(t)> = 0$, $<\xi_i(t) \; \xi_i(t + \tau)> = D\delta(t - \tau)$.

Let us analyze the Poincaré maps and the power spectra of the stable limit cycles Γ^+ in the phase-locked region under a small noise perturbation with intensity $D = 0.01$. This noise excitation causes the unstable manifolds to become visible [9.40-9.41]. Moreover, as a criterion for the existence of a homoclinic trajectory we can calculate the condition which gives rise to a positive Lyapunov exponent of a solution of the perturbed system (9.34) [9.42]. So, starting from the inside of the phase-locked region and identifying the transition of the maximal Lyapunov exponent from a negative to a positive value, it is possible to construct the curve l_h [9.43].

Figure 9.38 shows the computed results for increasing values of the parameter α inside the phase-locked region but in the vicinity of l_1. Using the calculated maximum Lyapunov exponent, the curve l_h was obtained and plotted in the diagram of Fig.9.34. Evidently, results of such computations depend, to some extent, on the noise intensity

D. However, the results derived from this calculation can be used for a qualitative interpretation of the dynamics.

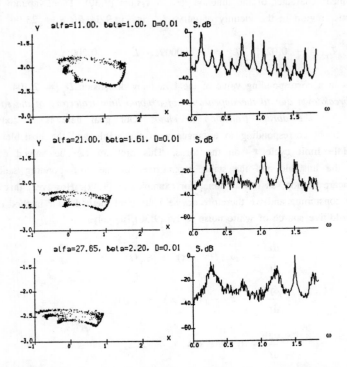

Fig.9.38. Poincaré sections and corresponding power spectra of the phase-locked cycle Γ^+ of the system (3) in the presence of noise. Visualization of homoclinic structures.

Hence, the torus $T^2(\alpha,\beta)$ does not exist in the region bounded by the curves l_h and l_1. It was destroyed because of the emergence of a stable homoclinic structure, which is not a part of the attractor. The stable limit cycle Γ^+ is the only attractor here. Its Poincaré section and power spectrum are shown in Fig.9.36. However, if we leave this region in the direction B' (see diagram of Fig.9.34), then the limit cycle Γ^+ merges with the limit cycle Γ^- and disappears when crossing the curve l_1 thereby giving birth to chaos abruptly in the region labelled $Chaos_2$. Fig.9.39 illustrates the phenomenon of the abrupt appearance of chaos in this region. The results of our computations illustrating the exit from the phase-locked region in the vicinity of corresponding points from Fig.9.38 are presented here. Hence, the abrupt transition to chaos, corresponding to the scenario B' of the Afraimovich-Shilnikov theorem on torus breakdown, also takes place in the system (9.30).

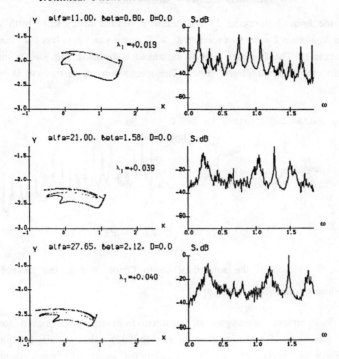

Fig.9.39. An illustration of the abrupt transition to chaos$_2$ in the system (9.30) (route B' in Fig.9.34,a).

C. *Torus breakdown due to the loss of smoothness: soft transition to chaos.* In the above scenarios A and B, the torus is destroyed before chaos is born. In case A, the torus is destroyed on the curve l_2 and the transition to chaos is connected with the period doubling bifurcations of a stable limit cycle, which no longer lie on the torus. In case B, the torus is destroyed on the curve l_h. The transition to chaos in this case occurs by crossing the curve l_1, when the stable limit cycle Γ^+ disappears, and the homoclinic structure forms a quasi-attractor. In case C we have a different situation. A direct transition $T^2(\alpha,\beta) \Rightarrow$ Chaos$_2$ takes place here! The loss of smoothness of the torus $T^2(\alpha,\beta)$ is a necessary condition for this direct transition.

As it is seen from the diagram in Fig.9.34, the bifurcation curve l_h starts near a point $\alpha \cong 5.2$, $\beta \cong 0.38$. For $\alpha \leq 5$ the torus exists. The torus $T^2(\alpha,\beta)$ loses its smoothness on its approach to the curve l_1 inside the 1:6 phase-locked region. The exit from the 1:6 phase-locked region through the curve l_1 leads to a soft transition to chaos here. Figure 9.40 illustrates this mechanism. The exit from the 1:6 phase-locked region near the homoclinic curve l_h, but below it, results in a "soft" birth of Chaos$_2$. Although the maximum Lyapunov exponent $L > 0$ in this case, it is less than the Lyapunov exponent in case B. The Poincaré map of the attractor shown in Fig.9.40 is displayed by an almost smooth invariant closed curve. The mechanism of destruction is the loss of smoothness here.

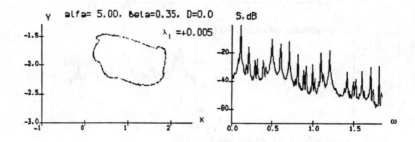

Fig.9.40. An illustration of the soft transition to Chaos$_2$ due to the loss of the torus $T^2(\mu)$ smoothness (route C in Fig.9.34,b).

As a final remark, although the transition from the phase-locked torus to an ergodic one (the route C' in the diagram of Fig.9.33) exists in the 1:6 phase-locked region in Fig.9.34, it is not easy to pinpoint the exact parameters (α,β) for this situation in practice. The reason is as follows. First, inside the 1:6 phase-locked region the multipliers of the limit cycle Γ^+ are complex-conjugate numbers practically everywhere along the boundary of the tongue near the curve l_1. The rotation of the invariant manifold of Γ^- in the vicinity of Γ^+ caused by the complex-conjugate multipliers led to the loss of smoothness of the torus. Second, in view of the location of the curve l_h, the critical curve l^2_{cr} representing the boundary between the ergodic torus and chaos is very near to the curve l_1, but is located outside the phase-locked region. Therefore, the probability of a transition from the phase-locked region to chaos, or from the same regime into another phase-locked region is very high (see the region near l^2_{cr} in Fig.9.34).

Concluding remarks. The main result of this paragraph is as follows. All the three mechanisms responsible for the breakdown of the two-dimensional torus are realized in the autonomous three-dimensional system (9.30) with piecewise-linear characteristic $f(\xi)$. The first mechanism is the destruction of the torus due to the loss of stability of the phase-locked limit cycle Γ^+ via period doubling on curve l_2 (Fig.9.34). The second mechanism is the destruction of the torus $T^2(\alpha,\beta)$ caused by the phenomenon of the homoclinic tangency of the stable and the unstable manifolds of the saddle limit cycle Γ^- on the torus (the curve l_h in Fig.9.34). The third mechanism is the destruction of the torus due to the loss of the smoothness on the curve l_1 below the point of intersection of the curves l_h and l_1 (see route C in Fig.9.34).

All the three routes to chaos are realized due to torus breakdown: (1) transition via a cascade of period-doubling bifurcations (route A), (2) abrupt transition to chaos in the homoclinic region via saddle-node bifurcation of the limit cycle Γ^+ (route B'), and (3) soft transition to chaos due to the loss of torus smoothness (route C).

These results give evidence that, at least from the experimental point of view, all conclusions of the Afraimovich-Shilnikov theorem on torus breakdown proved for smooth dynamical systems are also applicable to the piecewise-linear system (9.30). The non-smoothness of the function $f(\xi)$ in (9.30) generates some peculiarities in the system dynamics, which were identified here. However, these peculiarities do not have any significant influence on the basic aspects of the bifurcational phenomena.

CHAPTER 10

BIFURCATIONS OF DYNAMICAL SYSTEM IN THE PRESENCE OF NOISE

10.1. SOME METHODS OF STOCHASTIC CALCULUS

The presence of noise leads to topology concept to be impossible as applied when investigating bifurcations. Really, let a dynamical system have the only stable trivial solution $x = 0$. Phase trajectories would not converge to equilibrium state with noise having an arbitrary small intensity. Their stationary behavior will be defined by the *stationary probability density* $p_s(x)$ having its maximum at $x = 0$. As another example, let us consider a model system of the pitchfork bifurcation [10.1] as follows:

$$dx/dt = \varepsilon x - x^3. \qquad (10.1)$$

As $\varepsilon < 0$, the system (10.1) has only one equilibrium state in $x_0 = 0$ exhibiting a global stability. As $\varepsilon > 0$, there are two locally stable equilibria, $x_{1,2} = \pm (\varepsilon)^{1/2}$, and a saddle $x_0 = 0$. The situation is qualitatively changed by substituting an additive white noise into the right-hand part of (10.1). With positive ε, independently of initial conditions, the behavior of the system is defined by the stationary probability density $p_s(x)$ having its maxima at the points $x_{1,2} = \pm (\varepsilon)^{1/2}$ which correspond to the stable nodes of the unperturbed system (10.1), and a maximum at the point $x_0 = 0$, which corresponds to the saddle (Fig.10.1). Thus, the global stability of a system is induced by noise [10.2]. The bifurcation can be recorded in this case by varying the structure of the probability density, and just by varying the number of extrema $p_s(x)$.

Therefore, taking into account the noise influence leads to the transition from the study of limit sets in the phase space of a dynamical system to the consideration of an *ensemble of realizations* and their statistic characteristics. Such characteristics may include stationary probability density, correlation function, power spectrum, etc. There are two general approaches in the theory of stochastic processes: the first method is based on the consideration of *Langevin's equations* and the second one is built upon the solution of *kinetic* equations [10.3].

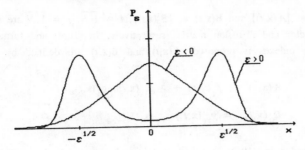

Fig.10.1. Stationary probability density of the system (1) excited by additive noise.

Langevin's equations, or *stochastic differential equations* (SDEs), are written in the form

$$d\mathbf{x}/dt = \mathbf{f}(\mathbf{x},t,\mu) + \mathbf{g}(\mathbf{x},t)\xi(t), \qquad (10.2)$$

where $\mathbf{x} = (x_1, x_2, ..., x_N)$ is the vector of state variables; $\mathbf{f}(\mathbf{x},t,\mu) = (f_1(\mathbf{x},t,\mu), ..., f_N(\mathbf{x},t,\mu))$; $\mathbf{g}(\mathbf{x},t)$ is the $N \times N$ matrix, $\mathbf{g}(\mathbf{x},t) = \{g_{ij}(\mathbf{x},t)\}$, $i = \overline{1,N}$, $j = \overline{1,N}$; $\mu = (\mu_1, \mu_2, ..., \mu_M)$; $\xi(t) = (\xi_1(t), ..., \xi_N(t))$. The quantities μ_k, $k = \overline{1,M}$ are parameters. The random functions $\xi_i(t)$ are called the Langevin's sources. The quantity N defines the dimension of the phase space of the appropriate dynamical system

$$d\mathbf{x}/dt = \mathbf{f}(\mathbf{x},t,\mu), \qquad (10.3)$$

and M is the dimension of parameter space. $\mathbf{x}(t)$ is a Markovian process while the random processes $\xi_i(t)$ are δ-correlated and are the Gaussian ones. In that case the transition from the SDE (10.2) to the *Fokker-Planck equation* (FPE) becomes possible. An ambiguity in interpretation of stochastic integrals appears [10.3]. Ito's and Stratonovich's interpretations are more useful leading to equivalent results in most cases. An alternative form for writing the SDE has been presented in Klimontovich's paper [10.4] and is based on the kinetic form of the FPE. This approach seems to be most valid when considering the natural noise. The treatment of stochastic integrals in the sense of Stratonovich, used in the present book, is more suitable for the study of the effect of external noise from the physical point of view.

Thus, with the Gaussian white noise, the SDE (10.2) corresponds to the FPE for the probability density $p(\mathbf{x},t)$ written in the form

$$\partial_t p(\mathbf{x},t) = -\partial_{x_i} A_i(\mathbf{x},t) p(\mathbf{x},t) + \frac{1}{2} \partial_{x_i x_j} B_{ij}(\mathbf{x},t) p(\mathbf{x},t), \qquad (10.4)$$

where $\mathbf{A}(\mathbf{x},t) = \{A_i(\mathbf{x},t)\}$ and $\mathbf{B}(\mathbf{x},t) = \{B_{ij}(\mathbf{x},t)\}$, $i = \overline{1,N}$, $j = \overline{1,N}$ are called the drift coefficients vector and diffusion matrix, respectively. In (10.4) and further, summation over repeating indexes is proposed. $\mathbf{A}(\mathbf{x},t)$ and $\mathbf{B}(\mathbf{x},t)$ are defined by the following relations

$$A_i(\mathbf{x},t) = f_i(\mathbf{x},t) + \frac{1}{2} g_{jk}(\mathbf{x},t) \partial_{x_j} g_{ik}(\mathbf{x},t),$$
$$B_{ij}(\mathbf{x},t) = g_{ik}(\mathbf{x},t) g_{jk}(\mathbf{x},t).$$
(10.5)

The noise included in (10.2) is called *additive* when the diffusion matrix is constant. In this case Ito's and Stratonovich's interpretations agree. The noise is called *multiplicative*, or parametric, when $\partial B_{ij}/\partial x_k \neq 0$ $(i,j,k = \overline{1,N})$.

Further, we will consider the so-called homogeneous stochastic processes with diffusion and drift coefficients which are independent of time. Having solved the system of SDE (10.2) or FPE (10.4), one can obtain the characteristics of the process $\mathbf{x}(t)$ and study their evolution when varying the control parameters of the system.

The effect of external noise on discrete time systems (or maps) can be investigated using two methods. Stochastic equations for a map have the form

$$\mathbf{x}_{n+1} = \mathbf{f}(\mathbf{x}_n) + \mathbf{G}(\mathbf{x}_n)\xi_n.$$
(10.6)

The evolution of probability density $p(\mathbf{x},n)$ is defined by the Chapman-Kolmogorov equation [10.5]

$$p(\mathbf{x},n+1) = \int D(\mathbf{y})^{-1} W\{\mathbf{G}^{-1}(\mathbf{y})[\mathbf{x} - f(\mathbf{y})]\} p(\mathbf{y},n) d\mathbf{y},$$
(10.7)

where $D(\mathbf{x}) = \det[G(\mathbf{x})]$, $W(\mathbf{x})$ is the distribution function of noise. Equation (10.7) is an analog of the FPE for discrete maps with noise.

The stationary probability density is used in the above example of pitchfork bifurcation as a characteristic of stochastic process for bifurcation analysis. However, not all the bifurcations are accompanied by the qualitative variations of invariant measure structure. In these cases, other characteristics of the process are used, for example, the power spectrum or the mean first passage time of a trajectory from a certain phase space region. The general property of nonlinear stochastic systems is that different characteristics suffer qualitative variations when varying the control parameter in a certain vicinity of the bifurcation point of the unperturbed system. Therefore, one needs to use the notion of bifurcation region in the presence of noise instead of bifurcation point or line. Nevertheless, to analyze qualitatively the effect of external noise on the dynamics of a nonlinear system, the bifurcation can be associated with the qualitative variation of statistical characteristics of the process. It should be noted in what sense one bifurcation or another is understood.

The above does not claim to be mathematically rigorous, of course. There exists no common idea of the bifurcation of a dynamical system perturbed by noise yet. Nontheless, this notion is accepted by many authors [10.1,10.2,10.5,10.6] and allows us to solve the problems of qualitative study of the dynamics of stochastic systems.

The analysis of the local bifurcation of dynamical systems is reduced, as a rule, to solving the appropriate algebraic equations. Effective algorithms and software have been developed to continue the bifurcational solutions in the parameter space [10.7]. The problem of parameter continuation of the bifurcational solutions is complicated essentially in the presence of noise since the FPE or the system of SDEs must be solved to obtain the statistical characteristics of the process. The result can be derived by numerically integrating the appropriate equations only except the simplest cases.

Stationary solution of the FPE and its structure. It is possible to analytically obtain the stationary solution (10.4) for one-dimensional case ($N = 1$) or with conditions of detailed balance fulfilled [10.3]. For high-dimensional nonlinear ($N \geq 3$) systems, these conditions are not fulfilled, as a rule. Then, approximations and numerical methods are used to solve the FPE [10.8]. Nevertheless, investigation of the simplest systems, which allow an exact or approximate analytical solution, is of great interest because it enables us to reveal the general noise effects.

Consider a one-dimensional stochastic system which is described by the FPE as follows

$$\partial_t p(x,t) = -\partial_x A(x)p(x,t) + \frac{1}{2}\partial_{xx} B(x)p(x,t),$$

$$p(x,0) = p_0(x), \quad p(\pm\infty,t) = 0. \tag{10.8}$$

The stationary solution of equation (10.8) under specified boundary conditions yields

$$p_s(x) = \frac{C}{B(x)} \exp\left[2\int \frac{A(x)}{B(x)} dx\right], \tag{10.9}$$

where C is a constant defined by the normalization condition, namely:

$$\int_{-\infty}^{\infty} p_s(x)dx = 1. \tag{10.10}$$

The structure of the stationary probability density is defined by the presence of extrema and their number. For the points of extrema x_m, we obtain from (10.8) and (10.9) the following relation:

$$\frac{1}{2}\partial_x B(x) - A(x) = 0. \tag{10.11}$$

The points of extrema x_m are found by solving (10.11). The calculation of the second derivative of $p_s(x)$ provides determination of their type. The variation of the number of extrema is possible by varying the parameters of the system what corresponds to a bifurcational transition.

To obtain the non-stationary solution of (10.8) or to solve the high-dimensional FPE, again, approximation and numerical methods are needed. The methods for FPE solving are reviewed in the Risken's monography [10.8] where a technique of numerical one- and two-dimensional FPEs is presented. The problem of solving the high-dimensional FPE with $N \geq 3$ exhibits, for the present, serious difficulties. Complications appear as well when decreasing the intensity of noise. This is connected with the need to decrease the difference grid step.

Simulation of colored noise. Numerical integration of stochastic differential equations. Process $\mathbf{x}(t)$ loses its Markovian properties when the Langevin's sources $\xi_i(t)$ in (10.2) are not δ-correlated. In this case, the FPE technique becomes useless. To overcome this difficulty, the following procedure is applied [10.6-10.10]. The random sources $\xi_i(t)$ are simulated using *Ornstein-Uhlenbeck processes* [10.3]

$$d\xi/dt = -M\xi + D\mathbf{n}(t), \qquad (10.12)$$

where $\mathbf{n}(t)$ is the vector of Gaussian white noise sources, and M, D are the $N \times N$ matrices, respectively. For the simplest one-dimensional Ornstein-Uhlenbeck process having the form

$$d\xi/dt = -\gamma\xi - \gamma(2\sigma)^{1/2} n(t); \qquad (10.13)$$

the process $\xi(t)$ has the following exponential *correlation function*:

$$<\xi(t)\xi(t + s)> = \sigma\gamma \, exp(-s\gamma). \qquad (10.14)$$

The parameter $\gamma = 1/\tau_c$ defines the correlation time τ_c of the process and, as $\gamma \to \infty$ ($\tau_c \to 0$), the limit transition to white noise occurs: $<\xi(t)\xi(t + s)> = \sigma\delta(s)$.

Thus, the system (10.2), that has the colored noise sources $\xi_i(t)$, turns out to be considered by extending the phase space within the theory of Markovian process. Analytical results are possible here for low-dimensional ($N = 1,2$) systems and for essential limitations on noise parameter [10.9-10.10]. Most of results in this field were obtained with the help of analogous and numerical simulation [10.11].

The algorithms of numerically integrating the SDE are reviewed in [10.12-10.14]. We have used the method proposed in [10.12] that is an analog of Runge-Kutt's fourth-order algorithm. The algorithm is defined by the following difference scheme

$$\begin{aligned}
x_{k+1} &= x_k + (K_1 + 2K_2 + 2K_3 + K_4)/6, \\
K_1 &= f(x_k, t_k) + g(x_k, t_k)\Delta W_k, \\
K_2 &= f(x_k + K_1/2, t_k + h/2)h + g(x_k+K_1/2, t_k+h/2)\Delta W_k, \\
K_3 &= f(x_k + K_2/2, t_k + h/2)h + g(x_k+K_2/2, t_k+h/2)\Delta W_k, \\
K_4 &= f(x_k + K_3, t_{k+1})h + g(x_k + K_3, t_{k+1})\Delta W_k, \\
\Delta W_k &= \sqrt{h}\,\xi_k,
\end{aligned} \qquad (10.15)$$

where h and ξ are the integration step and the Gaussian pseudorandom numbers vector, respectively. A mean-square convergence has been demonstrated for this method.

When examining the influence of colored noise, the linear SDEs (10.12) are to be solved additionally. In [10.15-10.16], an algorithm is presented providing the exact solution of (10.12) omitting the numerical integration procedure. In the one-dimensional case (10.13), this algorithm gives the following scheme

$$\xi_{k+1} = exp(-\gamma h)\xi_k + [\sigma\gamma(1 - exp(-2\gamma h))]^{1/2} n_k, \qquad (10.16)$$

where n_k is the Gaussian source of pseudorandom numbers.

As for the problem of numerical simulation of the SDE solutions, an important remark has to be made on realization length required to obtain the characteristics of stochastic process. Let, for example, the Brownian motion be simulated in a potential having several deep wells. With low noise intensity, the average time T_s of particle residence in each of the potential wells is great. It is obvious that calculated results are incorrect if the time of integration is smaller than T_s. For chaotic systems, the period of Poincare return is a characteristic time scale [10.17]. The difficulty is that the characteristic time scales of a system are not known in advance, as a rule. One of the methods is to originally define the characteristic times of the deterministic system T_d and then to simulate a stochastic system for times $T \gg T_d$. Another way is integration using different initial conditions from a certain region in the phase space.

Lyapunov characteristic exponents. The spectrum of Lyapunov characteristic exponents (LCEs) is of superior importance under characterization of a dynamical system. Its signature allows us to define the nature of the system's motion and, particularly, to diagnose chaotic regimes with confidence. The LCE concept is extended to encompass the stochastic regimes. In [10.18-10.19], the relation between LCE and Lyapunov exponents of p-th moment of the linear stochastic system in the form $dx/dt = A\xi(t)x$ is ascertained.

Consider the following one-dimensional stochastic system

$$dx/dt = f(x) + \xi(t), \tag{10.17}$$

where $\xi(t)$ is a white noise with intensity 2σ: $<\xi(t)\xi(t+s)> = 2\sigma\delta(s)$. By definition, the Lyapunov exponent λ is

$$\lambda = \lim_{t \to \infty} (T)^{-1} ln|\delta x|, \tag{10.18}$$

where δx is a distance between two close trajectories. Let $p_s(x)$ be a stationary solution of the FPE corresponding to the SDE (10.17). Due to ergodicity of process $x(t)$, the Lyapunov exponent may be found by averaging over the ensemble of the realizations of process $x(t)$ [10.1]:

$$\lambda = \int_{-\infty}^{\infty} df/dx \; p_s(x)dx. \tag{10.19}$$

Singularities in the dependence of the Lyapunov exponent λ on a parameter μ at the bifurcation points are typical for dynamical systems. The presence of noise leads to a smoothed dependence $\lambda(\mu)$ and, when new regimes are not induced by noise, to stabilization of the system. The value of Lyapunov exponent of stochastic system is decreased as compared with the unperturbed case. This effect of stabilization of the system by noise can be easily understood [10.1]. Let us propose a system having a globally stable equilibrium state x_0. With noise, trajectories of the process may considerably deflect from x_0. Here, the nonlinearity of the system becomes important and returns the trajectories into vicinity of x_0 decreasing λ thereby.

The algorithms of numerical computation of the LCE spectrum have been described in [10.20]. Different algorithms are discussed in detail in [10.21] (involving that for stochastic systems).

The application of cumulant analysis for the study of bifurcations in stochastic systems. As known, the random process $x(t)$ can be defined in two ways. The first way is to specify the *n-dimensional probability density* $p_n(x(t_1),x(t_2),...,x(t_n))$. The second one specifies the hierarchy of the *moments* or *cumulants* of a process. The cumulant description is, by a series of causes, more convenient and has been developed by Stratonovich and Malakhov [10.22-10.24]. Using, as characteristics for bifurcational analysis the cumulants and moments, calculated *directly* out of the process realization or the stationary probability density, is not a success, as a rule [10.6]. The matter is that the system acquires global stability under the influence of noise. The dependence of the process cumulants on parameters is smooth and shows no distinctions at bifurcation points.

For the homogeneous Markovian diffusion process defined by the FPE as follows:

$$\partial_t p(\mathbf{x},t) = -\partial_{x_i} A_i(\mathbf{x}) p(\mathbf{x},t) + \frac{1}{2} \partial_{x_i x_j} B_{ij}(\mathbf{x}) p(\mathbf{x},t) \tag{10.20}$$

the transition from the kinetic FPE (or a system of SDEs corresponding to it) to the equations describing the evolution of cumulants is possible. Due to the system's nonlinearity, the chain of cumulant equations is unclosed: the evolution equation for the n-th cumulant will involve higher order cumulants. To close the chain, model approximations are used with the simplest one being Gaussian. The cumulants of the first and second orders are assumed to differ from zero. This approximation is justified for small external noise intensities.

Let us introduce the following designations for the first and second order cumulants

$$M_i \equiv <x_i>, \quad D_i \equiv <x_i,x_i>, \quad u_{ij} \equiv <x_i,x_j>. \tag{10.21}$$

In the Gaussian approximation according to (10.20), we can obtain the following system of ordinary differential equations which describe the evolution of cumulants (10.21) [10.24]

$$\begin{aligned}
dM_i/dt &= <A_i(x)>, \\
dD_i/dt &= 2<x_i,A_i(x)> + <B_{ii}(x)>, \\
du_{ij}/dt &= 2<x_i,A_j(x)> + 2<x_j,A_i(x)> + <B_{ij}(x)>.
\end{aligned} \tag{10.22}$$

In (10.22), the averaging within the Gaussian approximation is assumed.

The effective Gaussian distributions, approximating the original stationary distribution, are defined by the equilibrium states of the cumulant equations (10.22). This dynamical system can be analyzed using the ordinary methods concerning the qualitative theory of differential equation. Thus, the idea of this method consists in the transition from the system of stochastic equations to that of the ordinary differential equations describing the evolution of cumulants within the Gaussian approximation [10.25]. In this chapter, a high-parametric bifurcational analysis of rather complicated high-dimensional dynamical systems, perturbed by noise, is performed by using the cumulant analysis.

The influence of noise on the bifurcations of a periodic solution is convenient to be investigated for discrete maps where the fixed points are equivalent to the periodic orbits of a flow system.

The Gaussian approximation is, of course, quite rough for chaotic regimes. However, as it will be shown below, the cumulant analysis enables one, as well as in this case, to obtain results being qualitatively adequate to the data of numerical simulation of original stochastic systems.

To illustrate the cumulant approach in use, we return to the model of pitchfork bifurcation mentioned above:

$$dx/dt = \varepsilon x - x^3 + \xi(t), \quad <\xi(t)\xi(t+s)> = \sigma\delta(s). \tag{10.23}$$

The Fokker-Plank equation, corresponding to the SDE (10.23) reads as follows:

$$\partial_t p(x,t) = -\partial_x(\varepsilon x - x^3)p(x,t) + \frac{\sigma}{2}\partial_{xx}p(x,t) \tag{10.24}$$

and has the stationary solution $p_s(x)$:

$$p_s(x) = C\,exp[\sigma^{-1}(\varepsilon x^2 - x^4/2)]. \tag{10.25}$$

The equations for the evolution of cumulants within the Gaussian approximation have the form

$$dM/dt = \varepsilon M - 3MD - M^3, \quad dD/dt = 2(\varepsilon D - 3D^2 - 3M^2 D) + \sigma. \tag{10.26}$$

The stationary values of a cumulant are defined by the appropriate system of algebraic equations

$$\varepsilon M - 3MD - M^3 = 0, \quad 2(\varepsilon D - 3D^2 - 3M^2 D) + \sigma = 0. \tag{10.27}$$

Solving (10.27) yields the following equilibrium states:

$$\begin{aligned}
M_0 &= 0, \quad D_0 = [\varepsilon + (\varepsilon^2 + 6\sigma)^{1/2}]/6, \\
M_{1,2} &= \pm(\varepsilon - 3D_1)^{1/2}, \quad D_1 = [\varepsilon - (\varepsilon^2 - 3\sigma)^{1/2}]/6, \\
M_{3,4} &= \pm(\varepsilon - 3D_2)^{1/2}, \quad D_2 = [\varepsilon + (\varepsilon^2 - 3\sigma)^{1/2}]/6.
\end{aligned} \tag{10.28}$$

The stability of an equilibrium state is defined by the sign of the real part of the eigenvalues s of the linearization matrix:

$$det\begin{bmatrix} \varepsilon - 3D - 3M^2 - s & -3M \\ -12MD & -12D + 2(\varepsilon - 3M^2) - s \end{bmatrix} = 0. \tag{10.29}$$

Firstly, the eigenvalues for the equilibrium state M_0, D_0 are

$$s_1^{(0)} = -2(\varepsilon^2 + 6\sigma)^{1/2}, \quad s_2^{(0)} = [\varepsilon - (\varepsilon^2 + 6\sigma)^{1/2}]/2. \tag{10.30}$$

As follows from (10.30), $s_{1,2}^{(0)} < 0$ for any ε and σ and, therefore, the equilibrium state M_0, D_0 is always stable. This is due to the stationary probability density (10.25) which is symmetrical for any ε, and $<x> = 0$.

Secondly, the eigenvalues for the equilibrium states $M_{1,2}$, D_1 and $M_{3,4}$, D_2 are defined by the expressions in the form

$$s_{1,2}^{(1,2)} = \pm [(\varepsilon^2 + 9\sigma)^{1/2} - (\varepsilon^2 - 3\sigma)^{1/2} + 2\varepsilon]. \qquad (10.31)$$

The eigenvalue s_1 is always negative. The sign of s_2 depends on ε and σ. Let us consider the case of a weak noise where $\sigma \ll 1$. Then, expand the right-hand part of (10.31) in σ in a Taylor's series and, restricting by the second order terms, obtain ($\varepsilon > 0$)

$$s_2 \approx 3\sigma/\varepsilon - 45\sigma^2/(4\varepsilon^3). \qquad (10.32)$$

The bifurcation corresponds to the case $s_2(\varepsilon_0, \sigma_0) = 0$ and is defined by the following expression

$$\varepsilon_0^2 = 45\sigma_0/12. \qquad (10.33)$$

Consider the bifurcation diagrams of system (10.26). In Fig.10.2, the steady-state values of cumulant D are shown as a function of parameter ε. The equilibrium states D_1, $M_{1,2}$ are stable and D_2, $M_{3,4}$ are unstable. They merge and disappear at the bifurcation point ε_0 via a saddle-node bifurcation. Thus, the bifurcation described corresponds to the feature of "fold" type [10.7]. As $\varepsilon_0^2 < 45\sigma_0/12$ (within the weak noise approximation), the system (10.26) has only one steady-state stable solution of M_0, D_0. As $\varepsilon_0^2 > 45\sigma_0/12$, there are four solutions being symmetrical relative to M: a pair of stable (D_1, $M_{1,2}$) and a pair of unstable (D_2, $M_{3,4}$) solutions. These pairs merge at the bifurcation point ε_0. The above bifurcation has codimension 1.

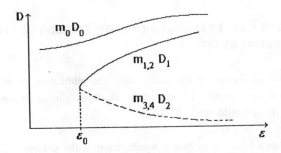

Fig.10.2. One-parametric bifurcation diagram for the system (10.26).

Thus, the qualitative variation of the structure of stationary probability density corresponds to the saddle-node bifurcation of the equilibrium states for the cumulant equation (10.26). A shift of bifurcation point appears. The bifurcational value of ε is defined (for $\sigma \ll 1$) by relation (10.33) for the system of cumulant equations (10.26) if the number of maxima $p_s(x)$ is doubled as $\varepsilon = 0$ independently of noise intensity σ.

It is of interest to take into account the great number of cumulants and to compare the bifurcation diagrams for the case of Gaussian approximation. Let us regard two cumulants more for the system (10.23): $k_3 = <x,x,x>$ and $k_4 = <x,x,x,x>$ what corresponds to the so-called *excess approximation* [10.24]. The evolution of cumulants is described by the following system of equations:

$$dM/dt = \varepsilon M - k_3 - 3MD - M^3,$$
$$dD/dt = 2(\varepsilon D - k_4 - 3Mk_3 - 3D^2 - 3M^2D) + \sigma,$$
$$dk_3/dt = 3(\varepsilon k_3 - 3Mk_4 - 9Dk_3 - 3M^2k_3 - 6MD^2), \quad (10.34)$$
$$dk_4/dt = 4(\varepsilon k_4 - 12Dk_4 - 3M^2k_4 - 9k_3^2 - 18MDk_3 - 6D^3).$$

The bifurcational analysis of system (10.34) has shown that the increase of the number of cumulants considered causes no variations in the qualitative bifurcation picture (Fig. 10.2).

It should be noted that the use of the excess approximation in some cases with high-dimensional systems comes against serious difficulties associated with the sudden increase of dimension of the system of cumulant equations. Furthermore, the increasing number of cumulants under study does not qualitatively affect the bifurcation picture, as a rule. Therefore, we shall confine ourselves below by the Gaussian approximation to analyze more complicated systems.

10.2 INFLUENCE OF EXTERNAL NOISE ON THE BIFURCATIONS OF EQUILIBRIUM STATE

Here, we consider general noise effects near the bifurcation points of equilibrium states with the simplest models as examples. Moreover, noise-induced transitions in bistable systems will be considered.

Codimension-two bifurcation (triple equilibrium) in the presence of external noise. The normal form for a specified bifurcation is as follows:

$$dx/dt = \alpha + \beta x - x^3, \quad (10.35)$$

where α and β are the control parameters. Equation (10.35) describes phenomena taking place in a wide scope of complicated systems involving the distributed ones [10.26-10.27]. As $\alpha = 0$, it presents the simplest model of basic bistable systems in synergetics.

The bifurcation diagram of the system (10.35) on the parameter plane (α,β) is shown in Fig.10.3. One or three robust equilibrium states may exist depending on the parameter values. These equilibrium states merge in pairs on lines S_1 and S_2. Point A corresponds to the merging of all the three equilibrium states and has codimension 2. There are three equilibrium states, two stable and one unstable (between them) in the system for the parameter values which lie inside the region 2 (see Fig.10.3) bounded by the curves S_1 and S_2. In the region 1, one globally stable equilibrium occurs.

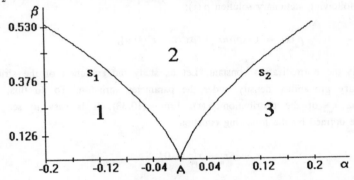

Fig.10.3. Bifurcation diagram of the system (10.35).

A bifurcation manifold, the triple equilibrium, is characterized by a "cusp" shown in Fig.10.4.

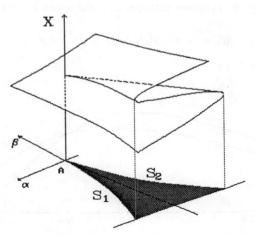

Fig.10.4. Bifurcation diagram of the system (10.35) illustrating a peculiarity of the "cusp" type.

Effects of additive white noise. The SDE is written in the form

$$dx/dt = \alpha + \beta x - x^3 + \xi(t), \quad <\xi(t)\xi(t+s)> = 2\sigma\delta(s). \tag{10.36}$$

The appropriate FPE reads as follows:

$$\partial_t p(x,t) = -\partial_x(\alpha + \beta x - x^3)p(x,t) + \sigma\partial_{xx} p(x,t). \tag{10.37}$$

It has the following stationary solution $p_s(x)$:

$$p_s(x) = C\exp[(\alpha x + \beta x^2/2 - x^4/4)/\sigma], \tag{10.38}$$

where C is the normalization constant. Let us study the evolution of the structure of the stationary probability density under the parameter variation. To do this, we find the extrema x_m of the distribution $p_s(x)$. From (10.38), it is easy to see that the extrema are defined by the following equation

$$\alpha + \beta x_m - x_m^3 = 0, \tag{10.39}$$

which exactly agrees with the equation for the equilibrium states of the unperturbed system (10.35). Thus, the bifurcational variations of the stationary probability density $p_s(x)$ proceed in a strict agreement with the bifurcation diagram shown in Figs 10.3 and 10.4.

Let us calculate the Lyapunov exponent λ for the system (10.36). Using the formula (10.19), we find

$$\lambda(\alpha,\beta) = \beta - 3\int x^2 p_s(x)dx = \beta - 3<x^2>.$$

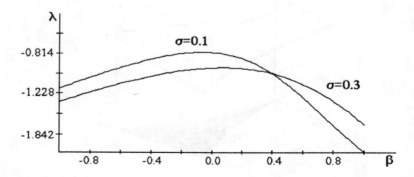

Fig. 10.5. Lyapunov exponent λ versus the parameter β as $\alpha = 0.1$ with various values of the noise intensity σ.

In Fig.10.5, λ is shown as a function of the parameter β as $\alpha = 0.1$ and for the noise intensity values $\sigma = 0.1$ and 0.3. From this figure it is obvious that the Lyapunov exponent dependence on β is smooth in the vicinity of the bifurcation point of the unperturbed system. The effect of stabilization of the system by noise is observed. With increasing the noise intensity, the Lyapunov exponent is decreased.

Now, the cumulant analysis of the system (10.36) is to be performed. Within the Gaussian approximation, the evolution of cumulants is defined by the following system of ordinary differential equations:

$$dM/dt = \alpha + \beta M - M(3D + M^2), \quad dD/dt = 2D(\beta - 3D^2 - M) + \sigma. \tag{10.40}$$

Calculations have demonstrated the bifurcation diagram of the system (10.40) to be qualitatively similar, for small σ, to the diagram obtained from equation (10.39).

The above results give evidence that the *additive white noise* does not change qualitatively the nature of behavior of the system (10.36) as compared with the original dynamical system (10.35).

Effects of multiplicative white noise. This case corresponds to the stochastic modulation of parameter β ($\beta' = \beta + \xi(t)$), namely:

$$dx/dt = \alpha + \beta x - x^3 + x\xi(t), \quad <\xi(t)\xi(t+s)> = 2\sigma\delta(s). \tag{10.41}$$

The appropriate FPE in the sense of Stratonovich

$$\partial_t p(x,t) = -\partial_x(\alpha+\beta x-x^3+\sigma x)p(x,t) + \sigma\partial_{xx} x^2 p(x,t), \tag{10.42}$$

has the stationary solution in the form

$$p_s(x) = C\exp[-(x^2 + \alpha)/(2\sigma)]x^{(\beta-\sigma)/\sigma}. \tag{10.43}$$

For $\beta < \sigma$, the integral $\int p_s(x)dx$ diverges and there exists no stationary solution of the FPE. The value $\beta = \sigma$ is bifurcational. For $\beta > \sigma$, the FPE has a stationary solution with the structure defined by extrema x_m that satisfies the following equation

$$\alpha + (\beta - \sigma)x_m - x_m^3 = 0. \tag{10.44}$$

The equations (10.44) and (10.39) do not agree. Hence, *the shift of bifurcation diagram appears under multiplicative influence*. The bifurcation surface in the three-dimensional parameter space (α,β,σ) is defined by the expression

$$\alpha = \pm \frac{2}{3\sqrt{3}} (\beta - \sigma)^{3/2} \tag{10.45}$$

and is shown in Fig.10.6. The analysis of (10.45) shows that the noise intensity becomes control parameter of the system. This yields the bifurcation lines of merging equilibrium states S_1 and S_2 (see Fig.10.4) which are transformed into planes, and the codimension-two point A which becomes a codimension-two line $\beta = \sigma$.

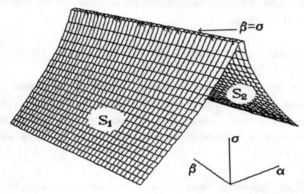

Fig.10.6. Central projection of bifurcation surface defined by Eq. (10.45) for the system (10.41).

A stochastic system can be exposed to the cumulant analysis like in the additive noise case whereby the results, qualitatively equivalent to (10.45), are obtained as $\sigma \ll 1$.

Common effects of additive and multiplicative white noise. In this case, the SDE is written in the form

$$\begin{aligned} dx/dt &= \alpha + \beta x - x^3 + x\xi(t) + \eta(t), \\ <\xi(t)\xi(t+s)> &= 2\sigma_1 \delta(s), \quad <\eta(t)\eta(t+s)> = 2\sigma_2 \delta(s). \end{aligned} \tag{10.46}$$

Consider the sources $\xi(t)$ and $\eta(t)$ to be statistically independent : $<\xi\eta> = 0$. The FPE is written as follows

$$\partial_t p(x,t) = -\partial_x (\alpha + \beta x - x^3 + \sigma_1 x) p(x,t) + \partial_{xx} (\sigma_1 x^2 + \sigma_2) p(x,t) \tag{10.47}$$

and has the following stationary solution

$$p_s(x) = C(\sigma_1 x^2 + \sigma_2)^{\gamma-1} \exp\left[F - \frac{x^2}{2\sigma_1} \right], \tag{10.48}$$

where $F = \dfrac{\alpha \, arctg(\sqrt{\sigma_1/\sigma_2})}{\sqrt{\sigma_1/\sigma_2}}$, $\gamma = (\beta/\sigma_1 + \sigma_2/\sigma_1^2 + 1)/2$.

In contrast to the case of multiplicative noise, the stationary probability density exists for any values of $\sigma_1 \neq 0$ and $\sigma_2 \neq 0$. It is easy to see, that the extrema of $p_s(x)$ are defined by the equation (10.44). Therefore, the *bifurcation picture*, for the common action of additive and multiplicative noise, *is similar to the case of multiplicative noise* except the behavior of FPE as $\sigma > \beta$.

Effects of additive colored noise. We simulate a *colored noise source* $z(t)$ by using one-dimensional *Ornstein-Uhlenbeck's process*

$$dx/dt = \alpha + \beta x - x^3 + z, \quad dz/dt = \tau_c^{-1}[-z + \sqrt{2\sigma}\xi(t)]. \qquad (10.49)$$

The two-dimensional Markovian process $\{x(t), z(t)\}$, defined by the SDE (10.49), corresponds to the FPE for the two-dimensional probability density $p(x,z,t)$:

$$\partial_t p = -\partial_x(\alpha + \beta x - x^3 + z)p + \tau_c^{-1}(\partial_z z p + \sigma \tau_c^{-1} \partial_{zz} p). \qquad (10.50)$$

As it was already noted, we have failed to obtain an exact analytical solution for the FPE (10.50). However, an approximate stationary solution can be obtained in the limit cases of very small or very large noise correlation times [10.9-10.10, 10.28-10.29]. For $\tau_c \ll 1$, one may proceed from (10.49) to an effective FPE for the one-dimensional probability density $p(x,t)$ as follows:

$$\partial_t p(x,t) = -\partial_x(\alpha+\beta x-x^3)p(x,t) + \sigma\partial_{xx}[1+\tau_c(\beta-3x^2)]p(x,t). \qquad (10.51)$$

This equation is valid for the weakly correlated noise. The extrema of the stationary probability are defined by the following equation

$$\alpha + (\beta + 3\sigma\tau_c)x_m - x_m^3 = 0. \qquad (10.52)$$

It follows from (10.52) that a bifurcation shift, that is proportional to the product $\sigma\tau_c$, appears here as compared with the case of δ-correlated action.

Effect of multiplicative colored noise. In this case, the two-component Markovian process $\{x(t), z(t)\}$ is described by the following system of SDEs

$$dx/dt = \alpha + \beta x - x^3 + zx, \quad dz/dt = \tau_c^{-1}[-z + \sqrt{2\sigma}\xi(t)]. \qquad (10.53)$$

For $\tau_c \ll 1$, we write the effective FPE for one-dimensional probability density $p(x,t)$, namely:

$$\partial_t p = -\partial_x[\alpha + \beta x - x^3 + \sigma v(x)]p + \sigma \partial_{xx} xv(x)p,$$
$$v(x) = x - \tau_c(\alpha + 2x^3). \tag{10.54}$$

The extrema of stationary solution of (10.54) are defined by the following equation

$$\alpha + (\beta - \sigma)x_m - x_m^3(1 - 6\sigma\tau_c) = 0. \tag{10.55}$$

The bifurcation surface is specified by the condition given below

$$\alpha = \pm \frac{2}{3\sqrt{3}}(\beta - \sigma)^{3/2}(1 + 3\sigma\tau_c). \tag{10.56}$$

The small τ_c deviation from zero results in a small shift of the bifurcation diagram proportionally to $\sigma\tau_c$. Note, that this shift proceeds in the *reverse direction* to that of the shift induced by multiplicative noise (confer (10.45) and (10.56)).

The analysis of influence of fluctuation on the bifurcations of equilibrium state in a simple system provides the following conclusions. The additive δ-correlated (white) noise with low intensity causes no qualitative distinctions in the bifurcation diagrams of perturbed systems as compared with the unperturbed ones. The variations in bifurcation diagram are induced by the multiplicative noise. The noise intensity gains the properties of a control parameter of the system. This is responsible for necessity of considering the perturbed system in an extended parameter space involving the intensity of noise.

For colored noise, the additional bifurcation shift appears to be proportional to the noise correlation time.

Analysis of the phenomenon of a "hole" formation in the two-dimensional probability density of a bistable system perturbed by colored noise. One of the most interesting phenomena associated with the influence of colored noise on bistable systems is the transition induced by noise and followed by a qualitative change in topology of the stationary probability density [10.30]. A SDE of the system has the following form

$$dx/dt = -df(x)/dx + z(t), \tag{10.57}$$

where $f(x)$ is the bistable potential and $z(t)$ is the external Gaussian noise with zero average and correlation function, namely:

$$<z(t)z(t + s)> = \sigma/\tau_c \exp(-s/\tau_c). \tag{10.58}$$

Parameters σ and τ_c determine the intensity and correlation time of noise, respectively. The process $z(t)$ is simulated by the following SDE

$$dz/dt = \tau_c^{-1}[-z + (2\sigma)^{1/2}\xi(t)], \quad <\xi(t)\xi(t+s)> = \delta(s) \quad (10.59)$$

which determines the Ornstein-Uhlenbeck process with the correlation function (10.58). The two-component Markovian process $\{x(t),z(t)\}$ (10.57), (10.59) corresponds to the FPE for the two-dimensional probability density $p(x,z)$ as follows:

$$\partial_t p = \partial_x[(df/dx - z)p] + (\tau_c)^{-1}\partial_z(zp) + \sigma/\tau_c^2 \partial_{xx} p. \quad (10.60)$$

The effect discussed here, consists mainly in the following [10.30]. For a correlation time τ_c smaller than some critical time τ_0, the structure of the two-dimensional stationary probability density $p_s(x,z)$ is traditional. It contains two local maxima corresponding to the two minima of the potential $f(x)$. Between the maxima of $p_s(x,z)$, in the vicinity of the origin, there is a characteristic minimum in the distribution. It is governed by the maximum of the potential function, $f(x)$, and associated with the saddle nature of the equilibrium state of the unforced system

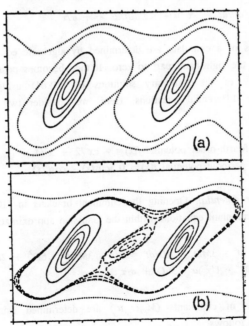

Fig.10.7. Contour lines of the two-dimensional stationary probability density of a bistable system excited by colored noise: a - $\tau_c < \tau_0$, b - $\tau_c > \tau_0$.

(10.57). As $\tau_c > \tau_0$, the *structure* of $p(x,z)$ is *qualitatively changed*. The phase space trajectories of the two-dimensional process on the (x,z) plane do not fall within the vicinity of the origin, what results in the appearance of the so-*called* "*hole*" in the two-dimensional distribution. This effect is illustrated in Fig.10.7 where the contour lines of stationary probability density $p_s(x,y)$ are shown for the values of $\tau_c < \tau_0$ (a) and $\tau_c > \tau_0$ (b).

In general case, it seems impossible to obtain the stationary solution of FPE (10.60). Approximate solutions can be obtained only in the case of essential limitations on the noise parameters σ and τ_c. The FPE (10.60) was solved numerically in [10.31-10.32] while in [10.30] the method for the analogous simulation of SDEs (10.57) and (10.59) was used.

To explore the bifurcational reasons of this effect, one may use the method of cumulant analysis [10.33]. Let us introduce the following designations for the cumulants: $M \equiv <x>, D \equiv <x^2> - M^2, u \equiv <x,z>$. The evolution of cumulants M, D and u, within the Gaussian approximation, is determined by the following system of ordinary differential equations

$$dM/dt = - <df/dx>, \quad dD/dt = - 2<x, df/x> + 2u,$$
$$du/dt = -\tau_c^{-1}u - <z, df/dx> + \sigma/\tau_c. \qquad (10.61)$$

The stationary values of a cumulant are determined by algebraic equations derived from (10.61) by equating the right-hand part to zero. It then becomes possible to conduct the analysis for stability of the stationary solutions of (10.61) using the conventional methods of ordinary differential equations. Here, we consider two specific shapes of potential $f(x)$, namely:

$$1 \text{ - the fourth-order potential } f(x) = - \varepsilon x^2/2 + x^4/4, \qquad (10.62)$$
$$2 \text{ - the periodic potential } f(x) = 0.5cos(x) - cos(2x). \qquad (10.63)$$

The fourth-order potential. Opening the cumulant brackets in (10.61) we yield the following system of cumulant equations within the Gaussian approximation

$$dM/dt = \varepsilon M - M^3 - 3MD, \quad dD/dt = 2(\varepsilon - 3M^2)D - 6D^2 + 2u,$$
$$du/dt = (\varepsilon - 3M^2 - 1/\tau_c)u - 3uD + \sigma/\tau_c. \qquad (10.64)$$

The stationary values of cumulants (M_0, D_0, u_0) are determined using the system of algebraic equations as follows

$$(\varepsilon - 3D_0)M_0 - M_0^3 = 0, \quad (\varepsilon - 3M_0^2)D_0 - 3D_0^2 + u_0 = 0,$$
$$(\varepsilon - 3M_0^2 - 1/\tau_c)u_0 - 3u_0 D_0 + \sigma/\tau_c = 0. \tag{10.65}$$

Of particular interest is the vicinity of the origin on the plane (x,z), where the noise-induced transition is realized; therefore, let us confine our bifurcational analysis of the stationary solutions with $M_0 = 0$ and $D_0 \ll 1$, which correspond to the characteristics of the ensemble of trajectories for the two-dimensional process $\{x(t), z(t)\}$ in the indicated region. The nature of stability of the stationary solutions is determined by the eigenvalues, s, of the linearization matrix of system (10.64):

$$\det \begin{bmatrix} \varepsilon - 3D_0 - s & 0 & 0 \\ 0 & 2\varepsilon - 12D_0 - s & 2 \\ 0 & -3u_0 & \varepsilon - 1/\tau_c - 3D_0 - s \end{bmatrix} = 0. \tag{10.66}$$

Consider first the case of weak noise as $\sigma \ll 1$. While doing this, we ignore the values D^2 and uD in (10.64):

$$dM/dt = \varepsilon M - M^3 - 3MD, \quad dD/dt = 2(\varepsilon - 3M^2)D + 2u,$$
$$du/dt = (\varepsilon - 3M^2 - 1/\tau_c)u + \sigma/\tau_c. \tag{10.67}$$

The stationary values of the cumulants are:

$$M_0 = 0, \quad D_0 = \sigma/[\varepsilon(\varepsilon\tau_c - 1)], \quad u_0 = -\sigma/(\varepsilon\tau_c - 1). \tag{10.68}$$

The eigenvalues of the linearization matrix are determined by the following relationships:

$$s_1 = \varepsilon - 3D_0, \quad s_2 = 2\varepsilon, \quad s_3 = \varepsilon - 1/\tau_c. \tag{10.69}$$

If the eigenvalue s_2 is positive, as $\varepsilon > 0$, then potential $f(x)$ has two minima. The sign of s_3 is determined by the condition that the diffusion D_0 (10.68) is positive: $D_0 > 0$ when $\tau_c > 1/\varepsilon$, hence $s_3 > 0$. However, the sign of the eigenvalue s_1 depends on the relationship between the parameters ε, σ and τ_c. Let us determine the value of the noise correlation time τ_c for which $s_1 \geq 0$. From (10.68) and (10.69), we obtain

$$\tau_c \geq \tau_0 = 1/\varepsilon + 3\sigma/\varepsilon^3. \tag{10.70}$$

For $\tau_c < \tau_0$, the stationary point (10.68) is a *saddle-node* with stable one-dimensional and unstable two-dimensional manifolds. At the point $\tau_c = \tau_0$, the eigenvalue s_1 changes its sign as $\tau_c > \tau_0$ and becomes positive. In consequence, the structure of the phase trajectories abruptly changes in the vicinity of equilibrium state of the system of cumulant equations (10.67). This system undergoes a bifurcation: the saddle-node transforms into an absolutely unstable node, or repeller, (all the three eigenvalues $s_{1,2,3} > 0$). As a result, the trajectories of the process $\{x(t), z(t)\}$ bypass the vicinity of the origin. This, in turn, offers the appearance of a hole in the stationary distribution $p(x,z)$. Linearization in the vicinity of the equilibrium state naturally cannot give the answer about the rigidity of the boundaries of the hole region on the plane (x,z). However, it can be asserted that near the origin, phase trajectories will be very scarce in the case when $\tau_c > \tau_0$. In addition it should be noted that the expression for the bifurcational value of correlation time (10.70), that was derived from the equation for the cumulants (10.67), rigorously corresponds to the relationship obtained by another method in [10.34].

A more exact bifurcational analysis can be conducted using numerical analysis of the stationary solutions of (10.64). The results of these calculations are presented in Fig. 10.8 which shows the bifurcation lines in the plane of parameters (τ_c, σ) for different parameter ε values. Note, as well, the satisfactory agreement of our curves with the results of analogous simulation presented in [10.30].

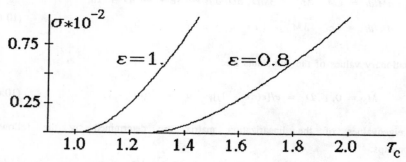

Fig. 10.8. Bifurcation diagram of the system (10.64) for two values of parameter ε.

Periodic potential. Opening the cumulant brackets in (10.63) and taking into account that, within Gaussian approximation

$$<sin(ax)> = exp(-a^2D/2)sin(aM),$$
$$<cos(ax)> = exp(-a^2D/2)cos(aM), \quad <x, sin(ax)> = M<sin(ax)> + aD<cos(ax)>,$$
$$<z, sin(ax)> = a<cos(ax)>u,$$

we obtain the following cumulant equations for the potential (10.63):

$$dM/dt = -exp(-D/2) \sin(M) + 2 \exp(-2D) \sin(2M),$$
$$dD/dt = 2\{u - 0.5 exp(-D/2)[M\sin(M) + D\cos(M)] - exp(-2D)[M\sin(2M) + 2D\cos(2M)]\} \quad (10.71)$$
$$du/dt = -u[1/\tau_c + 0.5exp(-D/2)\cos(M) - 4exp(-2D)\cos(2M)] + \sigma/\tau_c.$$

Let us consider, as in the case of fourth-order potential, the ensemble of trajectories of the process $\{x(t),z(t)\}$ traversing the vicinity of the origin. In the system (10.71), a "saddle-node - repeller" bifurcation is realized which is likely to be completely similar to (10.64). The numerically obtained bifurcation diagram on the plane (τ_c,σ) is shown in Fig.10.9. Here, we also witness a good agreement with the results of analogous simulation [10.30].

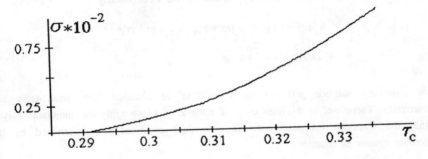

Fig.10.9. Bifurcation diagram for the system (10.71).

Thus, a bifurcation of the equilibrium state of the "saddle-node - repeller" type in the dynamical system of the cumulant equations corresponds to the transition induced by coloured noise in the bistable system. That is manifested by the appearance of a hole in the two-dimensional probability density.

Bifurcations of the equilibrium state in the Lorenz model in the presence of noise. The equilibrium states of the Lorenz system (3.8) can be found from the following equations
$$\sigma y - \sigma x = 0, \quad rx - y - xz = 0, \quad xy - bz = 0 \quad (10.72)$$

and their bifurcations are well known [10.38]. The origin $O(0,0,0)$ is a steady-state point for any parameter values. For $r < 1$, O is a stable node. For $r = 1$, a pitchfork bifurcation is observed: point O becomes unstable and two stable foci appear as $r > 1$, O_1 and O_2 ($\pm\sqrt{b(r-1)}, \pm\sqrt{b(r-1)}, r-1$). The above bifurcation is simulated using the simple system (10.23). O_1 and O_2 lose their stability as $r = r^* = \sigma(\sigma + b + 3)/(\sigma - b - 1)$.

In the presence of additive white noise, the SDEs of the Lorenz model are as follows [10.39]

$$dx/dt = \sigma y - \sigma x + \xi_1(t), \quad dy/dt = rx - y - xz + \xi_2(t),$$
$$dz/dt = -bz + xy + \xi_3(t). \tag{10.73}$$

where $\xi_i(t)$ are the statistically independent δ-correlated noise sources:

$$\langle \xi_i(t)\xi_j(t+s) \rangle = D\delta_{ij}\delta(s) \tag{10.74}$$

For our characteristic in the bifurcational analysis, we use the stationary probability density $p_s(x,y,z)$ which is the stationary solution of the FPE, namely:

$$\partial_t p = -\partial_x(\sigma y - \sigma x)p - \partial_y(rx - y - xz)p - \partial_z(-bz + xy)p +$$
$$+ D/2(\partial_{xx} + \partial_{yy} + \partial_{zz})p. \tag{10.75}$$

This stationary solution $p_s(x,y,z)$ is difficult to be obtained both analytically and numerically. Therefore, in the limit of weak noise $D \ll 1$, we apply the cumulant analysis within the Gaussian approximation. The cumulant equations are determined by the following system of equations:

$$dM_x/dt = \sigma(M_y - M_x), \quad dM_y/dt = M_x(r - M_z) - M_y - u_{xx},$$
$$dM_z/dt = -bM_z + M_x M_y + u_{xy}, \quad dD_x/dt = 2\sigma(u_{xy} - D_x) + D,$$
$$dD_y/dt = 2[-D_y + u_{xy}(r - M_z) - M_x u_{yz}] + D,$$
$$dD_z/dt = 2[-bD_z + M_y u_{xz} + M_x u_{yz}] + D, \tag{10.76}$$
$$du_{xy}/dt = -u_{xy}(1+\sigma) + D_x(r - M_z) - M_x u_{xz} + \sigma D_y,$$
$$du_{xz}/dt = -u_{xz}(b+\sigma) + M_y D_x + M_x u_{xy} + \sigma u_{yz},$$
$$du_{yz}/dt = -u_{yz}(b+1) + M_y u_{xy} + M_x D_y + u_{xz}(r - M_z) - M_x D_z.$$

In Fig.10.10, the bifurcation diagram of steady-state solutions of the system (10.76) is shown on the plane (r,D) ($\sigma = 10$, $b = 8/3$). The passing through the bifurcation line to higher values of parameter r corresponds to the bifurcation of the birth of equilibrium points $O_{1,2}$ or to that of doubling of the maxima number of the stationary probability density $p_s(x,y,z)$. As $D > 0$, the value of the bifurcational parameter r tends towards the value in the unperturbed case, $r_0 = 1$.

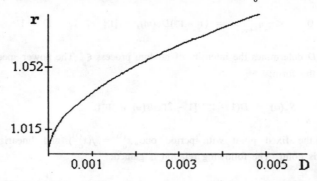

Fig. 10.10 Bifurcation diagram for the cumulant equations (10.76) for the Lorenz model.

10.5. PERIOD DOUBLING BIFURCATIONS IN THE PRESENCE OF NOISE

Below we examine in detail the influence of noncorrelated noise on period doubling bifurcations [10.37-10.43]. It has been found that the presence of noise leads to finite sequence of period doublings. A shift in parameter of both the points of period-doubling bifurcations and the critical point of transition to chaos appears. In the power spectrum, there are broad-band peaks at the subharmonics, preceding the bifurcations in the phase space [10.44]. The problems of the effect of noise with finite correlation time, comparison of theories based on one-dimensional models with the data obtained from numerical and full-scale experiments remain, however, still open.

Linear analysis and power spectrum [10.45]. Let us consider a family of one-dimensional maps

$$x_{n+1} = f(x_n) + \xi_n, \qquad (10.77)$$

where $f(x)$ is a function having a square-law maximum and ξ_n is the Gaussian colored noise source having a zero average. We will simulate these noise sources using an analog of Ornstein-Uhlenbeck process, the first order autoregressive process, and namely:

$$\xi_{n+1} = \Gamma\xi_n + \eta_n, \qquad (10.78)$$

where Γ is the parameter that determines the correlation time

$$\tau_c = -1/\log|\Gamma|,$$

and η_n denotes the white noise as follows:

$$<\eta_n> = 0, \quad <\eta_n \eta_{n+m}> = (1 - \Gamma^2)D\delta(m), \quad |\Gamma| < 1, \quad D \ll 1. \quad (10.79)$$

The parameter D determines the intensity of random process ξ_n. The power spectrum of the process ξ_n has the form:

$$S_\xi(\omega) = D(1 - \Gamma^2)/[1 - 2\Gamma\cos(\omega) + \Gamma^2]. \quad (10.80)$$

We take the fixed point with period one $x_0^{(1)} = f(x_0^{(1)})$ and linearize the map (10.77) near this point. The following relation is gained:

$$y_{n+1} = \mu_1 y_n + \xi_n, \quad (10.81)$$

where $y = x - x_0^{(1)}$. $\mu_1 = f'(x_0)$ is the characteristic multiplier for the fixed point $x_0^{(1)}$. For a fixed point of period 2^k, one can similarly write

$$y_{n+2^k} = \mu_k y_n + \xi_n, \quad (10.82)$$

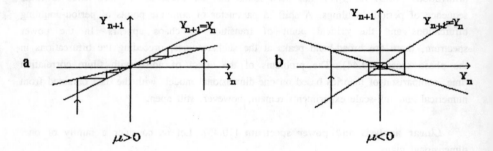

Fig.10.11. Phase portraits of the linearized map (10.81).

where y_n is now a small perturbation near the fixed point $x_0^{(k)} = f^{(k)}(x_0^{(k)})$ of period 2^k and μ_k is the characteristic multiplier of the fixed point $x_0^{(k)}$ of the k-iterated map. The phase portraits of the map (10.81) with no noise, $D = 0$, are shown in Fig.10.11. As $\mu > 0$ (Fig.10.11a), the fixed point at the origin is a stable node and, as $\mu < 0$ (Fig.10.11b), it is a stable focus. The value of multiplier $\mu = 0$ is bifurcational under "node - focus" transition. Note, that the above bifurcation is not recorded in the absence of perturbation.

The power spectrum of the linearized map (10.81) can be easily obtained as follows:

$$S_Y^1(\omega) = S_\xi(\omega)/(1 - 2\mu_1 cos(\omega) + \mu_1^2). \tag{10.83}$$

From this equation it follows that the power spectrum has a broad-band peak at the basic frequency $\omega = 0$ (or $\omega = 2\pi$) if the multiplier μ_1 is positive. For negative values of μ_1, the power spectrum has a broad-band peak at the subharmonic frequency $\omega = \pi$. Thus, the influence of noise gives rise to the fact that the power spectrum reflects a structure, which is typical for the fixed point with period two, long before the bifurcation point in the appropriate deterministic system occurs. It corresponds to noisy precursors of bifurcation as is described by Wiesenfeld [10.44]. In the same way we can obtain the power spectrum of the period-2^k fixed point:

$$S_Y^{(k)} = S_\xi(\omega)/[1 - 2\mu_k cos(2^k\omega) + \mu_k^2]. \tag{10.84}$$

If $\mu_k < 0$ then the power spectrum $S_Y^{(k)}$ has broad-band peaks at the subharmonics $\omega_k = (2n + 1)\pi/2^{k-1}$ ($n = 0,1,2,...$).

Let us consider the dependence of the intensity of these subharmonics $I^{(k)}(\Gamma)$ of the period-2^k fixed point on the parameter Γ of colored noise.

$$I^{(k)}(\Gamma) = D(1-\Gamma^2)/[1-2\Gamma cos(\omega_k)+\Gamma^2]/(1-2\mu_k+\mu_k^2). \tag{10.85}$$

It is easy to see that such a dependence takes its maximum value at the parameter $\Gamma=\Gamma_m^{(k)}$ determined by the following expression:

$$\Gamma_m^{(k)} = [1 - sin(\omega_k)]/cos(\omega_k). \tag{10.86}$$

Especially it gives for $k = 1$ (the period-2 fixed point) $\omega_1 = \pi$ and $\Gamma_m^{(1)} = -1$; for $k = 2$ $\omega_2 = \pi/2$ and $\Gamma_m^{(2)} = 0$; for $k = 3$ $\omega_3 = \pi/4$ (or $3\pi/4$) and $\Gamma_m^{(2)} \approx -0.41$ (or 0.41). Thus, the dependence of the subharmonic intensity of 2^k-cycles on the colored noise parameter Γ exhibits a resonance-like shape. In other words, the linear response of the system demonstrates a resonance-like sensitivity to the variations of the characteristic time scale of noise.

Bifurcational analysis of cumulant equation. We derive the cumulant equations within the Gaussian approximation for the two-dimensional map (x_n, ξ_n):

$$x_{n+1} = f(x_n) + \xi_n, \quad \xi_{n+1} = \Gamma\xi_n + \eta_n,$$
$$<\eta_n> = 0, \quad <\eta_n \eta_{n+m}> = (1 - \Gamma^2)D\delta(m). \tag{10.87}$$

First, a new variable, the deviation from the mean value is introduced [10.39]:

$\delta_n = x_n - \langle x_n \rangle$. Second, let us consider the limit of weak noise, $D \ll 1$. Then, the function $f(x)$ may be expanded in a Taylor's series in δ_n:

$$f(x) = f(\langle x \rangle + \delta) \approx f(\langle x \rangle) + f'(\langle x \rangle)\delta + 0.5f''(\langle x \rangle)\delta^2.$$

For the mean $\langle x_n \rangle$ value and δ_{n+1} from (10.87), we obtain

$$\langle x_{n+1} \rangle = \langle f(x_n) \rangle = f(\langle x_n \rangle) + 0.5f''(\langle x_n \rangle)\langle \delta_n^2 \rangle,$$
$$\delta_{n+1} = f'(\langle x_n \rangle)\delta_n + 0.5f''(\langle x_n \rangle)[\delta_n^2 - \langle \delta_n^2 \rangle] + \xi_n. \qquad (10.88)$$

From (10.88), we proceed to the equations for the cumulants of the deviation δ_n. Within the Gaussian approximation there are:

$$\begin{aligned}
\langle x_{n+1} \rangle &= f(\langle x_n \rangle) + 0.5f''(\langle x_n \rangle)\langle \delta_n^2 \rangle, \\
\langle \delta_{n+1}^2 \rangle &= [f'(\langle x_n \rangle)]^2 \langle \delta_n^2 \rangle + 2f'(\langle x_n \rangle)\langle \delta_n \xi_n \rangle + \langle \xi_n^2 \rangle, \\
\langle \delta_{n+1}\xi_{n+1} \rangle &= \Gamma[\langle \xi_n^2 \rangle + f'(\langle x_n \rangle)\langle \delta_n \xi_n \rangle], \\
\langle \xi_{n+1}^2 \rangle &= \Gamma^2 \langle \xi_n^2 \rangle + (1 - \Gamma^2)D.
\end{aligned} \qquad (10.89)$$

In the limit case of white noise, $\Gamma \to 0$, $\langle \delta_n \xi_n \rangle = 0$, a two-dimensional map is realized which is equivalent to that discussed in [10.39]:

$$\begin{aligned}
\langle x_{n+1} \rangle &= f(\langle x_n \rangle) + 0.5f''(\langle x_n \rangle)\langle \delta_n^2 \rangle \\
\langle \delta_{n+1}^2 \rangle &= [f'(\langle x_n \rangle)]^2 \langle \delta_n^2 \rangle + D.
\end{aligned} \qquad (10.90)$$

Furthermore, the evolution of stationary regimes of the system (10.89) is of interest under variation of the parameters. Therefore, the last equation of the system (10.89) may be omitted and the stationary value, which is equal to D, may be used in the remaining equations as $\langle \xi^2 \rangle$. By introducing new variables $X_n \equiv \langle x_n \rangle$, $Y_n \equiv \langle \delta_n^2 \rangle$, $Z_n \equiv \langle \delta_n \xi_n \rangle$, we finally obtain the following three-dimensional map:

$$\begin{aligned}
X_{n+1} &= f(X_n) + 0.5f''(X_n)Y_n, \\
Y_{n+1} &= [f'(X_n)]^2 Y_n + 2f'(X_n)Z_n + D, \\
Z_{n+1} &= \Gamma[D + f'(X_n)Z_n].
\end{aligned} \qquad (10.91)$$

For a particular form of $f(X) = 1 - aX^2$, we have

$$X_{n+1} = 1 - a(X_n^2 + Y_n), \quad Y_{n+1} = 4aX_n(aX_nY_n - Z_n) + D,$$
$$Z_{n+1} = \Gamma(D - 2aX_nZ_n). \tag{10.92}$$

The bifurcational mechanisms of this map are considered for the white noise in [10.39]. The qualitative bifurcation diagram is shown in Fig.10.12. Unlike the deterministic case where a soft doubling bifurcation is realized, the bifurcation is abrupt for the system of cumulant equations and does not correspond to a multiplier passing through "-1". Ordinary period-doubling bifurcations (through "-1") in the system (10.92) are possible for period-2^k *saddle cycles*. The transition to chaos in (10.92) is realized via the destruction of invariant curves formed on the base of period doubling cycles [10.39].

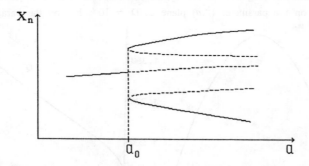

Fig.10.12. Qualitative bifurcation diagram for the system (10.92) in the vicinity of the first period doubling bifurcation.

Let us consider the bifurcation lines for the birth of the fixed point with period 2^1, 2^2 and 2^3 in the parameter planes (Γ, a) and (D, a). The results for the period-2^1 fixed point are shown in Fig.10.13. It is obvious from Fig.10.13a, that the dependence $a(\Gamma)$ takes its highest value at $\Gamma \to \Gamma_m^{(1)} = -1$ (cf. (10.86)). In Fig.10.13b, the birth point of period-2 cycle is shown as a function of noise intensity for $\Gamma = 0.5$. The bifurcation point is shifted towards higher values of the control parameter a. The bifurcation lines of period-2^2 fixed point are shown in Fig.10.14. The dependence $a(\Gamma)$ (Fig.10.14a) is characterized by a maximum at $\Gamma = \Gamma_m^{(2)} = 0$, i.e., it corresponds to the value of Γ at which the intensity of spectrum subharmonics in a linearized system has its maximum. The bifurcation line in the plane (D, a) (Fig.10.14b) replicates qualitatively the appropriate bifurcation line for the period-2 cycle (Fig.10.13b).

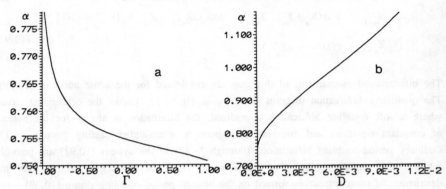

Fig.10.13. Bifurcation diagram of the system (10.92) for the first period doubling bifurcation: a - on the parameter (Γ,a) plane as $D = 10^{-5}$, b - on the parameter (D,a) plane as $\Gamma = -0.4$.

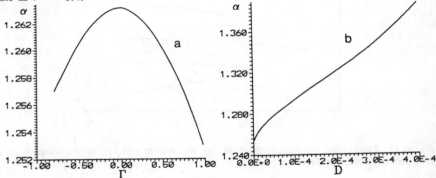

Fig.10.14. Bifurcation diagram of the system (10.92) for the second period doubling bifurcation: a - on the (Γ,a) parameter plane as $D = 10^{-5}$, b - on the (D,a) parameter plane as $\Gamma = -0.4$.

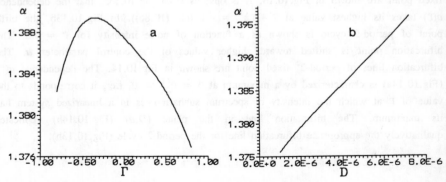

Fig.10.15. Bifurcation diagram of the system (10.92) for the third period doubling bifurcation: a - on the parameter (Γ,a) plane as $D = 10^{-6}$, b - on the parameter (D,a) plane as $\Gamma = -0.4$.

Consider now the bifurcation diagrams for the period-2^3 fixed point. The bifurcation line in the plane (Γ,a) is shown in Fig.10.15a. It takes its maximum at the parameter Γ value which corresponds approximately to $\Gamma_m^{(3)}$. The bifurcation line in the plane (D,a) is shown in Fig.10.15b.

Therefore, the bifurcation lines of birth of cycle with doubled period in the system of cumulant equations are defined by the presence of extrema corresponding to the values of the parameter Γ at which the intensities of subharmonics of the appropriate cycle have their maxima in the linearized system. Therefore, the correlation time of the noise becomes a control parameter which, when varied, provides bifurcation.

Colored noise effect at the threshold of transition to chaos. Now, we examine the influence of colored noise at the critical point of transition to chaos for a stochastic map (10.87). The point of transition to chaos can be identified with the first transition through zero of the largest Lyapunov exponent λ_1. Thus, to determine the critical point, the following equation must be solved:

$$\lambda_1(a_{cr},D,\Gamma) = 0. \qquad (10.93)$$

Under calculation, the noise parameters D and Γ were recorded and the critical point a_{cr} was found by varying the parameters D and Γ. The number of the map (10.87) iterations was determined by the accuracy of calculation of Lyapunov exponent ε (10^{-5}) and was, as a rule, $10^4 \div 10^5$. The results of calculation of the dependence $a_{cr}(\Gamma)$ for two values of

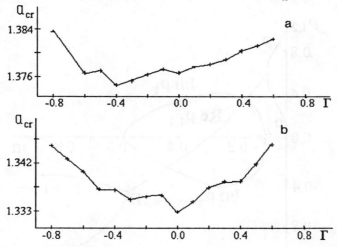

Fig.10.16. Critical parameter value a_{cr} versus the parameter Γ: a - $D = 0.0001$, b - $D = 0.001$.

noise intensity $D = 0.0001$ and $D = 0.001$ are shown in Fig 10.16. Observe from this figure that the dependence $a_{cr}(\Gamma)$ is nonmonotonous and has its peculiarities for values of Γ corresponding approximately to the characteristic frequencies of doubled period cycles. The dependence $a_{cr}(\Gamma)$ has its minimum at $\Gamma = 0$ (i.e., in the case of white noise). So, the increase of the noise correlation time provides the shift of the point of transition to chaos towards higher overcriticality values. Speaking otherwise, the system is stabilized by increasing the correlation time of external noise.

As the noise intensity D increases, the critical point a_{cr} is shifted towards lower values, i.e., the chaos is induced more earlier due to the increase of the noise intensity.

Numerical simulation and full-scale experiment using a differential system. The theoretical conclusions of our analysis of noisy map are experimentally verified as applied to the differential system (3.38). Figure 10.17 shows a typical evolution of the multipliers of a periodic solution near a parameter value at which a period-doubling bifurcation takes place in a fluid-mechanical system. In contrast to one-dimensional map, multipliers whose real part can pass through zero occur only in complex-conjugate pairs. Consequently, a bifurcation from a node to a focus, as some parameter is varied, will occur twice (the points A and B in Fig.10.17). Therefore it is not enough to set the multiplier equal to zero in the map. In connection with this, it is desirable to study the evolution of the spectra of the pertubed system (additive colored noise sources has been introduced in the first equation of the system (3.38)) as depending on parameter m.

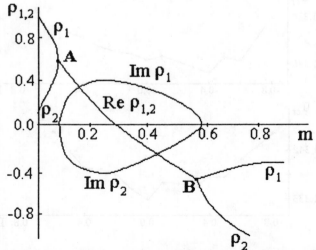

Fig.10.17. Characteristic multipliers $\rho_{1,2}$ versus the parameter m for the system (3.38) as $g = 0.3$.

Calculations show that a "node-focus" bifurcation (for which the parameter m passes through the point A, as in Fig.10.17) induce no broad-band peaks at the subharmonics in the spectrum. These peaks appear only for values of the parameter m that lie to the right of point B, where the multipliers are real and negative. The effect is more striking for the case, where the multiplier of the cycle, that is the largest in modulus, satisfies $0.5 < |\rho_1| < 1$. In order to elucidate the implication of Fig.10.18, we present power spectra for two values of parameter m: $m = 0.42$ (where multipliers are complex; see Fig.10.18a) and $m = 0.72$ (where the multipliers are real and negative; see Fig.10.18b). As it is clear from the plots, in the former case, the noise does not induce subharmonics in the spectrum. In the latter case, there are clearly discernible broad-band peaks near the frequencies $\omega_0/2$ and $3\omega_0/2$.

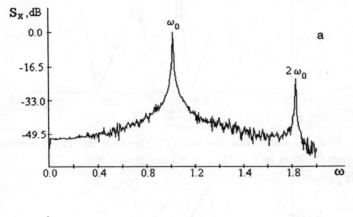

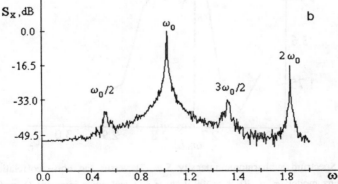

Fig.10.18. Power spectra for the system (3.38) as $g = 0.3$ with the intensity of additive white noise of $D = 0.0001$: a - $m = 0.42$, b - $m = 0.72$.

To reveal the effect of the noise correlation time a noise source was simulated using the Ornstein-Uhlenbeck process with the energy to remain constant when varying the correlation time. The results of calculation of the maximum of the power spectrum at the subharmonic frequency $3\omega_0/2$ as a function of correlation time τ_0 are shown in Fig.10.19 for the cycle T of the system (3.38) as $D = 10^{-4}$. The dependence has a striking maximum at $\tau_0 \approx (3\omega_0/2)^{-1}$. Similar dependence (see Fig.10.20) is obtained for the Lorenz system for the parameter values $\sigma = 10$, $r = 99.9$, $b = 8/3$, $D = 10^{-3}$. Thus, the intensities of the broad-band peaks of a subharmonic depend resonantly on the noise correlation time.

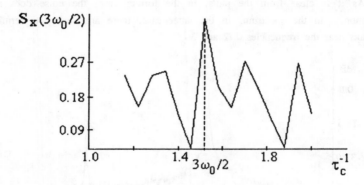

Fig.10.19. Spectrum subharmonic intensity $S_x(3\omega_0/2)$ versus the noise correlation time τ_c for the system (3.38) as $m = 0.72$, $g = 0.3$ and $D = 0.0001$.

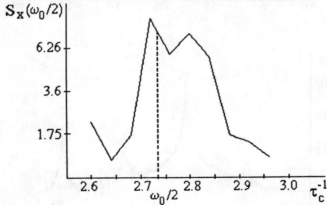

Fig.10.20. Spectrum subharmonic intensity $S_x(\omega_0/2)$ versus the noise correlation time τ_c for the Lorenz model: $\sigma = 10$, $r = 99.9$, $b = 8/3$ and noise intensity $D = 0.001$.

We also carried out a full-scale experiment using a radiophysical system that was modeled by the system (3.38) under conditions close to those used in the computer simulation. The results of measuring the power spectrum of the oscillations when varying the intensity and the correlation time of the noise reinforce our theoretical predictions and numerical data.

The above noise characteristics (intensity, correlation time) become control parameters for the system perturbed by noise. The effect of extention of the dimension of the control parameters space for the perturbed system, as compared to the original dynamical system, is of fundamental importance. From the general mathematical point of view, just this effect makes the phase transitions, induced by noise, possible and understandable. For example, a bifurcation of codimension higher than 1 can not be described by the bifurcation line crossing as analyzed one-parametrically. When considered two-parametrically, however, the bifurcation points of codimension 2 and more can appear on the bifurcation line of codimension 1.

CHAPTER 11

CHAOS STRUCTURE AND PROPERTIES IN THE PRESENCE OF NOISE

11.1. INTRODUCTION

Too modest place is assigned, as we see it, to the problem of fluctuation effects in the literature on dynamical chaos. This is accounted for, to some extent, by theoretical results obtained by Kifer and Sinai [11.1-11.4]. A proof of the theorem was constructed by Kifer [11.1]. According to this theorem, *small variations in the invariant measure* of a strange attractor are caused by its *small random perturbations*. By Ya.Sinai's expression, the intrinsic chaos of a system turns out to be stronger than that imposed from outside. These theoretical results are applicable, however, only to the model systems exhibiting hyperbolic attractors. Up to now, no real systems with hyperbolic attractors have been revealed. Chaotic attractors (and systems) can be classified according to Shilnikov as follows [11.5-11.6]:

1 - robust hyperbolic, or strange, attractors;

2 - Lorenz type attractors;

3 - quasihyperbolic attractors (quasiattractors).

The *Lorenz type attractors* are most close to the "real" strange ones. They have a property of hyperbolicity but they are *structurally unstable*. Realizable regimes of dynamical chaos are more often described by *quasihyperbolic attractors*. They are characterized by a large number of coexisting stable and unstable, regular and chaotic attractors, attracting and non-attracting homoclinic structures. When one varies parameters, these limit sets undergo an infinite number of different bifurcations. It is obvious that the appreciation of fluctuation effect in such systems is highly necessary.

From the viewpoint of the theory of stochastic process, the problem of studying the noisy chaos can be considered as Brownian motion of a chaotic system [11.7]. This problem is complicated by the fact that flow systems, exhibiting chaos, have the dimension of phase space $N \geq 3$. Thereby, analytical results are difficult to be obtained here. Numerical simulation is a basic investigation method which provides the main characteristics of the process.

The study of the influence of external noise on quasihyperbolic systems is of interest, as well, from the viewpoint of the qualitative theory of dynamical system. As known [11.8], in the vicinity of homoclinic trajectories, bifurcations take place which result in formation of denumerable sets of stable and saddle periodic trajectories. Thus, the exhaustive classification of motion types is impossible. Moreover, the invariant measure of a quasiattractor would have a denumerable set of singularities due to regular motions in the quasiattractor. The motion on the quasiattractor cannot be considered as ergodic complicating the consideration of such objects.

The definition of the notion of *"small noise"* is of principal importance in quasihyperbolic systems. If the stable limit cycle is an attractor in the system, then the noise intensity to the oscillation intensity ratio may be considered as a small parameter. The latter, when being introduced, remains in force when one varies the control parameters in the finite region of their values.

A quite different picture is observed in quasihyperbolic systems. The *weak noise perturbation*, as applied to one of the system's attractors, *is basically not small* relative to any other coexisting regime. Moreover, the highly remarkable variations of resultant dynamics can be caused, as compared with unperturbed system, by the noise with intensity as small as desired due to a denumerable set of bifurcations of attractors in the system as the control parameters are varied.

The rich spectrum content of a quasiattractor is defined by a denumerable set of periodic trajectories in its basins of attraction. Speaking otherwise, the regimes of *quasihyperbolic chaos* are described by a set of *time scales* corresponding to the peaks in the power spectrum at the characteristic frequencies of the system. The complicated phenomena of resonance type can be induced by varying the time scale of perturbations, for example, the correlation time.

The influence of external noise on the nonlinear dynamics of a system leads to effects of two types: a shift of the bifurcation diagram of the system (the presence of noise is qualitatively equivalent to a change in the control parameter of the system) and the appearance of new regimes that are absent from the unperturbed system (the so-called noise-induced transitions). For one-dimensional systems, the phenomenon of the last type is possible in the case of multiplicative noise perturbation [11.9]. In quasihyperbolic systems, $N \geq 3$, the noise-induced transitions are also feasible in the case of additive noise because of the noise-induced destruction of the separatrixes of hypersurface that separates regions with different types of motion in the phase space of the system.

Two situations are to be discerned: variation of parameters in the presence of external noise with the given intensity and variation of the noise intensity at the fixed parameters of the system. If we follow, for example, a sequence of period doubling bifurcations in the presence of noise with constant intensity, then the chain of doubling bifurcations observed would turn out to be finite and transition to chaos would occur for a lower value of the control parameter. The typical case of a shift of bifurcation diagram, followed by its fine-scale structure vanishing, is realized. The phenomenon described is not realized as the control parameters of the system are fixed when the noise intensity is increased. No cascade of the period doubling bifurcations can be induced by increasing the intensity of noise[1]!

At the same time, there is a number of other bifurcation phenomena which are qualitatively reproduced both by varying the parameters with no noise and by increasing the intensities of external fluctuations at the fixed parameter values. *The noise intensity* plays here the role of a *control parameter* in a certain sense [11.9]. The above considerations are useful to refer to when interpreting the results presented below.

11.2. REGIMES OF DYNAMICAL CHAOS UNDER THE INFLUENCE OF NOISE WITH FINITE INTENSITY

The variation of the intensity of external noise can result both in the stabilization of the system, i.e., in the decrease of the largest Lyapunov exponent, and in the increase of chaosity degree [11.10-11.11]. As a rule, if no bifurcational transitions are induced by noise, then the main effect is the stabilization of chaos by noise which is manifested by the decrease of the largest Lyapunov exponent. The latter can become even negative with a high noise intensity [11.12].

Consider in more detail the structure and characteristics of the Lorenz attractor ($r = 28$, $\sigma = 10$, $b = 8/3$) in the presence of δ-correlated additive noise with intensity D (see equations (10.73)). We begin with analyzing the structure of two-dimensional stationary probability density $P(X,Z)$. The results of numerical simulation are presented in Fig.11.1, where the central projections of $P(X,Z)$ are shown for $D = 0$, 1.0, 10.0, and in Fig.11.2, where the isolines of two-dimensional probability density are indicated for the appropriate values of the noise intensity. Fig.11.1a and Fig.11.2a correspond to the deterministic case $D = 0$. The surface of $P(X,Z)$ has two characteristic craters corresponding to saddle-foci: $X^0_{1,2} = \pm [b(r-1)]^{1/2}$, $Z^0_{1,2} = r - 1$ in whose vicinity chaotic oscillations take place. As the noise intensity grows (Figs 11.1b and 11.2b), the fine-scale structure of the attractor is smoothed. With a high noise intensity (Figs 11.1c and 11.2b), the craters corresponding to the saddle-foci disappear. The two-dimensional probability density $P(X,Z)$ acquires a homogeneous structure. The shape of the structure is defined by an unstable manifold of saddle-node at the origin. It should be noted, that the residence time of a phase trajectory on each of symmetric Lorenz

---

[1]This assertion is valid as applied both to single doubling bifurcations and to Andronov-Hopf bifurcations.

attractor parts is decreased with the increase of noise intensity.

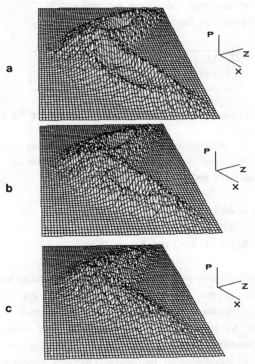

Fig.11.1. Central projections of the stationary two-dimensional probability density $P(X,Z)$ for the Lorenz model with noise as the parameters are $(\sigma,b,r) = (10,8/3,28)$: a - $D = 0$, b - $D = 1.0$, c - $D = 10.0$.

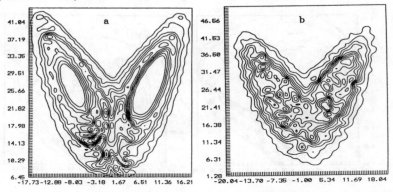

Fig.11.2. Isolines of the stationary two-dimensional probability density $P(X,Z)$ for the Lorenz noise model as the parameters are $(\sigma,b,r) = (10,8/3,28)$: a - $D = 0$, b - $D = 10.0$.

Then, we examine the largest Lyapunov exponent λ_1 as a function of the noise intensity. The results of numerical calculation are tabulated in Table 11.1 and show the decrease of chaosity degree with the growth of noise intensity.

Table 11.1.

D	0.0	0.1	1.0	10.0	20.0	30.0
λ_1	0.91	0.89	0.79	0.76	0.70	0.67

The largest Lyapunov exponent as a function of noise intensity for the Lorenz attractor

Let us consider the influence of external white noise on the chaotic attractor of a modified oscillator with inertial nonlinearity (MOIN). The stochastic equations for the MOIN in the case of additive noise are defined as follows:

$$dx/dt = mx + y - xz + \xi_1(t), \quad dy/dt = -x,$$
$$dz/dt = -gz + gI(x)x^2 + \xi_2(t), \qquad (11.1)$$
$$I(x) = 1, \, x > 0 \text{ and } I(x) = 0, \, x \leq 0.$$

We choose the parameter values $m = 1.10$ and $g = 0.3$ corresponding to a period-two band Feigenbaum's attractor in the absence of noise [11.13]. Fig.11.3 illustrates the variations of the two-dimensional probability density $P(X,Y)$ structure when increasing the noise intensity D. As seen from the figure, the attractor bands merge with the increase of noise intensity, followed by further smoothing of the distribution structure. The results of calculation of the largest Lyapunov exponent λ as a function of the noise intensity D are tabulated in Table 11.2 and show the decrease of λ_1. As $D \geq 0.1$, the Lyapunov exponent becomes negative.

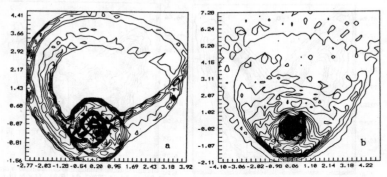

Fig.11.3. Isolines of the stationary two-dimensional probability density $P(X,Y)$ for the system (11.1) with the parameters $(m,g) = (1.1, 0.3)$: a - $D = 0.0$, b - $D = 0.01$.

Table 11.2.

D	0.0	0.001	0.1	0.2	0.3
λ_1	0.026	0.024	0.002	-0.04	-0.005

The largest Lyapunov exponent as a function of noise intensity for MOIN $m = 1.10$, $g = 0.3$.

Thus, the increase of noise intensity leads to the decrease of the chaosity degree.

For illustration, we present the data on construction of the two-dimensional probability density from the experimental realizations of a radiophysical MOIN model. The lay-out of experimental set is shown in Fig.11.4. A pair of realizations, $x(t)$ and $y(t)$, from MOIN influenced by noise oscillator, was digitized using an analog-digital converter (ADC) and set into computer. The external noise was applied to the MOIN in such a way as to correspond to the introduction of the Langevin's source only into the first equation of the mathematical model (11.1). The probability densities were constructed on the assumption of the process to be ergodic. Note in this connection that the fixed sampling rate is of no importance for calculating the stationary probability characteristics if a realization is sufficiently long.

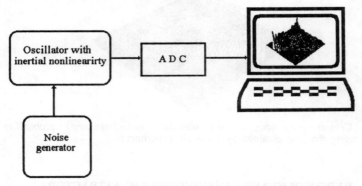

Fig.11.4. Lay-out of experimental set.

To study the noise influence on chaotic self-oscillations (under numerical simulation), the regimes of two-band attractor and developed chaos were chosen. The calculated results for the two-dimensional probability density are shown in Fig. 11.5. The figures illustrate the process of smoothing of the structure of the two-dimensional probability density as the intensity of external noise is increased.

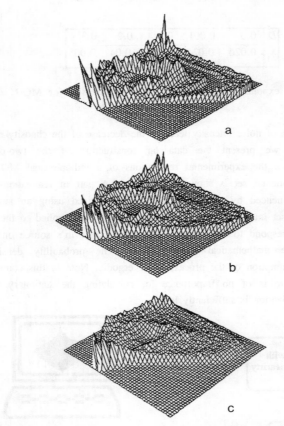

Fig.11.5. Central projections of the stationary two-dimensional probability density calculated using the data obtained by full-scale experiments.

11.3. HYPERBOLIC AND QUASIHYPERBOLIC ATTRACTORS UNDER THE INFLUENCE OF COLORED NOISE

Let us study the influence of correlated noise on the regimes of dynamical chaos with the Lorenz model as an example. As it was already mentioned, the phase transition "the Lorenz attractor - quasiattractor" is realized in the space of control parameters of this system [11.14].

For comparison, we consider two chaotic regimes in the Lorenz system: (1) $r = 28$, the Lorenz attractor, and (2) $r = 210.0$, quasiattractor. In Fig.11.6, the Poincaré maps are shown for the former (Fig.11.6a) and for the latter (Fig.11.6b) attractors in the absence of fluctuation ($D = 0$). The region of zero derivative is seen in Fig.11.6b showing the type of attractor. The appropriate power spectra $S_x(\omega)$ are presented in Fig.11.7. In the case of the Lorenz attractor (Fig.11.7a), the power spectrum has the well-known form and no sudden peaks. Such a structure represents the fact that the Lorenz attractor has no stable regular trajectories. The *spectrum* of *quasiattractor* (Fig.11.7b) is characterized by *sudden peaks* on the background of continuous noise spectrum component.

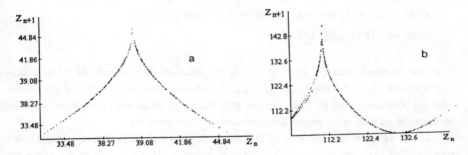

Fig.11.6. The Poincaré map of the Lorenz model attractors in the absence of noise: a - $(\sigma,b,r) = (10,8/3,28)$ (the case of Lorenz type attractor), b - $(\sigma,b,r) = (10,8/3,210)$ (the case of quasiattractor).

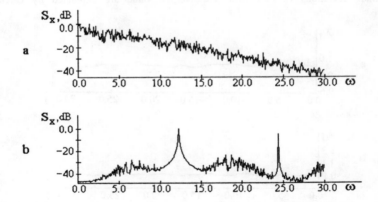

Fig.11.7. The power spectra of the Lorenz model in the absence of noise: a - $(\sigma,b,r) = (10,8/3,28)$ (the case of Lorenz type attractor), b - $(\sigma,b,r) = (10,8/3,210)$ (the case of quasiattractor).

In the case of *additive colored noise*, the stochastic equations for the Lorenz model are written in the form of (10.73) where random sources $\xi_i(t)$ are simulated by the SDE as follows:

$$d\xi_i/dt = \gamma\xi_i + (D\gamma)^{1/2}n_i(t), \quad <n_i(t)n_j(t + s)> = \delta_{ij}\delta(s), \tag{11.2}$$

and define the Ornstein-Uhlenbeck processes with the correlation time $\tau_c = 1/\gamma$. When analyzing the influence of multiplicative noise, the system's behavior is defined by the following SDE:

$$dx/dt = [\sigma + \xi_1(t)](y - x), \quad dy/dt = [r+\xi_2(t)]x - y - xz,$$
$$dz/dt = -[b+\xi_3(t)]z + xy.$$

It should be noted, that the form of writing the equations for the colored noise sources $\xi_i(t)$ provides their intensities to be equal to D and independent of the correlation time τ_c. This allows us to study, in the pure state, the phenomena associated with the influence of colored noise on the characteristics of chaotic regime.

Let us consider the influence of colored noise on the chaosity degree of the Lorenz attractor. We calculate the largest Lyapunov exponent λ_1 as a function of the parameter $\gamma = 1/\tau_c$ with constant noise intensity $D = 1.0$. The calculated results are presented in Fig.11.8 and show the largest Lyapunov exponent to be essentially *independent* of the noise correlation time both for additive (Fig.11.8a) and for multiplicative (Fig.11.8b) noise. Small deviations ($\pm 1\%$) from the average value are explained by calculation errors.

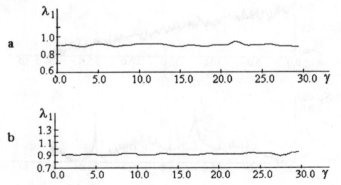

Fig.11.8. Largest Lyapunov exponent as a function of the noise correlation time in the regime of Lorenz type attractor; the parameters $(\sigma,b,r) = (10,8/3,28)$: a - additive noise $D = 1.0$, b - multiplicative noise $D = 0.1$.

The dependence of the largest Lyapunov exponent on the noise correlation time is qualitatively different for the regime of quasiattractor ($r = 210$). It is shown in Fig.11.9. λ_1 *depends strongly* on τ_c both for additive (Fig.11.9a) and for multiplicative (Fig.11.9b) noise influence. The deviations of the average λ_1 value are as great as

± 20% within the region of parameter γ under study, i.e., they are an order higher than that in the case of the Lorenz attractor.

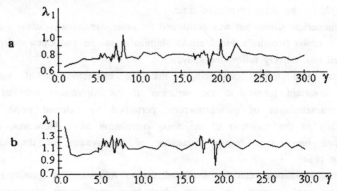

Fig.11.9. Largest Lyapunov exponent as a function of the noise correlation time in the regime of quasiattractor; the parameters are $(\sigma,b,r) = (10,8/3,210)$: a - additive noise $D = 1.0$, b - multiplicative noise $D = 0.1$.

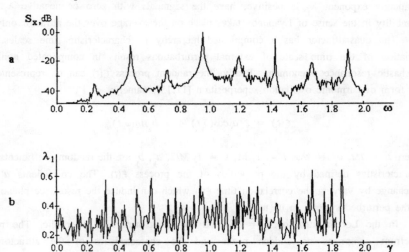

Fig.11.10. The power spectrum (a) and the dependence of the largest Lyapunov exponent on the noise correlation time for the system (11.1) as the parameters are $(m,g,D) = (1.1, 0.3, 0.001)$.

Further example is the effect of external colored noise on a quasiattractor in the MOIN system. The regime of the MOIN oscillations is defined, when influenced by additive colored noise, by equations (11.1) and (11.2). We examine a regime where $m = 1.1$, $g = 0.3$, and $D = 0.001$. In Fig.11.10, the power spectrum $S_x(\omega)$ in the absence of noise

(Fig.11.10a) and the largest Lyapunov exponent λ_1 as a function of the parameter $\gamma = 1/\tau_c$ (Fig.11.10b) are presented. As seen from the figure, the degree of chaosity *depends strongly* on the noise correlation time.

Similar numerical simulation was performed for other dynamical systems exhibiting a quasihyperbolic chaos (Rössler's and Marioka-Shimizu's systems and Chua's circuit). The results obtained qualitatively follow the above.

The data cited testify convincingly that the characteristics of the *Lorenz attractor* are invariant relative to the variation of the *correlation time* of external noise. The characteristics of *quasiattractors*, perturbed by external noise, show a *strong sensitivity* to the variation of the *noise correlation time*. Moreover, the noise correlation time becomes a control parameter of the system for the regimes of quasihyperbolic chaos.

The above phenomena can be accounted for as follows. The quasiattractors are characterized by stable periodic trajectories which have small basins of attraction. The small perturbations of these regimes lead to their destruction. Therefore, the phase trajectories of a quasiattractor, along with the hyperbolic regions where the largest Lyapunov exponent λ_1 is positive, have the segments with zero or negative λ_1. The instability in the sense of Lyapunov takes place on the average over the attractor only.

The quasiattractor has a complicated hierarchy of characteristic time scales. The variation of the time scale of external perturbations results in complicated nonlinear stochastic resonance phenomena. Really, the random process $\xi(t)$ can be represented in the form of harmonic oscillation superposition [11.15], namely:

$$\xi(t_i) = \sum_k a_k cos(\omega_k t_i) + \sum_k b_k sin(\omega_k t_i),$$

where $t_i = i\Delta t$, $\omega_k = k\Delta\omega$, $i = 1, M$, $k = 1, M/2$, a_k, b_k are the random coefficients with characteristics defined by the properties of the process $\xi(t)$. The coefficients a_k and b_k change by varying the correlation time $\xi(t)$ which can induce the resonance phenomena in the perturbed system, in its turn.

In the Lorenz attractor ($r = 28$), all the trajectories are hyperbolic. The power spectrum comprises no sudden peaks and the characteristics of the attractor are invariant relative to the variation of the correlation time of external noise.

11.4. BIFURCATIONS OF CHAOTIC ATTRACTORS IN THE PRESENCE OF NOISE

Let us explore the influence of external noise on the bifurcation of band merging of chaotic attractors in the MOIN system [11.13].

We consider the system of stochastic differential equations (11.1) and perform a numerical simulation at the characteristic points of the parameter plane with different noise intensities[2]. In Fig.11.11, the results of calculation of the largest Lyapunov exponent are indicated as a function of the noise intensity. It is seen that, depending on parameters, the noise increase provides both λ_1 increasing and its decreasing to zero. The noise can induce the transition to a regular motion, as well as to provide a higher chaosity degree. What is the reason of these principle distinctions in the response of the system to the noise action? As it was stated in detail, the response of the system to external noise is defined by the nature of the bifurcation diagram of the unperturbed system. If no internal bifurcations of attractor occur under the action of noise, the degree of mixing weakly depends on the fluctuation intensity. The noise is responsible for transfer of a phase trajectory into the basin of attraction of a cycle in the case of the proximity in parameters to the point of tangency bifurcations. The transition of "chaos-order" type is realized. In the cases, where the bifurcational transition in the chaos from an attractor with one structure to the attractors with other structure takes place, when affected by noise, the increase or decrease of degree of mixing is possible. One may argue the experimental results shown in Fig.11.11 to be a direct consequence of a shift of bifurcation diagram caused by the noise perturbation of dynamical system.

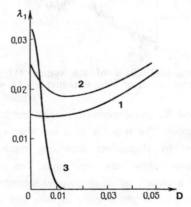

Fig.11.11. The dependence of largest Lyapunov exponent on the noise intensity for various values of parameter m: 1.09 (1); 1.10 (2); 1.16(3).

Let us consider in detail one of possible internal attractor bifurcations illustrating the above, and namely, that of band merging of the chaotic attractor due to

---

[2]The magnitudes of additive noise intensities, which correspond to real experiment conditions, were chosen from comparing the full-scale experiment and numerical simulation data when studying a sequence of period doubling bifurcations.

the influence of weak additive noise. Fig.11.12 shows the Poincaré maps and the power spectra of chaotic attractors of the system (11.1) for $m = 1.09$ and for two values of the noise intensity D. Noise with $D = 0.01$ induces the merging bifurcation of a period-four strange attractor band into a period-two band. From the physical point of view, low-intensity fluctuations are equivalent here to a slight increase in the parameter m. The intensity increase up to $D = 0.5$ produces a period-one chaotic attractor band (Fig.11.12,b). Noise induces a bifurcation that does not occur in the dynamical system (11.1) in the absence of noise.(Fig.11.12b)[3].

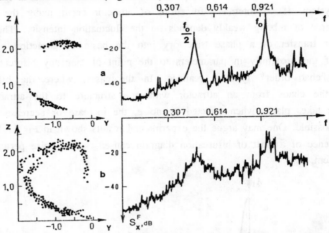

Fig.11.12. Regimes of dynamical chaos of the system (11.1) with the parameters $(m,g) = (1.09,0.3)$ for various values of noise intensity: a - $D = 0.01$, b - $D = 0.05$.

A property being typical for quasihyperbolic systems and clearly manifested in this experiment is to be highlighted. The response of a nonlinear system to isotropic noise perturbation is characterized by fluctuations decreasing transversely to the unstable manifold due to a predominant phase flow contraction. The noise effect makes itself evident in the expansion of a chaotic set along unstable separatrixes. As a result, when influenced by noise, the attractor in phase space expands just in the direction towards the unstable manifolds. In the vicinity of homoclinic bifurcation points, it causes the phenomena being qualitatively equivalent to the bifurcations of the attractor of deterministic system (the effect of attractor bands' merging in the example under study).

The influence of fluctuations on the band attractor bifurcations was experimentally

---

[3]The transition to a period-one attractor band is excluded when analyzing one-parametrically for $D = 0$ in the system (11.1). It becomes real, however, when we examine it two-parametrically. This again shows the effect of the bifurcation diagram shift in the whole.

investigated on a real MOIN as the level of external noise was controlled. The calculated results were entirely confirmed by experiments [11.13]. Furthermore, the above results are qualitatively reproduced, as well, by Henon model for which we succeeded in analyzing the chaotic band merging using the two-parametric approach.

The statistical properties of a dynamical system can be analyzed on the basis of the dependence of the autocorrelation function (ACF) $\Psi(\tau)$ on the time τ. For chaotic attractors, the ACF tends to zero as time elapses. In regimes with a well-developed stochastic nature, the ACF decreases exponentially, providing evidence that the system is close to being a finite Markov's chain [11.4]. The exponent is related to the correlation time τ_c, which should depend on the extent of mixing. For hyperbolic systems, τ_c is inversely proportional to the Kolmogorov-Sinai's entropy [11.16-11.17] and can be estimated from the following expression:

$$\tau_k = \left[\sum_{i=1}^{l} \lambda_i^+ \right]^{-1} \tag{11.3}$$

where λ_i^+ are the positive Lyapunov exponents for the chaotic trajectory. Is there any interrelation between the correlation time τ_c and the Kolmogorov-Sinai's entropy of chaotic attractors in dynamical systems with quasihyperbolic properties, and, if so, what is the nature of this interrelation? One could cite several examples which imply that the dependence (11.3) does not hold for quasiattractors.

Let us examine the interrelation between the correlation time τ_c and the value of the positive Lyapunov exponent λ^+ for the attractors of a Henon map in attractor bands' merging bifurcations taking into account fluctuations [11.18]. The Henon map has the form

$$x_{n+1} = 1 - ax_n^2 + y_n + \xi_1(n), \quad y_{n+1} = bx_n + \xi_2(n), \tag{11.4}$$

where $\xi_{1,2}(n)$ is the δ-correlated noise with intensity D. This map is a typical quasihyperbolic system with Feigenbaum's attractor. The envelope of the dependence $\lambda^+(a)$ satisfies the universal law

$$\lambda_1 = c(a - a^*)^\gamma, \quad \gamma = ln2/ln\delta, \tag{11.5}$$

in which, for $b = 0.3$, we have $a^* = 1.058$ and $c = 0.836$. The dependence (11.5) is well substantiated by experimentation up to the value $a \leq 1.16$.

If we assume that the correlation time is a function of a, then the relation (11.5) determines the functional dependence $\tau_c(a)$ which is shown by solid line in Fig.11.13. The circles in this figure refer to the results of direct calculations of $[\lambda^+(a)]^{-1}$.

Calculations of the dependence of the ACF on the parameter a reveal a clear discrepancy between the obtained results and the predictions. Up to the value $a \leq 1.155$ (the time at which a well-developed period-one band of an attractor is produced), the ACF remains approximately periodic, and the envelope does not tend toward zero. This

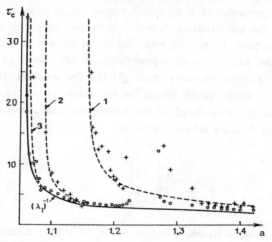

Fig.11.13. Correlation time τ_c as a function of the parameter a for Henon map in the absence of noise: $1 - a^* = 1.155$; $2 - a^* = 1.088$; $3 - a^* = 1.064$.

behavior stems from the complexity of the discrete sequence $\{x_n\}$, which contains both periodic and chaotic motion components. An attractor, consisting of a set of non-intersecting chaotic zones, can be described by a superposition of periodic oscillations (between zones) and noise [11.19], namely:

$$x_n = \sum_{\nu=1}^{p} A_\nu exp(j\omega_\nu n) + \xi(n). \tag{11.6}$$

A consequence of (11.6) is the particular nature of the distribution of the spectrum power density, which contains δ-function bursts at frequencies ω_ν against the background of a continuous noise spectrum. According to the Wiener-Khinchin theorem, the ACF having the form

$$\Psi_x(\tau) = \lim_{k \to \infty} \frac{1}{k} \sum_{n=1}^{k} [x_k - \bar{x}][x_{k+\tau} - \bar{x}] \tag{11.7}$$

is a superposition of periodic and δ-correlated components.

We use a procedure of eliminating the periodic motion components and determine the nature of the interrelation between $\Psi_x(\tau)$ and the Kolmogorov-Sinai's entropy λ^+ for individual chaotic zones of the attractor [11.19]. For this purpose, we subject a sequence of points belonging to some specific chaotic zone for statistical analysis depending on the number of periods of the attractor band. For example, in the case of a period-two band we need to store in the computer memory a sequence of every second points of iteration of the original mapping (11.4).

The calculated results are shown in Fig.11.13. The values of τ_c (shown with crosses) were calculated as functions of the parameter a, with account for the sequence of bands' merging bifurcations. The dependence $\tau_c(a)$ has sharp bursts at the points near 1.064, 1.088, and 1.155, where τ_c increases abruptly. In the intervals between these bursts, τ_c decreases exponentially, tending toward an approximating function $\tau_c = [\lambda^+]^{-1}$ (solid line). It was found out that this functional dependence $\tau_c(a)$ has discontinuities strictly at the bands' merging bifurcation points, where the chaotic zones of the attractor merge.

In the intervals between the bands' merging bifurcation points, the correlation time can be highly accurately approximated by the universal law

$$\tau_c = c_i(a - a_i^*)^{-\gamma} \tag{11.8}$$

where a_i^* are the critical points of the bands' merging bifurcations, and c_i are the corresponding constants. Calculations show that for period-eight, period-four, and period-two attractor bands these parameters have the following values: $a_8^* = 1.064$, $c_8 = 1.04$, $a_4^* = 1.088$, $c_4 = 1.07$, $a_2^* = 1.155$, and $c_2 = 1.79$. The approximation (11.8) is shown by dashed line in Fig.11.13 confirming the arguments mentioned above.

The results obtained here are interesting from several points of view.

1. It can be stated that the internal bifurcations of attractors have a fundamental influence on the statistical properties of chaos, which are determined not only by the Kolmogorov-Sinai's entropy but also - and to a greater extent - by the creation (or disappearance) of periodic motion components. The correlation time increases abruptly at the bands' merging bifurcation points.

2. Within chaotic zones of an attractor, the correlation time depends in a universal way on the parameter (see (11.8)).

3. Near the bands' merging bifurcation points, a typical picture is observed in the distribution of energy over the spectrum, which reflects the appearance of low-frequency components being characteristic of flicker noise.

The threshold of chaotic regime is shifted towards the lower parameter a values by introducing small additive perturbations (11.4). As $D = 0.01$, the critical value of a^* in (11.5) becomes as large as 1.034. Dependence $\lambda_1^D(a)$ maintains the form of (11.5) and is approximated by the formula

$$\lambda_1^D(a) = c(a - a^*)^k, \tag{11.9}$$

where k no longer agrees with the universal constant γ in (11.5). So, for $b = 0.3$ and $D = 0.01$, the constants of (11.7) take the values $c = 1.57$, $k = 0.82$, $a^* = 1.034$. The picture, being qualitatively identical to a case where $D = 0$, is observed in the evolution of regularities of ACF $\Psi_x(\tau)$. At the bands' merging bifurcation points, the correlation time has discontinuities followed by an exponential decrease of dependence $\tau_c(a)$ which tends to $\lambda_1^{-1}(a)$.

The results of numerical simulation are indicated in Fig.11.14 for $b = 0.3$ and $D = 0.01$. The designations are similar to the case of Fig.11.16 where $D = 0$. Two exponential $\tau_c(a)$ branches are shown which correspond to the period-two and period-one bands of Henon attractor. The plots presented are well described by the dependences of (11.9) type where the values of all constants are defined by noise intensity.

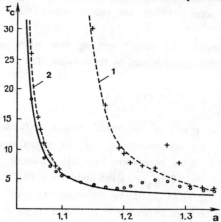

Fig.11.14. Correlation time τ_c as a function of the parameter a for Henon map in the presence of noise with intensity $D = 0.01$.

In the whole, one may say that the regularities in $\tau_c(a,D)$ behavior are more pronounced in the presence of fluctuations. Numerical data do not fit the approximating theoretical curves (11.9). Since the number of the chaotic attractor bands' merging bifurcations is restricted by the noise perturbation of the system, the number of discontinuities in the dependence $\tau_c(a)$ is limited by the finite fluctuation intensities (Fig.11.13) unlike the case where $D = 0$ (Fig.11.14).

The regularities stated are fulfilled not only for a Henon map but they are also typical for the systems comprising saddle-focus quasiattractor appearing via a cascade of Feigenbaum's doublings. Particularly, the above phenomena were also observed in a number of special experiments for a bifurcation of the period-four attractor band into the period-two attractor band in the oscillator (3.38) [11.16].

11.5. TRANSITIONS IN CHAOTIC SYSTEMS INDUCED BY NOISE

In the present paragraph, noise effects will be considered which lead to regimes which are not observed in the appropriate unperturbed systems. One of such examples has been discussed in Chapter 4 where the modulation intermittency was studied.

The noise induced transitions will be considered for coupled logistic maps, the Lorenz system, and a model of laser with nonlinear absorption as examples.

Chaos induced by noise in systems having robust homoclinic structures. The more interesting phenomena, causing the structures of attractors to be qualitatively rebuilt, are observed in the presence of homoclinic structures in the system. As an example, we consider the results of the numerical analysis of a stochastic discrete system that describes the interaction of two Feigenbaum's oscillators when influenced by additive δ-correlated noise with intensity D [11.20]:

$$x_{n+1} = 1 - ax_n^2 + \gamma(y_n - x_n) + \xi_1(n),$$
$$y_{n+1} = 1 - ay_n^2 + \gamma(x_n - y_n) + \xi_2(n).$$
(11.10)

The bifurcation diagram of unperturbed system (11.10) in the vicinity of resonance 2:5 was discussed earlier (see Fig.7.8). The bifurcation line l_h of Fig.7.8 corresponds to the appearance of a structurally unstable homoclinics where the unstable manifolds of a saddle five-cycle on the invariant curve of the map (7.7) are tangent to their stable manifolds. In the resonance region bounded by lines l_1 and l_0, one of the five-cycles of the system is always stable (both below and above the line l_h !). The LCE spectrum, corresponding to the stable five-cycle, consists of two negative exponents.

We shall examine the characteristics of the regimes under the action of noise with relatively low intensity $D \ll 1$. The signature of the LCE spectrum of a resonant stable cycle on the invariant curve in the region being close to the base of a synchronization tongue is not sensitive to noise of rather high intensity ($D \approx 0.1$), and that of the type "-","-" is retained. The characteristics of motion correspond to the classical situation of broadening of the spectrum oscillation lines under the action of noise. Near the line of non-robust homoclinics, the picture changes qualitatively. The low-intensity noise induces bifurcation in the signature of the LCE spectrum: "-", "-" $\Rightarrow$ "+", "-", indicating the development of the exponential instability of individual trajectories in the attractor. The results are shown in Fig.11.15,b, where the dependence $\lambda_1(D)$ of the largest Lyapunov exponent is presented for two points in the parameter plane, above and below the line l_h. It can be seen that in the region above the line l_h, which corresponds to the presence of robust homoclinic trajectories, the low-intensity noise $D \approx 0.005$ induces chaotic behavior in the system, that corresponds to the chaotic attractor regime (curve 1). Stochasticity below but close to line l_h arises at somewhat larger values of noise intensity (curve 2).

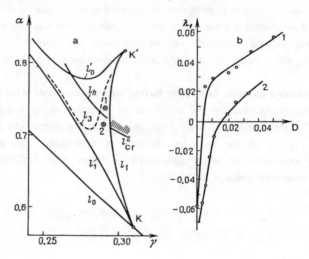

Fig.11.15. a - The bifurcation diagram of map (11.10): l_0 - the line of the invariant curve birth, l_1 - the lines of saddle-node bifurcations, l_3 - the line on which stable cycle multipliers become complex-conjugate, l_0' - the Hopf bifurcation line for the stable cycle, l_h - the line of structurally unstable homoclinics, l_{cr}^2 - a segment of the line of destruction of invariant curve outside the resonance; b - the dependence $\lambda_1(D)$ for the points 1 and 2 on the bifurcation diagram.

Let us examine the characteristics of chaotic regimes induced by fluctuations. Fig.11.16,a shows a phase portrait of the system (11.10) for $a=0.73$, $\gamma=0.292$, and $D = 0.001$ (see curve 1 in Fig.11.15,b). The small circles show a stable five-cycle for $D = 0$. Fig.11.16,b shows a stable cycle and a stochastic attractor for $D=0.01$. The set of characteristics that we have obtained during numerical simulation (the continuous spectrum and the decrease of the ACF, the presence of positive exponents in the LCE spectrum) is evidence in favour of generation of a stochastic regime induced by noise.

Similar calculations were carried out for various dynamical systems. It was found that the effects we are studying occur near robust homoclinic trajectories of saddle limiting cycles, near saddle points on invariant closed curves, and near saddle cycles on two-dimensional tori. Noise-induced transition to stochasticity has also been substantiated by the full-scale experiment on the MOIN.

These results are of general nature and indicate that regular attractors with robust homoclinic trajectories are not realized experimentally. What is actually seen under the action of relatively low-intensity fluctuations are chaotic regimes of oscillation, which by their characteristics, are not distinguished from the regimes of dynamical chaos[4].

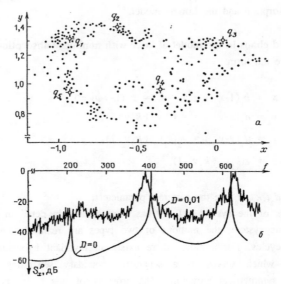

Fig.11.16. a - The Poincaré map of system (11.1) for $D = 0$ (points q_1-q_5) and $D = 0.01$ (chaotic set); b - The corresponding power spectra of the processes.

The exponential instability of the system is an experimental criterion to diagnose the structure of its phase trajectories in cases where the homoclinic points are difficult, for any reason, to be calculated directly but their existence is assumed. In particular, the experimental and numerical investigations of the dynamics of a non-autonomous oscillator in the resonance regions have provided the results which are qualitatively similar to the above, when influenced by external noise. These data can be considered as a consequence of the robust saddle cycle homoclinics on the resonant torus which was stated theoretically.

Invariant hypersurface destruction in the phase space under the influence of fluctuations. Interaction of attractors. Dynamical systems, which possess some symmetry, are characterized by invariant hypersurfaces being present in the phase space. The

---

[4]The realization of transition to chaos via Landau-Hopf scenario seems to be excluded due to homoclinic structures of cycles appearing on low-dimensional tori.

hypersurface cannot be crossed by any phase trajectory of deterministic system. Such *crossings* become *possible* in the *presence* of *fluctuations*. They induce the *attractors' interaction phenomena* and can lead to chaotic behavior.

As an example, consider a dynamical system that simulates the dynamics of a laser with nonlinear absorption and the Lorenz model.

Noise induced chaos in the model of laser with nonlinear absorption. The equations for the model have the form

$$\begin{aligned}
x' &= b_1(1-x-v_1z^2), \quad y' = b_2(1-y-av_2z^2),\\
v_1' &= d_1(x-v_1) - v_1(k_1v_1+k_2v_2-1),\\
v_2' &= d_1(y-v_2) - v_2(k_1v_1+k_2v_2-1),\\
z' &= z(k_1v_1+k_2v_2-1),
\end{aligned} \qquad (11.11)$$

where b_1, b_2, d_1, d_2, k_1, k_2 are the system's parameters.

The dynamics of the system (11.11) has been studied in detail in [11.11]. In the phase space of this system, the motions of two types are realized; namely, stationary points and limit cycles. Furthermore, there exists an invariant hypersurface $z = 0$ in the phase space which serves as a separatrix boundary separating the basins of attraction of two symmetrical attractors. The crossing of this boundary by any phase trajectory is excluded. The regimes of dynamical chaos are absent in the system (11.11). Introducing the additive noise into the model's equations leads to a qualitative change of the motion. The noise of low intensity $D = 0.001$ actually provides no variation of the motion nature leading only to a wider spectrum line of Fig.11.17 (curve 1). Sharp peaks remain in the spectrum which correspond to the frequency of limit cycle and its harmonics. With the growth of noise intensity up to $D = 0.01$, the noise-induced transition takes place which is diagnosed both by the phase portrait (trajectories cross the invariant hypersurface $z = 0$) and by the shape of the power spectrum of Fig.11.17

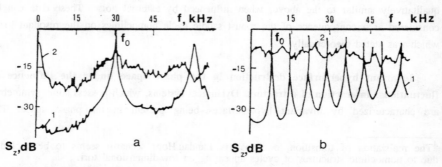

Fig.11.17. The power spectra of the system (11.11): $1 - D = 0.001$, $2 - D = 0.01$.

(curve 2). The spectrum takes the shape that is typical for chaotic regime. The presence of a low frequency component is evidence of intermittency between the attractor parts symmetric with respect to $z = 0$. The calculations of the LCE spectrum have corroborated the presence of the noise induced dynamical chaos in the system.

Noise-induced transition in the Lorenz system. The noise-induced transition in the regions of regular motion regimes has been studied in [11.24]. Here we study the noise-induced transition in the region of existence of a quasiattractor for the parameter values of $\sigma = 10$, $r = 210$, and $b = 8/3$. The stochastic differential equations for the Lorenz system are defined by the system (10.73).

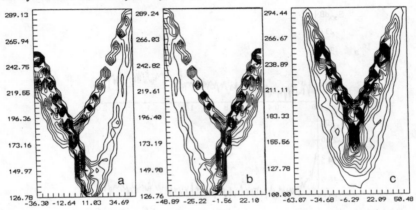

Fig.11.18. Isolines of the stationary two-dimensional probability density $P(X,Z)$ for the Lorenz model with noise, at the parameters $(\sigma,b,r) = (10,8/3,210)$, noise intensity D and initial point : a - $D = 0$, $(X_0,Y_0,Z_0) = (40,1,100)$, b - $D = 0$, $(X_0,Y_0,Z_0) = (40,1,100)$, c - $D = 1.0$, $(X_0,Y_0,Z_0) = (40,1,100)$.

Fig.11.18 shows the isolines of the two-dimensional stationary probability density $P(X,Z)$ obtained by numerical integration of (10.73): in the absence of noise (Fig.11.18,a,b) and with the noise of intensity $D = 1.0$ (Fig.11.18,c). In the absence of noise under various initial conditions two symmetrical attractors are generated in the system (Fig.11.18,a,b). For $D \neq 0$, the analysis of the structure of the probability density shows that the noise causes the two symmetrical attractors to join into a single chaotic manifold. For low-intensity noise, the phase trajectory remains for a long time in each of the attractors and undergoes rapid transitions between them. Thus, the intermittency of the "chaos-chaos" type is induced in the system by the external noise. To confirm the conclusion that intermittency is present, we calculate the power spectrum $S_x(\omega)$ of the process. The low-frequency region of the power spectrum is shown in Fig.11.19,a in the absence of noise. In the presence of noise the power spectrum evolves into the region of low frequencies, as is indicated in Fig.11.19,b ($D=0.2$). The appearance of the low-frequency components in the power spectrum is due to the existence of two time scales in the process. The former is associated with the prolonged residence of a phase trajectory in each of the symmetrical attractors, and the latter is due to transitions between them. The average residence time T_s, that the trajectory spends in each of the merged attractors, is related to the half-width of the low-frequency spectrum component $\Delta\Omega$ by the obvious relation $T_s = \alpha 2\pi/\Delta\Omega$, where α is the proportionality factor. As the intensity of noise increases, the residence time becomes less: $T_s \propto \exp(\beta/D)$, as is indicated by the smoothing of the low-frequency region of the power spectrum (Fig.11.19,c).

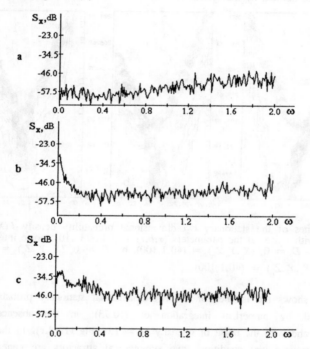

Fig.11.19. The low-frequency region of the power spectrum with the parameters $(\sigma,b,r) = (10,8/3,210)$ and noise intensity: a - $D = 0.0$, b - $D = 0.2$, c - $D = 0.8$.

Thus, a noise-induced transition is observed in the Lorenz system in the region of parameters corresponding to a quasiattractor. As a result, the attractors interact in the form of a "chaos-chaos" intermittency [11.25]. We note that this effect is still found when the action of noise is multiplicative and when the sources $\xi_i(t)$ have a finite correlation time.

11.6. STATISTICAL PROPERTIES OF INTERMITTENCY IN QUASIHYPERBOLIC SYSTEMS IN THE PRESENCE OF NOISE

The internal bifurcations of regular and chaotic limit sets lead to the phenomenon of *intermittency* manifesting itself in a random alternation of different phases of the system's motion [11.23]. The intermittency *induced by noise* can appear in real physical systems under fluctuation [11.24].

According to generation of the bifurcation mechanism, three intermittency types are distinguished [11.25]. Furthermore, the intermittency can be classified by the type of interacting attractors. For example, the intermittency of "cycle-chaos", "chaos-chaos", and "torus-chaos" types are possible.

The phenomenon of intermittency is quite typical for quasihyperbolic systems. It was observed experimentally in different physical systems [11.26]. The intermittency exhibits a number of remarkable statistical properties one of which is a characteristic shape f^a of the power spectrum in the region of low frequencies [11.25].

In many papers, which have reported this phenomenon in the presence of noise [11.27-11.31], the first type intermittency, as influenced by *additive white noise* is considered. They leave out of attention such important phenomenon characteristics as the distribution of laminar phase duration and the power spectrum. Noise influence on "chaos-chaos" intermittency was almost not investigated.

In the present paragraph, the above problems are reviewed and a technique is proposed based on the mean first passage time theory (MFPT) that enables one to formulate a unified approach to study the properties of intermittency in the presence of external noise [11.32-11.34].

Let us propose the regions Θ_1 and Θ_2 separated by hypersurface Γ to be available in the N-dimensional phase space of a certain dynamical system. Phase trajectories are present in each of them for a long time and undergo transitions from one region to another through the boundary γ being a part of hypersurface Γ, i.e., the intermittency between the regions Θ_1 and Θ_2 takes place. The arbitrary number of interacting regions can be considered in principle. For example, the "cycle-chaos" intermittency is characterized by two regions: one of them is in the vicinity of the limit cycle having disappeared and the other corresponds to the basin of attraction of a chaotic limit set. Note, that the separatrix hypersurfaces serve often as boundaries of interacting regions in quasihyperbolic systems which separate the basins of attraction of different limit sets in the phase space.

The *time statistics* of trajectory residence in each of these regions (the laminar phase duration in the case of "cycle-chaos" intermittency) and the *power spectra* are of basic interest when studying the statistical properties of the intermittency phenomenon. While defining these characteristics, a problem of achieving the boundary γ by a phase trajectory, that emerges from a certain point $\mathbf{x} = x(x_1,...,x_N)$, arises naturally. The *theory of boundary achievement* by a Markovian process (or mean first passage time theory MFPT) can be used to solve this problem taking into account the fluctuation effect which has been proposed first by L.S.Pontryagin [11.34].

Let the system be described by Fokker-Planck equation for N-dimensional probability density $P(\mathbf{x},t)$

$$\frac{\partial P}{\partial t} = \frac{1}{2}\sum_{i,j=1}^{N}\frac{\partial^2}{\partial x_i \partial x_j}[B_{ij}(\mathbf{x})P(\mathbf{x},t)] - \sum_{i=1}^{N}\frac{\partial}{\partial x_i}[A_i(\mathbf{x})P(\mathbf{x},t)]. \qquad (11.12)$$

where $A_i(\mathbf{x})$ and $B_{ij}(\mathbf{x})$ are the drift coefficients' vector and the diffusion matrix elements, respectively. The probability of the first boundary achievement from point $\mathbf{x} \in \Theta_1$ $p(\mathbf{x},t)$ obeys the first Pontryagin's equation:

$$\frac{\partial p}{\partial t} = \frac{1}{2}\sum_{i,j=1}^{N}B_{ij}(\mathbf{x})\frac{\partial^2}{\partial x_i \partial x_j}p(\mathbf{x},t) + \sum_{i=1}^{N}A_i(\mathbf{x})\frac{\partial}{\partial x_i}p(\mathbf{x},t), \qquad (11.13)$$

that has the following initial condition

$$p(0,\mathbf{x}) = 0, \ \mathbf{x} \in \Theta_1\backslash\Gamma \qquad (11.14)$$

and boundary conditions

$$p(\mathbf{x},t) = 1, \ \mathbf{x} \in \gamma \text{ and } p(\mathbf{x},t) = 0, \ \mathbf{x} \in \Gamma\backslash\gamma. \qquad (11.15)$$

The equation for the probability density of the time of the first boundary achievement $w(t,\mathbf{x})$ has the similar form:

$$\frac{\partial w}{\partial t} = \frac{1}{2}\sum_{i,j=1}^{N}B_{ij}(\mathbf{x})\frac{\partial^2}{\partial x_i \partial x_j}w(\mathbf{x},t) - \sum_{i=1}^{N}A_i(\mathbf{x})\frac{\partial}{\partial x_i}w(t,\mathbf{x}) \qquad (11.16)$$

with initial and boundary conditions

$$w(0,\mathbf{x}) = 0, \ \mathbf{x} \in \Theta_1\backslash\Gamma,$$

$$w(\mathbf{x},t) = \delta(t), \ \mathbf{x} \in \gamma \text{ and } w(\mathbf{x},t) = 0, \ \mathbf{x} \in \Gamma\backslash\gamma. \qquad (11.17)$$

From (11.16) we can derive the equation for the moments of the time of the first boundary achievement $T_n(x)$:

$$\frac{1}{2}\sum_{i,j=1}^{N} B_{ij}(\mathbf{x})\frac{\partial^2}{\partial x_i \partial x_j}T_n(t,\mathbf{x}) - \sum_{i=1}^{N} A_i(\mathbf{x})\frac{\partial}{\partial x_i}T_n(t,\mathbf{x}) = -nT_{n-1},$$

$$T_n(\mathbf{x}) = \int_0^\infty t^n w(t,\mathbf{x})dt, \; T_0 \equiv 1, \; T_n(\mathbf{x})|_{\mathbf{x}\in\gamma} = 0. \quad (11.18)$$

As $n = 1$, $T(\mathbf{x})$ defines the mean time of the first boundary achievement from the starting point $\mathbf{x}$. Averaging over the ensemble of trajectories returning to the region Θ_1 provides the *mean residence time* $<T>$ for the phase trajectory in the region Θ_1

$$<T> = \int T_1(\mathbf{x})P_B(\mathbf{x})d\mathbf{x}, \quad (11.19)$$

where $P_B(\mathbf{x})$ is the probability density of returning trajectories.

To assess the form of the power spectrum, a signal is idealized as a two-state process $u(t)$ being equal to zero in the region Θ_1 and to a unity outside this region, $u(t) = 0$, $\mathbf{x} \in \Theta_1$, $u(t) = 1$, $\mathbf{x} \in \backslash \Theta_1$. It is sufficient to know the characteristic distribution functions of durations and of the pauses or, as in our case, that of the time of the first boundary achievement from the regions Θ_1 and Θ_2, to calculate the power spectrum of the pulse process $u(t)$. The characteristic distribution function of the time of the first boundary achievement is found by applying Laplace transform to the equation (11.16) as follows:

$$\frac{1}{2}\sum_{i,j=1}^{N} B_{ij}(\mathbf{x})\frac{\partial^2}{\partial x_i \partial x_j}W(s,\mathbf{x}) - \sum_{i=1}^{N} A_i(\mathbf{x})\frac{\partial}{\partial x_i}W(s,\mathbf{x}) = sW(s,\mathbf{x}),$$

$$W(s,\mathbf{x}) = \int_0^\infty w(t,\mathbf{x})e^{-st}dt, \; W(s,\mathbf{x})|_{\mathbf{x}\in\gamma} = 1. \quad (11.20)$$

By averaging the solution of (11.20) over the ensemble of returning trajectories and by substituting $s \to i\omega$, we derive the *characteristic distribution function* of the time of the *first boundary achievement* $W(i\omega)$. Further, the expression for the power spectrum can be obtained. For example, as the time of the phase trajectory's residence in the region Θ_1 highly exceed that of its return (intermittency "cycle-chaos"), the expression for the power spectrum has the form [11.25]:

$$S_u(\omega) = 1/2\{[1 - W(i\omega)]^{-1} + [1 - W(-i\omega)]^{-1}\}. \quad (11.21)$$

Thus, using (11.13)-(11.21), one can define the main characteristics of intermittency process.

Noise effect on "cycle-chaos" intermittency. This type of intermittency is simulated by a discrete map having the form:

$$x_{n+1} = \varepsilon + x_n + ax_n^2, \quad |\varepsilon| \ll 1, \; |x| \ll 1. \tag{11.22}$$

The plot of the map (11.22) before and after transition is shown in Fig.11.20. As $\varepsilon < 0$, two fixed points exist one of them being saddle. As $\varepsilon = 0$, these points merge and, as $\varepsilon < 0$, they vanish. With $\varepsilon > 0$, $\varepsilon \ll 1$, a laminar channel is formed where the motion is almost regular and which is bounded by a certain l value. The phase trajectories return to the laminar channel when achieving the boundary $x = l$. For simplicity, we consider trajectories x_n, returning to the laminar channel $[-l, l]$, to have the uniform distribution $P_B(x)$; namely:

$$P_B(x) = 1/2l. \tag{11.23}$$

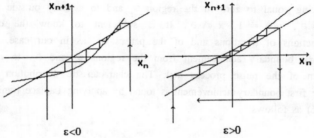

Fig.11.20. The phase portraits of the map (11.22).

The transition from a discrete map to the differential system is possible with specified bounded ε and x [11.27]. Supplementing the additive Gaussian noise $\xi(t)$ gives the following SDE

$$dx/dt = \varepsilon + ax^2 + \xi(t), \quad \langle \xi(t) \rangle = 0. \tag{11.24}$$

In case of white noise, $\langle \xi(t)\xi(t+s) \rangle = 2D\delta(s)$. The equation for the probability density of the time of the first boundary achievement $x = l w(t, x)$ has the form:

$$\frac{\partial w}{\partial t} = D \frac{\partial^2 w}{\partial x^2} + (\varepsilon + ax^2) \frac{\partial w}{\partial x}$$

$$w(0, x) = 0, \quad w(t, l) = \delta(t), \quad \left.\frac{\partial w}{\partial x}\right|_{x=-l} = 0. \tag{11.25}$$

The absorption boundary appears at the point $x = l$ and the boundary of reflection type is observed at the point $x = -l$. Equation (11.20) is written as follows, in our case:

$$D\frac{\partial^2 W(s,x)}{\partial x^2} + (\varepsilon + ax^2)\frac{\partial W(s,x)}{\partial x} = sW(s,x),$$

$$W(s,l) = 1, \quad \left.\frac{\partial W}{\partial x}\right|_{x=-l} = 0. \tag{11.26}$$

For the mean time of the first $x = l$ boundary achievement, we obtain the following equation

$$D\frac{\partial^2 T(x)}{\partial x^2} + (\varepsilon + ax^2)\frac{\partial T(x)}{\partial x} = -1$$

$$T(l) = 0, \quad \left.\frac{\partial T}{\partial x}\right|_{x=-l} = 0. \tag{11.27}$$

By averaging $T(x)$ over the ensemble of realization trajectories which return into the laminar channel $[-l, l]$ we find the average duration of the laminar phase τ_1:

$$\tau_1 = \int_{-l}^{l} T(x)P_B(x)dx = 1/2l \int_{-l}^{l} T(x)dx. \tag{11.28}$$

The solution of (11.27) for the given boundary conditions has the form

$$T(x) = 1/D \int_{x}^{l} exp[\phi(z)] \int_{-l}^{l} exp[\phi(y)]dydz,$$

$$\phi(y) = 1/D \int (\varepsilon + ay^2)dy = (\varepsilon y + ay^3/3)/D. \tag{11.29}$$

Unfortunately, we failed to obtain the analytical solutions of (11.25) and (11.26) in general case. No integrals in (11.29) are taken, as well. Therefore, let us consider a number of simplifying approximations.

1. *Deterministic case*, $D = 0$. For $T(x)$ from (11.27) and (11.28), the expression is derived that is known in the literature (see, for example, [11.29]) for the mean duration of the laminar phase τ_1

$$\tau_1 = 2l/\sqrt{a\varepsilon} \; arctg\sqrt{a/\varepsilon} \; l. \tag{11.30}$$

2. *Linear laminar channel*, $ax^2 = 0$. In this case, the complete analysis is a success. As $D = 0$, we obtain from (11.27) and (11.28) the following expression for the mean duration of the laminar phase

$$\tau_1 = l/\varepsilon. \tag{11.31}$$

For the characteristic function $W(i\omega)$ from (11.26), we find by elementary transformation that

$$W(i\omega) = [1 - exp(-i\omega\tau_1)]/i\omega\tau_1. \tag{11.32}$$

From (11.32), the reverse Fourier transform is easy to be obtained, i.e., the probability density of distribution of the duration of the laminar phase $w(t)$, namely:

$$w(\tau) = 1, \quad t \le \tau, \quad w(t) = 0, \quad t > \tau. \tag{11.33}$$

Thus, the probability density $w(t)$ is uniform. Substituting (11.32) in (11.21) yields the following power spectrum for the process

$$S(\omega) = \frac{\omega\tau_1 [sin(\omega\tau_1) - \omega\tau_1]}{2cos(\omega\tau_1) + 2\omega\tau_1 sin(\omega\tau_1) - (\omega\tau_1)^2 - 2}. \tag{11.34}$$

The form of the power spectrum for $l = 10^{-2}$, $\varepsilon = 10^{-4}$ is presented in Fig.11.21. The spectrum comprises a sequence of maxima with the decreasing envelope. Such a spectrum structure is accounted for by the presence of characteristic time scale in the process due to the mean duration of laminar phase. Note, that the form of this spectrum depends essentially on the way in which the trajectory returns into the laminar channel, i.e., on the structure of the probability density $P_B(x)$.

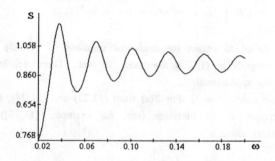

Fig.11.21. The power spectrum (11.34) for $l = 0.01$, $\varepsilon = 0.0001$.

As $D \ne 0$, we derive for the mean time of the first boundary achievement the following expression

$$D \frac{\partial^2 T(x)}{\partial x^2} + \varepsilon \frac{\partial T(x)}{\partial x} = -1,$$

$$T(l) = 0, \quad \frac{\partial T}{\partial x}\bigg|_{x=-1} = 0, \qquad (11.35)$$

with the solution having the form

$$T(x) = D/\varepsilon^2 \{exp(-2\varepsilon l/D) - exp[-\varepsilon(x-l)/D]\} + (l-x)/\varepsilon. \qquad (11.36)$$

By averaging over the ensemble of trajectories returning into the laminar channel, we obtain the following expression for the mean duration of the laminar phase

$$\tau_1 = D^2/(2\varepsilon^3 l)[exp(-2\varepsilon l/D)-1] + D/\varepsilon^2 exp(-2\varepsilon l/D) + l/\varepsilon. \qquad (11.37)$$

As $D \to 0$, $\tau_1 \to l/\varepsilon$, i.e., with the noise intensity tending to zero, the mean duration of laminar phase tends to its deterministic value (11.31). It is decreased with increasing noise intensity what is illustrated in Fig.11.22. Then, let us pass to analysis of the noise effect using the shape of the power spectrum. To do this, one needs to determine the characteristic distribution function of the laminar phase duration $W(i\omega)$.

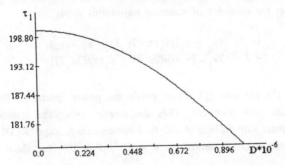

Fig.11.22. The mean laminar phase length as a function of the noise intensity (11.37) for $l = 0.01$, $\varepsilon = 0.0001$.

In our case, the equation (11.26) has the form

$$D \frac{\partial^2 W(s,x)}{\partial x^2} + \varepsilon \frac{\partial W(s,x)}{\partial x} = sW(s,x),$$

$$W(s,l) = 1, \quad \frac{\partial W}{\partial x}\bigg|_{x=-1} = 0. \qquad (11.38)$$

The general solution of (11.38) is defined by the following expression

$$W(s,x) = C_1 exp(\gamma_1 x) + C_2 exp(\gamma_2 x), \quad (11.39)$$

where $\gamma_{1,2}$, in their turn, are determined by the following expression

$$\gamma_{1,2} = \frac{-\varepsilon \pm \sqrt{\varepsilon + 4Ds}}{2D}, \quad (11.40)$$

and constants $C_{1,2}$ are found from the boundary conditions using the following system of equations

$$C_1 exp(\gamma_1 l) + C_2 exp(\gamma_2 l) = 1,$$
$$\gamma_1 C_1 exp(-\gamma_1 l) + \gamma_2 C_2 exp(-\gamma_2 l) = 0. \quad (11.41)$$

The result of simple transformations for $W(s,x)$ is as follows:

$$W(s,x) = \frac{\gamma_2 exp(\gamma_1 x - \gamma_2 l) - \gamma_1 exp(\gamma_2 x - \gamma_1 l)}{\gamma_2 exp[(\gamma_1 - \gamma_2)l] - \gamma_1 exp[(\gamma_2 - \gamma_1)l]}. \quad (11.42)$$

The averaging over the ensemble of returning trajectories yields

$$W(s) = \frac{1}{2\gamma_1 l} + \frac{(\gamma_2 - \gamma_1)[\gamma_1 exp(2\gamma_2 l) - \gamma_1 - \gamma_2]}{2\gamma_1 \gamma_2 [\gamma_2 exp(2\gamma_1 l) - \gamma_1 exp(2\gamma_2 l)]}. \quad (11.43)$$

Let us substitute (11.43) into (11.21) to obtain the power spectrum. The computations and final result are very awkward. They are omitted here. The transformations were performed on computer with the help of analytical mathematics package "REDUCE". The form of the power spectrum is presented as $\varepsilon = 10^{-4}$, $l = 10^{-2}$ for three noise intensity

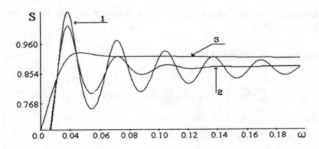

Fig.11.23. The power spectra of intermittency: $1 - D = 10^{-8}$, $2 - D = 10^{-7}$, $3 - D = 10^{-6}$.

values $D = 10^{-8}$, 10^{-7}, 10^{-6} in Fig.11.23 (curves 1-3), respectively. The spectrum becomes smooth as the noise intensity is increased: the maxima are smeared at the frequencies corresponding to the characteristic time scale of the process. Furthermore, the maxima in the power spectrum are shifted towards higher frequencies since the mean duration of laminar phase is decreased as the noise intensity grows. As $D = 10^{-6}$ (curve 3), the power spectrum is actually uniform.

As already noted, we failed in finding the analytical solutions for equations (4.15)-(4.17). They can be solved, however, numerically using the finite-difference methods. The calculated results for $\varepsilon = 10^{-3}$, $l = 10^{-1}$ are cited in Fig.11.24. Here, the probability density of the time of the first boundary achievement $w(\tau)$ is shown for three noise intensities: $D = 10^{-5}$ (curve 1), $D = 10^{-4}$ (curve 2) and $D = 5 \times 10^{-4}$ (curve 3). With low noise intensity (curve 1), $w(t)$ exhibits two maxima which correspond to the minimum and the maximum laminar phase durations, respectively. The structure of $w(\tau)$ is simplified with the increase of noise intensity and, as the latter becomes high, it is described by the reverse power law $\tau^{-\mu}$. The mean duration of laminar phase is obviously decreased.

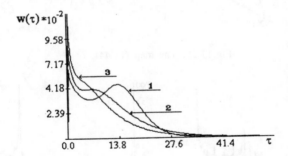

Fig.11.24. The probability density of distribution of residence times: 1 - $D = 10^{-5}$, 2 - $D = 10^{-4}$, 3 - $D = 5 \times 10^{-4}$.

Noise effect on "chaos-chaos" intermittency. The intermittency of "chaos-chaos" type was considered in [11.35-11.37]. In [11.38], a simple model for investigating the properties of this phenomenon was proposed which represents the Feigenbaum's map used twice:

$$x_{n+1} = 1 - a + 2a^2 x_n^2 - a^3 x_n^4 . \tag{11.44}$$

The plot of the map is shown in Fig.11.25. The fixed points x_1 and x_3 are always unstable; points x_2 and x_4 lose their stabilities via the bifurcation of period doubling (multiplier $\rho = -1$) while increasing the parameter a. As a grows, the multipliers of the two fixed points x_2 and x_4 pass through the critical value $\rho^* = -1.601$.

Simultaneously, two independent Feigenbaum's attractors are born, as $a = a^* = 1.4011551$, with the basins of attraction separated by the "separatrix" x_3. With further a increase, both attractors evolve via a series of reverse doubling bifurcations to the regimes which correspond to the map of a segment into itself and have the distribution obeying the "arcsinus" law [11.19]. As $a = a^{**} \approx 1.543689$, the bifurcation of *merging of attractors* takes place. Before this bifurcation, as $a^* < a < a^{**}$, the probability density $p(x)$ of the process appears as two non-intersecting functions (Fig.11.26a). As the threshold is exceeded, $a > a^{**}$, the attractors merge and the probability density takes the form shown in Fig.11.26b.

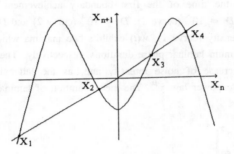

Fig.11.25. The map (11.44).

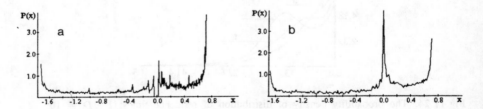

Fig.11.26. The probability density $p(x)$ for the map (11.44): a - $a = 1.5436$, b - $a = 1.5437$.

Let us consider the ACF and the power spectrum of the process before and after bifurcation of attractors' merging. The results of numerical simulation are presented in Fig.11.27. Below the threshold, $a < a^{**}$, the correlation function $R(k)$ is close to δ-function (Fig.11.27a, curve 1). The power spectrum is uniform (Fig.11.27b, curve 1). When exceeding the threshold, $a > a^*$, the statistical properties of the system change abruptly. The correlation function becomes slowly decreasing (Fig.11.27a, curve 2) and the power spectrum evolves towards lower frequencies (Fig.11.27b, curve 2). Thus, the characteristic time scale of the process is substantially increased. This effect is accounted for by the intermittency appearing between the attractors merged. The phase trajectories are for a long time in the basin of attraction of each attractor and undergo transitions through the boundary x_3. The parameter a increase leads to a smaller residence time of phase trajectories on each of the attractors. The calculations of the mean "laminar" phase duration, defined as the mean time of the first boundary achievement $x = x_3$ as a function of parameter a, demonstrate scaling that has the form

$$T \propto (a - a^{**})^{-\beta}, \quad \beta = 0.5, \qquad (11.45)$$

typical for many critical phenomena and, in particular, for the "cycle-chaos" intermittency.

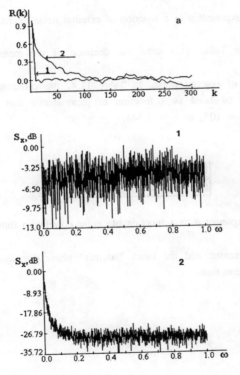

Fig.11.27. a - autocorrelation functions of map (11.44); b - corresponding power spectra: 1 - $a = 1.5436$, 2 - $a = 1.5437$.

To investigate the effect of external noise, let us introduce into the right-hand part of (11.44) a source of colored noise ξ_n simulated by the linear stochastic map (10.78). The presence of noise results in the transition to a merged attractor which is possible, as $a < a^{**}$. Theoretically, the probability of transition from the basin of attraction of one attractor to that of another exists, as $a < a^{**}$, even with low noise intensity D. The form of dependence (11.45) is maintained. However, the critical exponent β depends on noise intensity and correlation time.

Now, we consider the case of white noise. In Table 11.3, the calculation data for β as a function of noise intensity D are presented as $a = 1.544$.

Table 11.3.

D	0.0	10^{-5}	5×10^{-5}	10^{-4}	5×10^{-4}
β	0.51	0.49	0.35	0.34	0.17

Critical exponent β as a function of external noise intensity.

The results indicated in Table 11.3 show the decrease of the mean "laminar" phase duration.

Let us the effect of colored noise. In Table 11.4, the calculated results for the critical exponent β are tabulated as a function of parameter Γ that defines the noise correlation time with $D = 10^{-5}$, as $a = 1.544$.

Table 11.4.

Γ	0.0	0.2	0.7	0.8
β	0.49	0.499	0.50	0.51

Critical exponent β as a function of noise correlation time.

It is seen that β is increased and the mean "laminar" phase duration grows with the increase of noise correlation time.

CHAPTER 12
RECONSTRUCTION OF DYNAMICAL SYSTEMS FROM EXPERIMENTAL DATA

12.1. INTRODUCTION

One of the most important investigation problems in the field of chaotic dynamics is the one dealing with *reconstruction of attractors*, their characteristics and, at last, the model equations of system from the data of measurement of scalar time dependence (realization). Solving of such problems is carried out in two steps. At the first step the number of active degrees of freedom of the system under investigation is determined from the time series, i.e., its dimension, whose lower boundary can be obtained on the basis of computation of *correlation integral* [12.1, 12.2]. At the second step the model equations, approximately describing the current regime of the investigated system, are reconstructed from the realization [12.3-12.8]. As a result, it becomes possible to predict the evolution of the system from the finite length of the realization.

A serious hindrance in solving of such a problem is unavoidable presence of experimental noise in the realization. Thereupon, the role of fluctuations depends essentially on the properties of the system under study and its concrete control parameters.

It is quite clear that the problem of prediction is of a special interest and is important as applied to chaotic systems. In this case, the *influence of fluctuations* may play a principle role [12.9-12.11]. Thus, for instance, it was found, that in the system with *quasiattractors* small fluctuations can bring about abrupt changes of the resultant regime of oscillations in comparison with the regime of unperturbed system [12.11-12.13]. As a result, it is not excluded that the reconstruction of the system from the realization would strongly depend on the intensity and the statistical properties of noise.

Different from *ideal hyperbolic* systems with chaotic behavior, the real systems with quasiattractors are characterized by the phenomena of *homoclinic tangency* (non-robust homoclinics) [12.14-12.17]. This circumstance results in birth (disappearance) of denumerable set of cycles and their bifurcations. In this case, the systems with quasiattractors exhibit a *strong sensitivity to noise influence*.

In this respect it is of interest to study *homoclinic phenomena* while solving the problem of model equations reconstruction from the experimental realization under noise perturbation [12.13].

12.2. METHODS AND ALGORITHMS

As convenient examples for investigation of homoclinic phenomena in systems with quasiattractors let us consider the cascade of inverse period doubling bifurcations behind the critical point in Feigenbaum's scenario [12.18] and the mechanism of invariant curve destruction due to appearance of non-robust homoclinic trajectory [12.19, 12.20]. As a model of Feigenbaum's transition let us take the two-dimensional Henon map:

$$x_{n+1} = 1 - ax_n + y_n,$$
$$y_{n+1} = bx_n. \quad (12.1)$$

Let us investigate the second scenario using as an example two coupled logistic maps (see Chapter 7):

$$x_{n+1} = 1 - \alpha x_n^2 + \gamma(y_n - x_n),$$
$$y_{n+1} = 1 - \alpha y_n^2 + \gamma(x_n - y_n). \quad (12.2)$$

The portions of bifurcation diagrams in the region of control parameters values of the dynamical systems (DS) of interest (12.1) and (12.2) are shown in Fig. 12.1 and 12.2, respectively. Bifurcation curves Γ_i (i=0,1,2) in Fig 12.1 correspond to the phenomenon of homoclinic tangency of *stable* and *unstable manifolds* of saddle stationary points with period 2^i of the map (12.1). To the right of the line Γ_i robust *homoclinic intersection* of stable and unstable manifolds of corresponding saddle cycles of (12.1) takes place. There are robust *homoclinic structures* available [12.18]. The bifurcation line l_h on the diagram, shown in Fig. 12.2, corresponds to the phenomenon of non-robust homoclinic tangency of unstable and stable manifolds of saddle cycles on the resonant invariant curve of the system (12.2). Above this line a robust homoclinic structure is realized in system (12.2) [12.19, 12.20]. Let us note here an important for further investigations circumstance. In system (12.1) the homoclinic structures to the right of lines Γ_i are *included in chaotic attractor*. In system (12.2) the intersection of line l_h results in appearance of an *"invisible"* homoclinic structure (here the attractor is a resonant cycle on the invariant curve).

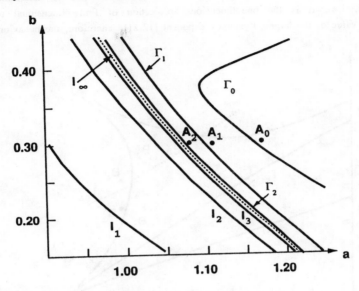

Fig. 12.1. Bifurcation diagram for map (12.1) on a,b plane:

l_i - period doubling bifurcation curves for period-2^i cycles,

l_∞ - critical curve of strange attractors arising,

Γ_i - curves of homoclinic tangency of stable and unstable manifolds for period-2^i cycles.

A_0, A_1, A_2 - points with $b=0.3$ and $a=1.17$, 1.10 and 1.07, respectively.

The investigation procedure is as follows. As a realization, we shall consider the sequence of coordinate values $\{x_i\}$, $i=1,...,N$, of the system (12.1) or (12.2) obtained for fixed values of control parameters. We shall select the values of parameters in accordance with bifurcation diagrams shown in Fig. 12.1 and Fig. 12.2 to know the system's attractor type in a given point of the parameter plane. To explore the influence of fluctuations let us add to the right-hand parts of maps an *additive normally distributed noise* $\xi(i)$ with zero average and intensity D. Let us consider the realization $\{\tilde{x}_i\}$ obtained as a result of iteration and summed with noise together with the original one. The comparison of results of DS reconstruction from $\{x_i\}$ and $\{\tilde{x}_i\}$ at different points of control parameters plane allows one to study the influence of homoclinic phenomena and noise level.

Let us characterize in brief the method of construction of DS model equations from the realization. Suppose we have *time series* $\{x_i,\ i=1,...,N\}$, obtained as a result of measurements. We shall also suppose that the realization $\{x_i\}$ is deterministically engendered, i.e., it is the one-dimensional projection of finite-dimensional attractor with finite value of the largest Lyapunov exponent [12.21]. Therefore, the behavior of

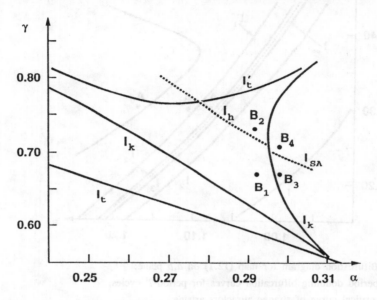

Fig. 12.2. Bifurcation diagram for system (12.2) on α,γ plane:
l_t - curve of invariant ergodic curve arising from the cycle of period 2,
l_k - curves of saddle-node bifurcation for period-10 cycles,
l_h - curve of homoclinic tangency of stable and unstable manifolds for the saddle period-10 cycle,
l_t' - curve of invariant ergodic curve arising from the period-10 cycle,
l_{SA} - curve of a strange attractor arising,
B_1, B_2 - points with $\gamma=0.292$ and $\alpha=0.67$, 0.73,
B_3, B_4 - points with $\gamma=0.298$ and $\alpha=0.67$, 0.708.

an unknown DS can be described with the help of finite-dimensional map $f: R^k \to R^k$, where k is the dimension of the phase space of the initial DS.

In order to have a possibility to construct the *model approximating* this time series it is necessary to know the approximate number of DS's degrees of freedom, i.e., the dimension of space into which the attractor corresponding to $\{x_i\}$ can be embedded (*embedding dimension* $m \geq k$). There are various methods known to estimate m [12.22-12.27]. Let us suppose that we have determined the value of m. Then, according to *Takens theorem*, we can restore m-dimensional trajectory with the help of *delay method* [12.28]:

$$y_i = \{x_i, x_{i+1}, ..., x_{i+m-1}\}, \quad i=1,..., N_0 = N-m+1. \tag{12.3}$$

Next task is to construct a map $g: R^m \to R^m$, approximating an unknown map f. There are two methods to derive g, namely, local and global ones. The local method does not need the map g to be smooth. In this case, actually, a certain algorithm is being developed, that allows to predict from the given segment of the time series its future behavior [12.3-12.5]. In order to calculate $y_{i+1} = g(y_i)$, one must find a certain number of nearest neighbors for point y_{i+1}. Thus, the functional dependence $g(y)$ is being changed from point to point and is not smooth. The global method supposes approximation of map g as a whole using all the points y_i and allows one to obtain the analytical form of writing the reconstructed map for the whole of the attractor [12.6-12.8]. Both methods have their advantages and shortcomings, that are known well enough in the literature. For our purposes, the most convenient is the method of *global reconstruction*, because of simplicity of analytical expressions and small dimensions of attractors of the chosen maps (12.1) and (12.2).

Let us search for map g in the form of m-dimensional polynomial of degree d, i.e.,

$$g(z) = \begin{bmatrix} g_1(z_1,...,z_m) \\ ..... \\ g_1(z_1,...,z_m) \end{bmatrix}, \tag{12.4}$$

$$g_j(z_1,...,z_m) = \sum_{i_1+...+i_m \leq d} a_{j,i_1,...,i_m} \prod_{k=1}^{m} z_k^{i_k},$$

$$j=1,...,m.$$

The substitution $z = y_i$ in (12.4) leads to the system of linear equations for determination of the unknown coefficients $a_{j,i_1,...,i_m}$ (whose number for each j is $M = (m+d)!/(m!d!)$)

$$g(y_i) = y_{i+1}, \quad i=1,...,N_0, \tag{12.5}$$

or in the coordinate form

$$g_j(x_i, x_{i+1},...,x_{i+m-1}) = x_{i+j}, \quad j=1,...,m, \quad i=1,...,N_0. \tag{12.6}$$

This system is decomposed into m independent subsystems of N_0 equations with M unknown quantities. Each subsystem can be solved by using the least squares method, i.e., by minimizing the following expression:

$$\sum_{i=1}^{N_0} [x_{i+j} - g_j(x_i,\ldots,x_{i+m-1})]^2, \quad j = 1,\ldots,m. \tag{12.7}$$

12.3. RESULTS OF MAP (12.1) RECONSTRUCTION

Let us consider, for example, the dynamics of system (12.1) with the values of control parameters $a=1.10$, $b=0.3$ (point A_1 in the bifurcation diagram shown in Fig. 12.1). At point A_1 located to the right of bifurcation curve of chaotic attractor birth l_∞ and between the lines Γ_1 and Γ_0, all the saddle cycles with period 2^i

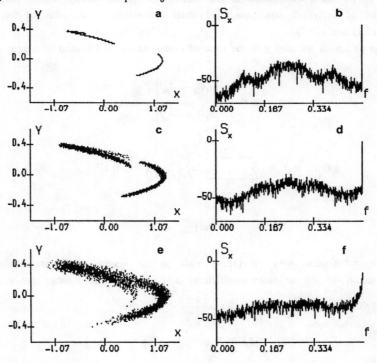

Fig. 12.3. Phase portraits and power spectra for the map (12.1) at point A_1 in Fig. 12.1 for $D=0.0$ (a, b), $D=0.01$ (c, d) and $D=0.03$ (e, f).

($i=1,2,...$) have robust homoclinic trajectories. The cycle with period 1 ($i=0$) is an exception, and the rough intersection of its stable and unstable separatrixes starts to the right of line Γ_0. This determines the structure of the chaotic attractor of system (12.1) at point A_1, representing two non-intersecting chaotic regions on the phase plane. Attractor (12.1) at point A_1 and power spectrum $S_x(f)$ are shown in Fig. 12.3(a),(b).

Let us add to the right-hand parts of the discrete equations of system (12.1) sources of white noise with intensity D and by usual iterations of map, let us construct attractor for point A_1 at the values of $D=0.01$ (Fig. 12.3 (c),(d)) and $D=0.03$ (Fig. 12.3 (e),(f)) on the phase plane. For noise intensity $D=0.01$ a noisy attractor is realized, consisting, as before, of two non-intersecting regions (Fig. 12.3(c)). An increase of noise intensity induces the qualitatively similar phenomena of appearance of homoclinic orbit of saddle cycle with period 1. This results in the merging of chaotic regions of the attractors into a single structure (Fig. 12.3(e)) [12.18]. Accordingly, the changes are observed in the frequencies' distribution in the power spectrum. With the increase of noise intensity, spectra $S_x(f)$ become more uniform (cf. Fig. 12.3(b) and (f)).

And now let us reconstruct the model maps by the method described above. We shall use the discrete sequences x_i ($D=0.0$) and $\tilde{x}_i$ ($D=0.01$ and 0.03) as the realizations. The right-hand parts of map (12.1) are represented by the polynomials of degree $d \leq 2$. Therefore, for $D=0.0$ this method of reconstruction allows one to restore with absolute precision the form of maps from just one coordinate x. As a result, the following map is obtained

$$x'_{n+1} = y'_n,$$
$$y'_{n+1} = 1 - a(y'_n)^2 + bx'_n, \quad (12.8)$$

where the new coordinates are related to the old ones according to the following formulas:

$$x_n = y'_n, \quad y_n = bx'_n. \quad (12.9)$$

Using iterations of the reconstructed two-dimensional maps, let us construct the corresponding phase portraits of the restored attractors whose coordinates were transformed according to (12.9), and power spectra. The results of reconstruction are shown in Fig. 12.4 and they are the evidence of two important phenomena. First, reconstruction results in an essentially complete noise filtration of the realization! In fact, the results shown in Fig. 12.4 are essentially identical and correspond to the original data shown in Fig. 12.3(a),(b)[1]. Second, the noise-induced homoclinic phenomenon disappears. The reconstruction from $\tilde{x}_i$, corresponding to a single noisy attractor shown in Fig. 12.2(e), for instance, results in restoration of the attractor with the original structure from the two chaotic regions (see Fig. 12.4(e) and Fig. 12.3(a)).

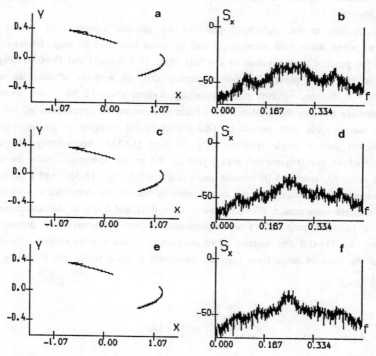

Fig. 12.4. Phase portraits and power spectra for reconstructed maps at point A_1 in Fig. 12.1 for $D=0.0$ (a, b), $D=0.01$ (c, d) and $D=0.03$ (e, f).

The numerical simulation have been conducted at points of bifurcation diagram shown in Fig. 12.1, corresponding to attractors with one-band ($a=1.17$, $b=0.3$, point A_0) and four-band ($a=1.07$, $b=0.3$, point A_2) structure. The results obtained are as follows: *filtration of the noise component* in the realization takes place, and the reconstructed attractors are topologically equivalent to the attractors of unperturbed system for the corresponding parameters values.

---

[1]Quantitative correspondence of original and reconstructed attractors' characteristics will be discussed below.

12.4. RECONSTRUCTION OF SYSTEM (12.2)

The bifurcation diagram of unperturbed system (12.2) in the vicinity of resonance 2:5 is shown in Fig. 12.2. Within the zone of synchronization, bounded by the curves of saddle-node bifurcations of resonant cycles l_k and by torus birth curve l'_t, there are a stable (T_0) and saddle (T_1) cycles with period 10. The closure of unstable separatrixes of the saddle points of cycle T_1 onto the stable points of cycle T_0 in the region of resonance form the "invisible" invariant curve. On line l_h a non-rough homoclinic tangency of stable and unstable separatrixes takes place and above it (in the region of resonance) a robust invisible homoclinic structure and a stable cycle T_0 coexist, the latter being the only attractor here [12.19, 12.20].

Map (12.2), like the map (12.1), is given by polynomials of the second degree. Therefore, for $D=0.0$, a precise reconstruction is realized, which results in the map in the following form:

$$\begin{aligned}
x'_{n+1} &= y'_n \\
y'_{n+1} &= (1+2\gamma-\alpha/\gamma) + \alpha(1/\gamma-1){y'_n}^2 + 2\alpha(\alpha/\gamma-\gamma){x'_n}^2 - \\
&\quad - (\alpha^3/\gamma){x'_n}^4 - 2\alpha^2{x'_n}^3 - (2\alpha^2/\gamma){x'_n}^2 y'_n - \\
&\quad - 2\alpha x'_n y'_n + 2\alpha x'_n + 2(\alpha-\gamma)y'_n ,
\end{aligned} \qquad (12.10)$$

where the new coordinates are related to the old ones according to the following formulas:

$$x_n = x'_n, \ y_n = x'_n + y'_n/\gamma - 1/\gamma + (\alpha/\gamma)x'_n . \qquad (12.11)$$

The transformation (12.11) after reconstruction for $D \geq 0$ allows one to obtain the phase portraits of attractors comparable with the original ones.

Let us select the two characteristic points in the space of parameters: B_1 ($\alpha = 0.67, \gamma = 0.292$) and B_2 ($\alpha = 0.73, \gamma = 0.292$). Point B_1 is located below the line l_h, point B_2 is correspondingly above it. The reconstruction of unperturbed system (12.2) from realization x_i in both cases results in model systems with the attractors in the form of period-10 cycles. The illustration of the above for point B_2 is shown in Fig. 12.5.

A more complicated picture is observed under the noise perturbation. When reconstructing the model system from the noisy realization $\tilde{x}_i$ at point B_1 an attractor is being restored in the form of cycle with period 10 (cycle T_0) (see Fig. 12.6). In the absence of homoclinic phenomena the algorithm of reconstruction again manifests the property of *noise filtration* and provides reconstruction of the regime similar to that of the unperturbed system (cf. the data in Fig. 12.6 with those in Fig. 12.5).

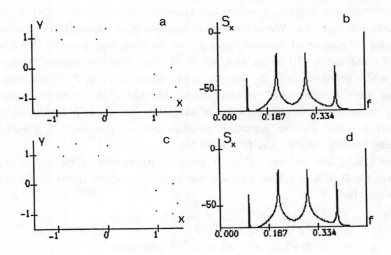

Fig. 12.5. Phase portraits and power spectra for system (12.2) at point B_2 in Fig. 12.2 for $D=0.0$ (a, b) and for the reconstructed map (c, d).

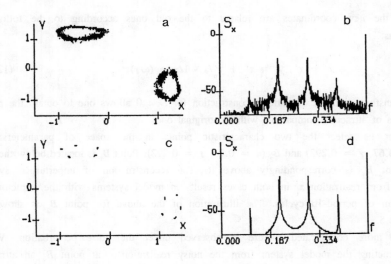

Fig. 12.6. Phase portraits and power spectra for system (12.2) at point B_1 in Fig. 12.2 for $D=0.05$ (a, b) and for the reconstructed map (c, d).

Above the homoclinic curve l_h there is *another* situation which is illustrated in Fig. 12.7. In this case the reconstruction results in simulation of the system, whose attractor *significantly differs* from the attractor of the unperturbed system! The structure of the reconstructed attractor is close to the invariant curve, with exhibits the loss of smoothness due to nonlinearity. The noise influence is qualitatively equivalent to the shift of control parameters into the region outside resonance (into the region of ergodic motion with incommensurate frequencies). To check this supposition the reconstruction of model systems and the corresponding attractors in the region of ergodic invariant curve and torus-chaos in system (12.2) (points B_3 and B_4 in the diagram shown in Fig. 12.2) was carried out. The results obtained are shown in Fig. 12.8 and Fig. 12.9 and are qualitatively equivalent to those shown in Fig. 12.7. At point B_3 from the noisy realization we reconstructed a smooth invariant curve, while at point B_4 we reconstructed an attractor whose structure is similar to the destroyed invariant curve.

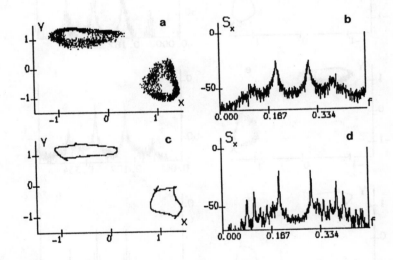

Fig. 12.7. Phase portraits and power spectra for system (12.2) at point B_2 in Fig. 12.2 for $D=0.05$ (a, b) and for the reconstructed map (c, d).

Thus, the application of reconstruction algorithm to the processing of the noisy realization in the presence of invisible homoclinic structure produces another results (comparing with the regimes when homoclinic structures are included in the attractor). In this case the reconstruction results in restoration of attractors, whose structure practically *repeats the geometry of unstable manifolds*. Thereupon, the filtration of non-correlated experimental noise is observed distinctly.

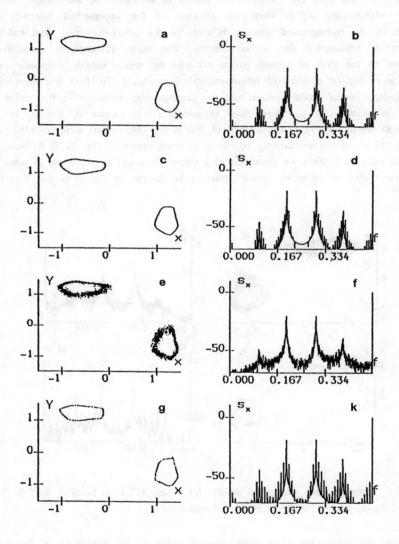

Fig. 12.8. Phase portraits and power spectra for system (12.2) at point B_3 in Fig. 12.2 for $D=0.0$ (a, b), $D=0.05$ (e, f) and for the corresponding reconstructed maps as $D=0.0$ (c, d), $D=0.05$ (g, k).

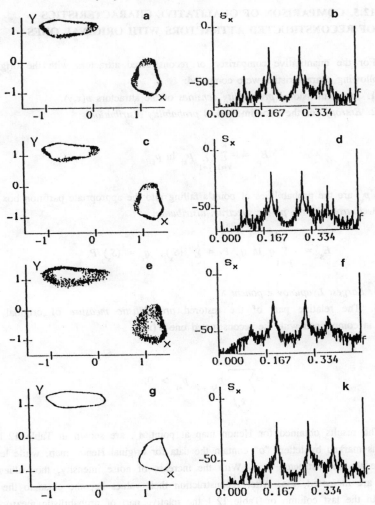

Fig. 12.9. Phase portraits and power spectra for system (12.2) at point B_4 in Fig. 12.2 for $D=0.0$ (a, b), $D=0.05$ (e, f) and for the corresponding reconstructed maps as $D=0.0$ (c, d), $D=0.05$ (g, k).

12.5. COMPARISON OF QUALITATIVE CHARACTERISTICS OF RECONSTRUCTED ATTRACTORS WITH ORIGINAL ONES

For the quantitative comparison of reconstructed attractors with the original ones the following characteristics were computed:

1. Two-dimensional *probability densities* of the attractors $p(x,y)$.
2. *Entropy* for the two-dimensional *probability distribution*

$$E_p = -\sum_{k=1}^{K} \sum_{j=1}^{J} p_{kj} \ln p_{kj}, \qquad (12.12)$$

where p_{kj} are the probabilities of point's falling into the appropriate partition box.

3. *Entropy* of the *power spectrum distribution*

$$E_s = -\sum_{j=1}^{J} q_j \ln q_j, \quad P = \sum_{j=1}^{J} (S_x)_j, \quad q_i = (S_x)_j/P. \qquad (12.13)$$

4. *Largest Lyapunov exponent* λ.

5. The relative part of the restored *probabilistic measure* of original attractor $p(x,y)$ in comparison with the reconstructed one $p^r(x,y)$

$$\rho = \sum_{k,j} p_{kj}, \quad p_{kj}^r > 0. \qquad (12.14)$$

The results obtained for Henon map at point A_1, are shown in Table 12.1, where columns marked by letters "o" contain the data for original Henon map, while letters "r" indicate the reconstructed map. With the increase of noise intensity, the values E_p, E_s and λ are increased, too. After reconstruction their values become close to the original ones. In the last column of Table 12.1 the relative part of probabilistic measure of the original attractor as $D = 0.0$ is shown in comparison with the reconstructed one as $D \geq 0$. As it was found out, for $D = 0.01$ 91% of measure is restored, while for $D = 0.03$ it was 86% only. This phenomenon is practically difficult to be noted using the phase portraits (Fig. 12.4), but can be observed when one carefully examines the power spectra shown in Fig. 12.4. For $D > 0$, the peak is more pronounced at frequency $f_0/4$, than for $D = 0.0$. Thus, noise filtration in the process of reconstruction is accompanied by the loss of some (scarcely probable) part of the attractor, which in this case does not practically affect the value of computed characteristics.

Table 12.1.

D	E_p		E_s		λ		ρ
	o	r	o	r	o	r	
0.0	3.82	3.82	0.62	0.62	0.180	0.180	1.0
0.01	4.87	3.74	0.86	0.52	0.183	0.181	0.91
0.03	5.90	3.82	4.22	0.51	0.199	0.181	0.86

Henon map at $a = 1.1$, $b = 0.3$.

In Fig. 12.10 the two-dimensional probability densities $p(x,y)$ for Henon map at points A_1 and A_0 as $D > 0$ and the corresponding densities $p^r(x,y)$, obtained as a result of reconstruction, are shown. From the distributions corresponding to the single-band attractors, different but qualitatively equivalent distributions of the two-band (at point A_1) and one-band (at point A_0) attractors are restored. The homoclinic intersection of manifolds of period-1 cycle at point A_0 results in the formation of the noisy distribution between two chaotic zones maximum. At point A_1 the phenomenon of

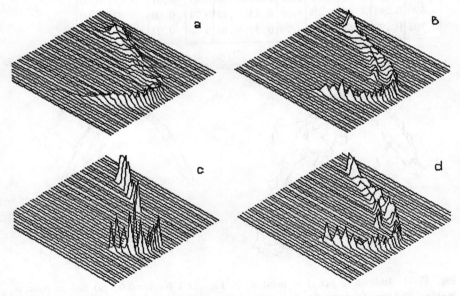

Fig. 12.10. Probability distribution densities at the attractor for map (12.1) at point A_1 in Fig. 12.1. for $D=0.03$ (a), at point A_0 in Fig. 12.1 for $D=0.01$ (b) and for the corresponding reconstructed maps (c, d).

bands merging is induced by noise, what corresponds to the minimum of distribution density in the region between the bands. This difference is well illustrated in Fig. 12.11, were the isolines $\log_{10} p(x,y)$ are shown at points A_1 and A_0, as $D > 0$. Only two level lines are constructed. The first line (l_1) corresponds to the minimum value of $p(x,y)$ (i.e., to the case when only one point falls into the box) and outlines the outer contour of the attractor. The second line (l_2) corresponds to the value of $0.2p_{max}$ and bounds the most probable regions of the attractor. In Fig. 12.11(a) line l_2 draws two disconnected regions, while in Fig. 12.11(b) the region bounded by line l_2 includes practically the whole attractor. This difference provides for possibility of qualitatively correct reconstruction.

The data for the system of two coupled logistic maps are represented in Table 12.2, where the letters "r" indicate reconstructed map. The entropy values E_p and E_s at all

Table 12.2.

α	γ	E_P			E_S			λ		
		D=0	D>0	r	D=0	D>0	r	D=0	D>0	r
0.670	0.292	2.30	5.81	2.34	0.23	0.25	0.21	-0.08	-0.05	-0.07
0.730	0.292	2.31	6.75	5.01	0.35	0.50	0.41	-0.07	0.03	0.03
0.670	0.298	4.62	6.08	4.60	0.23	0.28	0.21	0.00	-0.03	0.00
0.708	0.298	5.67	6.63	6.08	0.35	0.46	0.40	0.02	0.01	0.04

Two coupled logistic maps.

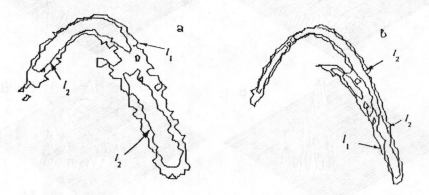

Fig. 12.11. Isolines $\lg p(x,y)$ at point A_1 in Fig. 12.1 for $D=0.03$ (a) and at point A_0 in Fig. 12.1 for $D=0.01$ (b): l_1 corresponds to the least value of $p(x,y)$, l_2 corresponds to the value $0.2p$.

points, as $D = 0.01$ are increased, however, after the reconstruction they come back to the initial values only where the *smooth torus* exists (resonant torus at point B_1 and ergodic torus at point B_3). At points B_2 and B_4, where the torus is not smooth, E_s and E_p remain increased. The largest Lyapunov exponent behaves in much in the same way. At points B_1 and B_3, it returns to negative and zero values. At point B_2, the exponent becomes positive, although in the initial regime it takes a negative value, and at point B_4 - remains positive, but increases in comparison with the initial one. In this case it has some sense to compute ρ values only for points B_3, B_4. At point B_3 only 74% of probabilistic measure is reconstructed, while at point B_4 there is only 59% of it. The increase of λ points out that in the reconstructed part of the attractor a stronger exponential divergence of trajectories takes place, than that in the lost part. This phenomenon is related to the fact that in the course of reconstruction the part of the attractor is

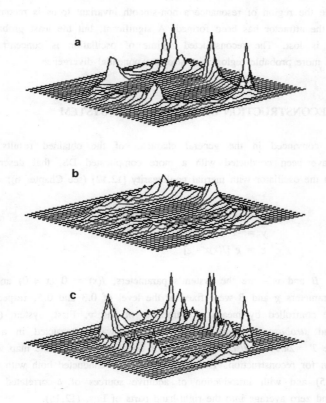

Fig. 12.12. Probability distribution density on the attractor for system (12.2) at point B_4 in Fig. 12.2 for $D=0.0$ (a), $D=0.05$ (b) and for the reconstructed map as $D=0.05$ (c).

restored, which is located in the vicinity of the invariant torus destroyed due to the loss of smoothness, on which the major part of probabilistic measure is concentrated, and characterized by the higher degree of exponential divergence. The above said is shown in Fig. 12.12, that illustrates the initial, noisy and reconstructed probability densities at point B_4.

Thus, it is evident that the results of reconstruction depending on properties of *homoclinic structures* differ significantly. If the homoclinic structure is *included* in the attractor, algorithm filters the noise and restores the initial attractor to the extent of quantitative characteristics. With this, an insignificant part of the attractor can be lost. If the homoclinic structure is *not included* in the attractor, then it hinders restoration of the initial attractor. In the region of resonance a hidden homoclinic structure is reconstructed, which is visualized under the influence of noise. Outside the region of resonance a non-smooth invariant torus is reconstructed, on whose basis the attractor has been formed. A significant, but the least probable part of the attractor is lost. The reconstructed regime of oscillation is concentrated in the narrower and more probable region with strong exponential divergence.

12.6. RECONSTRUCTION OF DIFFERENTIAL SYSTEM

To get convinced in the general character of the obtained results, numerical simulation have been conducted with a more complicated DS, that describes forced oscillations in the oscillator with inertial nonlinearity [12.12] (see Chapter 6):

$$\begin{aligned} \dot{x} &= mx + y - xz + B\sin(\nu t) \\ \dot{y} &= -x \\ \dot{z} &= g\,[I(x)x^2 - z] \end{aligned} \quad (12.15)$$

where m, g, B and ν - are the system's parameters, $I(x) = 0$ ($x \le 0$) and $I(x) = 1$ ($x < 0$). Parameters g and B were fixed at the level of 0.3 and 0.5, respectively. The regimes were controlled by means of parameter m and ν. First, system (12.15) was integrated and *stroboscopic map* of the trajectory was constructed in a period of external force $T = 2\pi/\nu$. A *discrete sequence* of coordinate $\{x_i^s\}$ of the map was *used as the realization* for reconstruction. The experiments were conducted both with unperturbed system (12.15) and with introduction of additive sources of δ-correlated noise with intensity D and zero average into the right-hand parts of Eqs. (12.15).

We have taken an advantage of qualitatively equivalent behavior of systems (12.2) and (12.15) in the vicinity of resonances, investigated in detail in [12.20]. Similarly to the experiments, described in Sec.12.4, the characteristic points in the space of parameters m, v in the bifurcation diagram of DS (12.15) were selected, which are shown in Fig. 12.13. The right-hand parts of system (12.15) are given by more complicated functions, than merely polynomials, as it happened to be in the previous cases. Naturally, the stroboscopic map would be approximated by polynomials with a

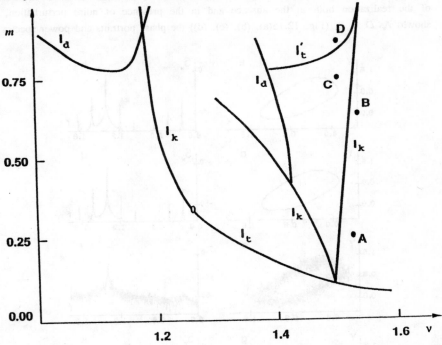

Fig. 12.13. Bifurcation diagram for the model (12.15) on the m, v plane as $g=0.3$, $B=0.5$:
  l_t, l'_t - torus arising curves,
  l_k - saddle-node bifurcation curves,
  l_d - period-doubling bifurcation curves,
  A, B - points with $v=1.525$ and $m=0.25$, 0.65, respectively,
  C, D - points with $v=1.516$ and $m=0.75$, 0.85, respectively.

greater error, even in the absence of noise.

The reconstruction of system (12.15) map has been conducted in the regime, that corresponded to the attractor in the form of smooth torus in the region of ergodic torus ($m = 0.25$, $v = 1.525$, point A in Fig. 12.13) as $D = 0$ and $D = 0.05$. Both for initial and perturbed system (12.15), a *smooth invariant curve* is reconstructed, qualitatively equivalent to the attractor of system (12.15), that is shown in Fig. 12.14. To achieve greater similarity, the phase portraits in this case and later on are constructed using a single coordinate x with delay.

A torus-chaos regime is worse restored. In Fig. 12.15, the results of computation of the initial system at point B in Fig. 12.13 ($m = 0.65$, $\nu = 1.525$), and reconstruction of the realization both in the absence and in the presence of noise perturbation, are shown. As $D = 0$, (Fig. 12.15(a), (b), (c), (d)) the phase portraits and power spectra of

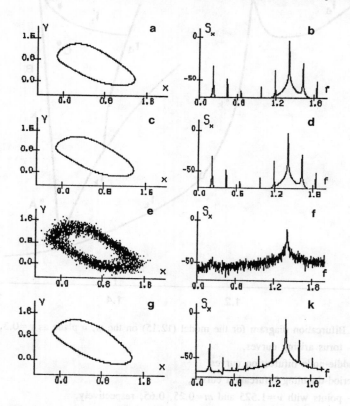

Fig. 12.14. Phase portraits and power spectra for the stroboscopic map of system (12.15) at point A in Fig. 12.13 for $D=0.0$ (a, b) and $D=0.05$ (e, f) and for the corresponding reconstructed maps as $D=0.0$ (c, d) and $D=0.05$ (g, k).

the initial and reconstructed attractors are highly different. Detailed examination of the two-dimensional distributions of probability densities for these two regimes (Fig. 12.16) sheds light on the nature of the non-correspondence obtained. It is evident that again the most probable part of the initial attractor is reconstructed, which is similar in its shape to a non-smooth torus. However, three spiral lines are added, whose structure reminds the shape of the unstable manifolds of saddle equilibrium point, on whose basis the torus attractor has originated and developed [12.29, 12.30].

Thus, in the process of reconstruction, the distortion of the initial attractor structure is the consequence of the loss of the least probable part and visualization of some hidden one of its phase portrait. Addition of noise with intensity $D = 0.05$ significantly affects the attractor's shape (its fractal structure is smeared), although

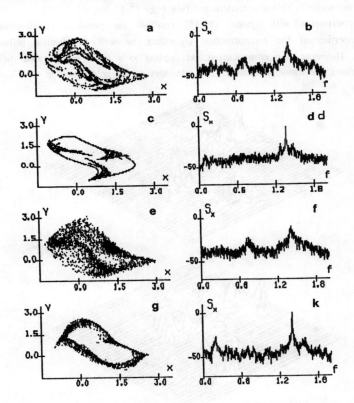

Fig. 12.15. Phase portraits and power spectra for the stroboscopic map of system (12.15) at point B in Fig. 12.13 for $D=0.0$ (a, b) and $D=0.05$ (e, f) and for the corresponding reconstructed maps at $D=0.0$ (c, d) and $D=0.05$ (g, k).

no qualitative changes are brought about, what is testified by power spectrum plot (Fig. 12.15 (e), (f)). As a result of reconstruction, an attractor is obtained, that is more similar to the initial one, than for $D = 0$ (Fig. 12.15(g), (k)). This seems to be explained by the fact that noise increases the frequency of attendance of the less probable attractors regions.

The similar picture is observed in the region of resonance both in the absence and in the presence of invisible homoclinics in qualitative correspondence with the data of Sec. 12.4. In the first case, the resonant cycle is reconstructed from the noisy realization, while in the second case, chaotic attractor is reconstructed. In Fig. 12.17 the characteristic results are shown to illustrate the above said. *A noisy cycle in the presence of homoclinics is not reconstructed* (see Fig. 12.17 (c), (d)).

The experiments with system (12.15) confirm the previous conclusions about filtering properties of the reconstruction algorithm, as well as specific influence of homoclinics. However, the computations, as applied to a more complicated differential system, revealed one more peculiarity. The attractors gained in the course of

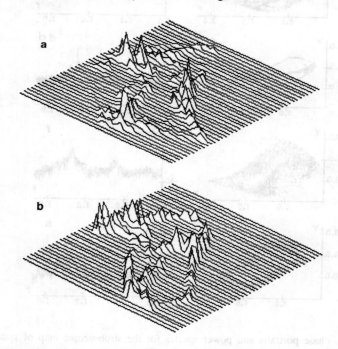

Fig. 12.16. Probability distribution densities on the attractor for stroboscopic map of system (12.15) at point B in Fig. 12.13 for $D=0.0$ (a) and for the corresponding reconstructed map (b).

reconstruction significantly differ from the initial ones in their shape, although their qualitative structure is equivalent. This important circumstance brings an idea, that in general case of more complicated systems, one can state with confidence only about reconstruction of attractors, equivalent in the sense of probabilistic measure.

The results of the investigations presented here allow one to conclude about the possibility to reconstruct the model equations relative to simple dynamical systems from the realization under the influence of uncorrelated noise. The reconstruction is understood in the sense, that the attractors of restored model system appear to be equivalent to the attractors of the initial system. The algorithms of reconstruction provide for *filtering of additive noise*.

It has also been found that the possibility of reconstruction essentially *depends on the character of homoclinic trajectories of the initial system*. If the homoclinic

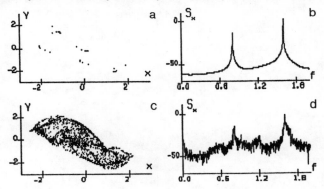

Fig. 12.17. Phase portraits and power spectra of reconstructed stroboscopic maps of system (12.15) at point C in Fig. 12.13 for $D=0.05$ (a, b) and at point D in Fig. 12.13 for $D=0.05$ (c,d).

structure is included in the attractor, then the reconstruction results in restoration of the attractors of initial system. The influence of *small noise* in this case induces *small perturbation* of the attractor's probabilistic measure and due to the filtering properties of algorithm is practically eliminated in the process of reconstruction.

A different situation takes place in the case of *homoclinic structures* which are *not included* in the attractor of the initial system. In the absence of noise, the probabilistic measure is concentrated on the attractor, whose structure is not determined by homoclinics. *Noise perturbation induces new regimes*. Then, the structure of attractors essentially depends on the homoclinic structure. As a result, the distribution density of the perturbed system attractor is being formed under the influence of homoclinics. In the final analysis, in this case of reconstruction the model system is restored, whose attractor reproduces the invisible homoclinic structure! Depending upon the intensity of noise perturbation we can have entirely different results of reconstruction.

And the last. The experiments conducted have shown that the problem of reconstruction is solved in the probabilistic sense. The most probable part of the attractor is restored; the least probable part is lost. We cannot claim for sure that this is the final conclusion. It is quite possible that with the increase of duration of time series and with further improvement of approximation techniques the results of reconstruction will be more complete and detailed.

BIBLIOGRAPHY

CHAPTER 1

[1.1] A.M. Lyapunov - *The Complete Works*, **1,2**, Acad. Press. Moscow (1954-1956).

[1.2] B.P. Demidovich - *Lectures on Mathematical Theory of Stability*, Nauka Publishers, Moscow (1967).

[1.3] V.S. Anishchenko - *Dynamical Chaos - Basic Concepts*, Teubner-Texte zur Physik, **Bd.14**, Leipzig (1987).

[1.4] T.S. Parker, L.O. Chua - *Practical Numerical Algorithms for Chaotic Systems*, Springer-Verlag (1989).

[1.5] V.S. Anishchenko - *Complicated Oscillations in simple Systems: Appearance Routes, Structure and Properties of Dynamical Chaos in Radiophysical Systems*, Nauka Publishers, Moscow (1990).

[1.6] V.F. Bylov et.al.- *Theory of Lyapunov Exponents and its Applications to Questions of Stability Theory*, Nauka Publishers, Moscow (1966).

[1.7] V.L. Oseledez "Multiplicative Ergodic Theorem: Lyapunov Characteristic Exponents for Dynamical Systems", Works of Mosk. Math. Sci. **19**, p.179 (1968).

[1.8] Yu.B. Pesin.- "Lyapunov Characteristic Exponents and Ergodic Theory", Usp. Mat. Nauk, **32**, No.4, pp.55-112 (1977).

[1.9] V.A. Andronov, A.A. Vitt, S.E. Haykin - *Theory of Oscillations*, Nauka Publishers, Moscow (1981).

[1.10] H. Haken - *Advanced Synergetics*. Springer-Verlag (1983).

[1.11] Yu.I. Neimark - *Method of Return Mapping in the Nonlinear Oscillations Theory*, Nauka Publishers, Moscow (1972).

[1.12] A.A. Andronov, L.S. Pontryagin - "The Rough Systems", Dokl. Acad.Nauk USSR **14**, p.5 (1937).

[1.13] A.A. Andronov et.al.- *Bifurcation Theory of Dynamical System on the Plane*, Nauka Publishers, Moscow (1967).

[1.14] V.V. Nemytzky, V.V. Stepanov - *Qualitative Theory of Differential Equations*, Fizmatgiz Publishers, Moscow (1947).

[1.15] N.N. Bautin - *Behavior of Dynamical System near the Boundaries of Stability Region*, Nauka Publishers, Moscow (1984).

[1.16] L.E. Marsden, M. McCracken - *The Hopf Bifurcation and its Applications*, Springer-Verlag, New York (1976).

[1.17] G. Iooss, D.D. Joseph - *Elementary Stability and Bifurcation Theory*, Springer-Verlag, New York, Heidelberg, Berlin (1980).

[1.18] B.D. Hassard, N.D. Kazarinoff, Y.-H. Wan - *Theory and Applications of Hopf Bifurcation*, Cambridge Univ. Press (1981).

[1.19] V.I. Arnold - *Complementary Chapters of Ordinary Differential Equation Theory*, Nauka Publishers, Moscow (1978).

[1.20] A.D. Bazykin, Yu.A. Kuznetzov, A.I. Khibnik - *The Bifurcational Diagrams of the Dynamical Systems on the Plane*, Center of Biolog.Res., Pushchino (1985).

[1.21] T. Poston, I. Steward - *Catastrophe Theory and its Applications*, Pitman, London, San Francisco, Melbourne (1978).

[1.22] V.I. Arnold - *Catastrophe Theory*, Znanie Publishers, Moscow (1983).

[1.23] N.N. Bautin, E.M. Leontovich - *The Qualitative Research Methods for Dynamical Systems on the Plane*, Nauka Publishers, Moscow (1976).

[1.24] A.I. Khibnik - *Periodic Solutions of the Differential Equations System*, Algorithms and Programs in FORTRAN 5, Center of Biolog.Res., Pushchino (1979).

[1.25] L.P. Shilnikov - "The Bifurcations Theory and Lorenz model", Appl.2 In: J. Marsden, H. McCracken - *The Hopf Bifurcation and its Applications*, Russ. Transl. Mir Publishers, Moscow (1980).

CHAPTER 2

[2.1] V.S. Anishchenko - *Dynamical Chaos - Basic Concepts*, Teubner-Texte zur Physik, **Bd.14**, Leipzig (1987).

[2.2] T.S. Parker, L.O. Chua - *Practical Numerical Algorithms for Chaotic Systems*, Springer-Verlag (1989)

[2.3] M. Holodniok, A. Klic, M. Kubicek, M. Marek - *Metody Analyzy Nelinearnich Dynamickych Modelu*. Academia, Praha (1986).

[2.4] V.S. Anishchenko - *Complicated Oscillations in simple Systems: Appearance Routes, Structure and Properties of Dynamical Chaos in Radiophysical Systems*, Nauka Publishers, Moscow (1990).

[2.5] V.S. Anishchenko - *Dynamical Chaos in Physical Systems - Experimental Investigation of Self-Oscillating Circuits*, Teubner-Texte zur Physik, **Bd.22**, Leipzig (1989).

[2.6] M. Henon - "On the Numerical Computation of Poincaré Maps", Physica **5D**, No.2,3, pp.412-414 (1982).

[2.7] M.I. Rabinovich, A.L. Fabrikant - "Stochastic automodulation of waves in non-equilibrium media", J.Theor. and Exp. Phys. **77**, No.2(8), pp.617-629 (1979).

[2.8] C. Hayashi, H. Kawakami - "Bifurcations and the Generation of Chaotic States in the Solutions of Non-linear Differential Equations", Theor. and Appl.Mech., 4 Intern. Symp. **B.1**, pp.537-542, Sofia (1981)

[2.9] A.I. Khibnik - *Periodic Solutions of the Differential Equations System*, Algorithms and Programs in FORTRAN 5, Center of Biolog.Res., Pushchino (1979).

[2.10] V.I. Oseledec - "Multiplicative ergodic theorem. Characteristic Lyapunov exponents of dynamical systems", Works of Math. Sci. of Moscow **19**, pp.179-210 (1968).

[2.11] Yu.B. Pesin - "Characteristic Lyapunov exponents and the smooth ergodic theory", Usp. Math. Nauk **32**, No.4, pp.55-112 (1977).

[2.12] I. Shimada, T. Nagashima - "A Numerical Approach to Ergodic Problem of Dissipative Dynamical Systems", Progr.Theor.Phys. **61**, No.6, pp.1605-1616 (1979).

[2.13] G. Benettin, L. Galgani, J.M. Strelcyn - "Kolmogorov Entropy and Numerical Experiments", Phys.Rev. **41A**, No.6, pp.2338-2442 (1976).

[2.14] Yu.I. Neimark, L.P. Shilnikov - "On one Case of the Periodic Motions Birth", Izv. Vuzov. Radiophysika **8**, No.2, pp.330-340 (1965).

[2.15] L.P. Shilnikov - "A case of the existence of a countable number of periodic motions", Soviet. Math. Docl. **6**, pp.163-166 (1965); "A contribution to the problem of the structure of an extended neighborhood of a rough equilibrium of saddle-focus type", Math. USSR Sbornik **10**, pp.91-102 (1970).

[2.16] A. Arneodo, P. Collet, C. Tresser - "Oscillator with chaotic behavior: An illustration of a theorem by Shilnikov", J. Stat. Phys. **27**, pp.171-182 (1982).

[2.17] Yu.A. Kuznetzov - *One-dimensional Separatrixes of the Ordinary Differential Equations System depending on the Parameter*, The Matters on the Computers Software **8**, Pushchino, Scientific Center of Biological Researches of AS USSR (1983).

[2.18] J.D. Farmer - "Information Dimension and the Probabilistic Structure of Chaos", Z. Naturforsh. **337A**, H.11, s.1304-1325 (1982).

[2.19] J.D. Farmer, E.Ott, J.A.Yorke - "The Dimension of Chaotic Attractors", Physica **7D**, No.1-3, P.153-180 (1983).

[2.20] L.S. Young - "Dimension, Entropy and Lyapunov Exponents", Ergod. Theory and Dyn. Systems **2**, pp.109-124 (1982).

[2.21] F. Hunt, F. Sullivan - "Efficient Algorithms for Computing Fractal Dimensions", In: *Dimensions and entropies in Chaotic Systems*, G. Mayer-Kress, Editor, pp.74-81, Springer-Verlag, N.Y. (1986).

[2.22] *Dimensions and Entropies in Chaotic Systems*, G. Mayer-Kress, Editor, Springer-

Verlag, N.Y.(1986).

[2.23] P. Grassberger - "On the Hausdorf Dimension of Fractal Attractors", J. Stat.Phys. **26**, No.1, pp.173-179 (1981).

[2.24] P. Grassberger - "Generalized Dimension of Strange Attractors", Phys.Lett. **97A**, No.6, pp.227-231 (1983).

[2.25] P. Grassberger, I.Procaccia - "Measuring the Strangeness of Strange Attractors", Physica **9D**, No.1,2, pp.189-208 (1983).

[2.26] F. Takens - "Detecting Strange Attractors in Turbulence", Lect. Notes in Math. **898**, pp.366-381, Springer-Verlag, Warwick (1980).

[2.27] J.L. Kaplan, J.A. Yorke - "Chaotic Behavior of Multi-Dimensional Difference Equations", Lect. Notes in Math. **730**, pp.204-227 (1979).

[2.28] J.D. Farmer - "Chaotic Attractor of an Infinite-Dimensional Dynamical System", Physica **4D**, No.3, pp.366-393 (1982).

[2.29] D.A. Russel, J.D. Hanson, E. Ott - "Dimension of Strange Attractors", Phys.Rev.Lett. **45**, No.14, pp.1175-1178 (1980).

[2.30] V.S. Anishchenko, M.A. Safonova - "Comparative Analysis of Various Dimensions of a Chaotic Attractor", Sov.Tech.Phys.Lett. **15(6)**, pp.472-473 (1989).

CHAPTER 3

[3.1] V.S. Anishchenko, V.V. Astakhov, T.E. Letchford, M.A. Safonova - "On bifurcations in three-dimensional two-parameter self-contained oscillation system with a strange attractor", Izv. Vuzov - Radiofizika **26**, No.2, pp.169-176 (1983).

[3.2] V.S. Anishchenko - *Dynamical Chaos in Physical Systems. Experimental Investigation of Self-Oscillating Circuits*, Teubner-Texte zur Physik, **Bd.22**, Leipzig (1989).

[3.3] O.E. Rossler - "An Equation for Continuous Chaos", Phys.Lett. **57A**, No.5, pp.397-398 (1976).

[3.4] A.S. Pikovsky, M.I. Rabinovich - "A simple oscillator with stochastic behavior", Sov.Phys.Dokl. **239**, No.2, pp.301-304 (1978).

[3.5] E.N. Lorenz - "Deterministic Nonperiodic Flow", J.Atmos.Sci. **20**, No.2, pp.130-141 (1963).

[3.6] V.I. Babitsky, P.S. Landa - "Auto-Oscillation Systems with inertial Self-Excitation", Z.Angew.Math. und Mech. **64**, No.8, pp.329-339 (1984).

[3.7] K.E. Theodorchik - "Auto-Oscillation Systems with inertial non-linearity", Zh. Tekhn. Fiz. **16**, No.7, pp.845-854 (1946).

[3.8] V.S. Anishchenko, V.V. Astakhov, T.E. Letchford - "Multifrequency and stochastic oscillations in a generator with inertial non-linearity", Radiotekhnika i elektronika **27**, No.10, pp.1972-1978 (1982).

[3.9] N.N. Bautin - "On arising of a limit cycle from equilibrium point of focus type", Zh.Exp. i Teor.Fiz. **8**, No.6, pp.648-654 (1938).

[3.10] T. Poston, I. Stewart. *Catastrophe Theory and its Applications*, Pitman (1978).

[3.11] V.S. Anishchenko, V.V. Astakhov, T.E. Letchford, M.A. Safonova. "On structure of quasyhyperbolic chaos in an inertial generator", Izv. Vuzov - Radiofizika **26**, No.7, pp.832-842 (1983).

CHAPTER 4

[4.1] V.S. Anishchenko - *Dynamical Chaos in Physical Systems. Experimental Investigation of Self-Oscillating Circuits*, Teubner-Texte zur Physik, **Bd.22**, Leipzig (1989).

[4.2] V.S. Anishchenko - *Complicated oscillations in simple systems: Appearance, routes, structure and properties of dynamical chaos in radiophysical systems*, Nauka Publishers, Moscow (1990).

[4.3] V.S. Anishchenko, V.V. Astakhov - "Experimental study of strange attractor initiation mechanism and structure in generator with inertial non-linearity", Radiotekhnika i elektronika **28**, No.6, pp.1109-1115 (1983).

[4.4] F.M. Izrailev, M.I. Rabinovich, A.D. Ugodnikov - "Approximate Description of Three-Dimensional Dissipative Systems", Phys.Lett. **86A**, No.6-7, pp.321-325 (1981).

[4.5] V.S. Anishchenko, V.V. Astakhov, T.E. Letchford, M.A. Safonova. "On bifurcations in three-dimensional two-parameter self-contained oscillation system with a strange attractor", Izv. Vuzov - Radiofizika **26**, No.2, pp.169-176 (1983).

[4.6] J. Testa, J. Perez, C. Jeffries - "Evidence for Universal Chaotic Behavior in Driven Non-linear Oscillator", Phys.Rev.Lett. **48A**, No.11, pp.714-717 (1982).

[4.7] B.A. Huberman, J. Rudnik - "Scaling Behavior of Chaotic Flows", Phys.Rev.Lett. **45**, No.3, pp.154-156 (1980).

[4.8] F.T. Arecci, F. Lisi - "Mechanism Generating 1/f Noise in Non-linear Systems", Phys.Rev.Lett. **49**, No.2, pp.94-98 (1982).

[4.9] F.T. Arecci, R. Badii, A. Politi - "Low frequency Phenomena in dynamical Systems with many attractors", Phys.Rev. **29A**, pp.1006-1009 (1984).

[4.10] V.S. Anishchenko - "Interaction of strange attractors and chaos-chaos intermittency", Sov. Tech. Phys. Lett. **10**, pp.266-270 (1984).

[4.11] V.I. Babitsky, P.S. Landa - "Auto-Oscillation System with inertial Self-

Excitation", J. Angew. Math. and Mech **64**, No.8, s.329-339 (1984).

[4.12] C. Sparrow. *The Lorenz Equations: Bifurcations, Chaos and Strange Attractor*, Appl.Math.Sci., Springer Verlag (1982).

[4.13] V.S. Anishchenko, T.E. Letchford, M.A. Safonova - "Influence of dissipative non-linearity on bifurcations in chaotic systems", Radiotekhnika i elektronika **29**, No.7, pp.1355-1361 (1984).

CHAPTER 5

[5.1] V.S. Anishchenko - *Dynamical Chaos in Physical Systems: Experimental Investigation of Self-Oscillating Circuits*, Teubner-Texte zur Physik, **Bd.22**, Leipzig (1989).

[5.2] A. Arneodo, P. Collet, C. Tresser - "Oscillators with Chaotic Behavior: an Illustration of a Theorem of Shilnikov", J.Stat.Phys. **27**, No.1, pp.171-182 (1982).

[5.3] P. Gaspard, G. Nikolis - "What can we learn from homoclinic Orbits in Chaotic Dynamics?", J.Stat.Phys. **31**, No.3, pp.449-518 (1983).

[5.4] P. Glendinning, C. Sparrow - "Local and Global Behavior near Homoclinic Orbits", J.Stat.Phys. **35**, No.5/6, pp.645-696 (1984).

[5.5] V.S. Anishchenko, M.A. Safonova - "Bifurcations of attractors in the presence of fluctuations", Sov.Phys.Tech.Phys. **33**, pp.391-397 (1988).

[5.6] D.L. Hitzl, F. Zele - "An Exploration of the Henon Attractor", J.Stat.Phys. **26**, No.4, pp.683-695 (1981).

[5.7] S.V. Gonchenko, L.P. Shilnikov - "On dynamical Systems with structurally unstable homoclinic curves", Sov.Math.Dokl. **33**, pp.234-239 (1986).

[5.8] L.P. Shilnikov. - "Bifurcation Theory and Turbulence", in: *Nonlinear and Turbulence Processes*, **2**, pp.1627-1635, Gordon and Breach, Harwood Academic Publishers, N.Y (1984).

[5.9] V.S. Anishchenko, M.A. Safonova - "Comparative Analysis of Various Dimensions of a Chaotic Attractors", Sov. Tech. Phys. Lett., **15(6)**, (1989).

CHAPTER 6

[6.1] E.N. Lorenz - "Deterministic Nonperiodic Flow", J.Atmos.Sci. **20**, No.2, pp.130-141 (1963).

[6.2] D. Ruelle, F. Takens - "On the Nature of Turbulence", Commun. Math. Phys. **20**,

No.2, pp.167-192 (1971).

[6.3] D. Ruelle - "The Lorenz Attractor and the Problem of Turbulence", Lect. Notes in Math. **565**, Springer, Berlin, pp.146-158 (1976).

[6.4] J.M. Guckenheimer, Ph. Holmes - *Non-linear Oscillations, Dynamical Systems and Bifurcation on Vector Fields*, Springer-Verlag, N.Y. (1983).

[6.5] A.L. Lichtenberg, M.A. Lieberman - *Regular and Stochastic Motion*, Springer-Verlag, N.Y. (1983).

[6.6] P. Berge, Y. Pomeau, C.H. Vidal - *Order within Chaos. (Towards Deterministic Approach to Turbulence)*, John Willey and Sons, N.Y. (1984).

[6.7] C. Sparrow - "Bifurcation, Chaos and Strange Attractors", Appl. Math. Sci. **41**, Springer-Verlag, Berlin (1982).

[6.8] V.S. Afraimovich, L.P. Shilnikov - "The ring principle and the problem on interaction of two autooscillation system", Prikl. matem. i mech. **42**, No.4, pp.618-627 (1977).

[6.9] V.S. Afraimovich, L.P. Shilnikov - "Invariant two-dimensional tori, their destruction and stochasticity", Methods of qualitative theory of differential equations, pp.3-26, Gorky University publisher, Gorky (1983).

[6.10] M.J. Feigenbaum, L.P. Kadanoff, S.J. Shenker - "Quasiperiodicity in Dissipative Systems: a Renormalization Group Analysis", Physica **5D**, No.2, pp.370-386 (1982).

[6.11] D. Rand, S. Ostland, J. Sethna, E. Siggia - "Universal Transition from Quasiperiodicity to Chaos in Dissipative Systems", Physica **8D**, No.3, pp.303-342 (1983).

[6.12] K. Kaneko. *Collapse of Tori and Genesis of Chaos in Dissipative Systems*, World Scientific, Singapore (1986).

[6.13] V.S. Anishchenko - *Dynamical Chaos in Physical Systems: Experimental Investigation of Self-Oscillating Circuits*, Teubner-Texte zur Physik, **Bd22**, Leipzig (1989).

[6.14] V.S. Anishchenko, V.V. Astakhov - "Bifurcation phenomena in an autostochastic oscillator responding to a regular external signal", Sov.Phys.Tech.Phys. **28**, pp.1326-1231 (1983).

[6.15] V.S. Anishchenko, T.E. Letchford, M.A. Safonova - "Phase Locking Effects and Bifurcations of Phase Locking and Quasiperiodic Oscillations in a Nonautonomous Generator", Izv. Vuzov, Radiofizika **28**, No.9, pp.1112-1125 (1985).

[6.16] V. Franceschini - "Bifurcations of Tori and Phase Locking in a Dissipative System of Differential Equations", Physica **6D**, No.3, pp.285-304 (1983).

[6.17] K. Kaneko - "Oscillation and Doubling of Torus", Progr.Theor.Phys. **72**, No.2, pp.202-215 (1984).

[6.18] S.P. Kuznetsov - "On periodical external perturbation of system demonstrating the transition to chaos due to period doubling bifurcations", Pisma v Zh. Eksp. i Teor. Fiz. **39**, No.3, pp.113-116 (1984).

[6.19] V.S. Anishchenko - "Destruction of quasiperiodic oscillations leading to chaos in dissipative systems", Sov.Phys.Tech.Phys. **31(2)**, pp.137-144 (1986).

[6.20] B.I. Shraiman - Transition from Quasiperiodicity to Chaos: a Perturbative Renormalization-Group Approach", Phys.Rev. **29A**, No.6, pp.3464-3466 (1984).

[6.21] S.J. Shenker - "Scaling Behavior in a Map of a Circle into Itself: Empirical Results", Physica **5D**, pp.405-411 (1982).

[6.22] J.D. Farmer, F. Sattiggia - "Renormalization of Quasiperiodic Transition to Chaos for Arbitrary Winding Numbers", Phys.Rev. **31A**, No.5, pp.3520-3522 (1985).

[6.23] V.S. Anishchenko, M.A. Safonova - "Mechanisms of destruction of invariant curve in a model map of plane", Radiotechnika i Elektronika **32**, No.6, pp.1207-1216 (1987).

[6.24] V.S. Anishchenko, T.E. Letchford, D.M. Sonechkin - "Universal properties of the soft transition to chaos via two-frequency oscillations", Sov.Phys.Tech.Phys., **33**, pp.517-522 (1988).

CHAPTER 7

[7.1] V.S. Anishchenko, T.E. Letchford, M.A. Safonova - "Stochasticity and destruction of quasiperiodic motion due to doubling in a system of coupled generators", Radiophys. & Quantum Electr. **27**, pp.381-90 (1984).

[7.2] V.S. Anishchenko - "Destruction of quasiperiodic oscillations leading to chaos dissipative systems", Sov.Phys.Tech.Phys. **31(2)**, pp.137-144 (1986).

[7.3] V.I. Arnold - "Loss of stability of oscillations near the resonances", In: *Nonlinear waves*, A.V.Gaponov-Grechov editor, pp.116-130, Nauka Publisher, Moscow (1979).

[7.4] K. Kaneko - *Collapse of Tori and Genesis of Chaos in Dissipative systems*, World Scientific, Singapore (1986).

[7.5] J.-M. Yuan, M. Tung, D.N. Fend, L.M. Narducci - "Instability and Irregular Behavior of Coupled Logistic Equations", Phys.Rev. **28A1**, No.3, pp.1662-1666 (1983).

[7.6] S.P. Kuznetsov - "On critical behavior of one-dimensional chains", Sov.Tech. Phys.Lett. **9**, No.2, pp.94-98 (1983).

[7.7] V.S. Anishchenko, M.A. Safonova - "Mechanisms of destruction of invariant curve in the model plane map", Radiotechnika i elektronika **32**, No.6, pp.1207-1216 (1987).

[7.8] S. Newhouse, D. Ruelle, F. Takens - "Occurrence of Strange Axiom A Attractor Near Quasi-Periodic Flows on T^m, $m=3$", Commun. Math. Phys. **64**, pp.35-40 (1978).

[7.9] H.T. Moon, P. Huerre, L.G. Redekopp - "Three-Frequency Motion and Chaos in the Ginzburg-Landau Equations", Phys.Rev.Lett **49**, No.7, pp.458-460 (1982).

[7.10] J.P. Gollub, S.V. Benson - "Many Routes to Turbulent Convection", J. Fluid. Mech. **100**, No.3, pp.449-470 (1980).

[7.11] V.S. Anishchenko, T.E. Letchford, M.A. Safonova - "Critical Phenomena under harmonic modulation of two-frequency oscillations", Sov.Tech.Phys.Lett. **11**, No.9, pp.536-541 (1985).

[7.12] V.S. Anishchenko, T.E. Letchford - "Breakdown of three-frequency oscillations and onset of chaos in a biharmonically excited oscillator", Sov.Phys.Tech.Phys. **31**, pp.1347-1349, (1986).

CHAPTER 8

[8.1] Ch. Hayashi. - *Nonlinear oscillations in physical systems*, McCraw-hill book company (1964).

[8.2] P. Berge, Y. Pomeau, Ch. Vidal - *L'ordre dans le chaos. Versune approche deterministe de la turbulence*, Hermann ed. (1988).

[8.3] V.S. Afraimovich, N.N. Verichev, M.I. Rabinovich - "Stochastically synchronized oscillations in dissipative systems", Izv. Vuzov. Radiofizika **29**, No.9, pp.105-1060 (1986).

[8.4] N.N. Verichev, A.G. Maksimov - "About synchronization of stochastic oscillations of parametrically excited nonlinear oscillators", Izv. Vuzov. Radiofizika **32**, No.8, pp.962-965, (1989).

[8.5] A.P. Volkovskii, N.F. Rul'kov - "Experimental investigations of bifurcations on the onset of stochastical synchronization", Sov.Tech.Phys.Lett. **15**, No.7, pp.5-10, (1989).

[8.6] V.V. Astakhov, B.P. Bezruchko, Yu.V. Guljaev, S.P. Seleznev - "Multistable regimes of dissipativelly coupled Feigenbaum systems", Sov.Tech.Phys.Lett. **15**, No.3, pp.60-65 (1989).

[8.7] Z.M. Pecora, T.J. Carroll - "Synchronization in chaotic systems", Phys.Lett. **64**, No.8, pp.821-824 (1990).

[8.8] V.S. Anishchenko, V.V. Astakhov - "Experimental investigation of the mechanism of arising and the structure of a strange attractor in a generator with inertia nonlinearity", Radiotechnika i Elektronika **28**, No.6, pp.1109-1115 (1983).

[8.9] V.S. Anishchenko - *Dynamical Chaos in Physical Systems: Experimental*

Investigation of Self-Oscillating Circuits, Teubner-Texte zur Physik, **Bd.22**, Leipzig (1989).

[8.10] V.S. Anishchenko, T.E. Vadivasova, D.E. Postnov, M.A. Safonova - "Synchronization of chaos", Int.J.Bifurcation and Chaos **2**, No.3, pp.633-644 (1992).

[8.11] V.S. Anishchenko, D.E. Postnov - "Selforganization in Chaos. A New method of experimental diagnostics", Sov.Tech.Phys.Lett. **15**, No.24, pp.28-32 (1990).

[8.12] C. Tresser - "About some theorems of L.P. Shilnikov", Ann.de J I.H.P **40**, pp.441-461 (1984).

[8.13] O.E. Rossler - "Continuous chaos - Four prototype equations", Ann.N.Y.Acad.Sci. **316**, pp.376-392 (1979).

[8.14] W. Shuxian - "Chua's Circuit Family", Proc. IEEE **75**, No.8, pp.1022-1032 (1987).

[8.15] Yu.L. Klimontovich - *Turbulent motion. Structure of chaos*. Kluwer, Academic Publishers, Dordrecht (1991).

CHAPTER 9

[9.1] L.P. Shilnikov - "A case of the existence of a countable number of periodic motions", Sov.Math.Doklady **6**, pp.163-166, Jan.-Feb.(1965).

[9.2] L.P. Shilnikov - "A contribution to the problem of the structure of an extended neighborhood of rough equilibrium state of saddle-focus type", Math. USSR-Sbornik **10**, pp.91-102, Jan.-Feb.(1970).

[9.3] L.O. Chua, M. Komuro, T. Matsumoto - "The double scroll family", Parts I and II, IEEE trans. Circuits Syst. **CAS-33**, pp.1073-1118 (1986).

[9.4] L.O. Chua - "Global unfolding of Chua's circuit", IEICE trans. Fundamentals (Special Section on Neural Nets, Chaos and Numerics) **E76-A**, pp.704-734 (1993).

[9.5] L.O. Chua - "The Genesis of Chua's circuit", Archiv für Elektronik and Ubertragungs-technik **46**, pp.250-257 (1992).

[9.6] R.N. Madan (editor) - *Chua's circuit: A Paradinm for Chaos*, World Scientific Publishing Co.Pte.Ltd. (1993).

[9.7] L.P. Shilnikov - "Bifurcation theory and turbulence", In: *Nonlinear and Turbulence Processes*, **2**, Gordon and Breach, Harwood Acad.Publ., pp.1627-1635 (1984).

[9.8] H.G. Schuster - *Deterministic Chaos*, Physik-Verlag, Weinheim (1984).

[9.9] V.S. Anishchenko - "Interation of strange attractors and chaos-chaos intermittency", Sov.Tech.Phys.Lett. **10**, pp.266-270 (1984).

[9.10] F.T. Areci, R. Badii, A. Politi - "Low-frequency phenomena in dynamical systems with many attractors", Phys.Rev. **A29**, pp.1006-1009 (1984).

[9.11] H. Ishii, H. Fudjisaka, M. Inoue - "Breakdown of chaos symmetry and intermittency in the double-well potential system", Phys.Lett. **A116**, pp.257-263 (1986).

[9.12] J.E. Hirsch, B.A. Hubermann, D.J. Scalapino - "Theory of intermittency", Phys.Rev. **A25**, pp.519-525 (1981).

[9.13] P.S. Landa, R.L. Stratanovich - "On the theory of intermittency", Izv. Vuzov. Radiofizika **30**, pp.65-69 (1987).

[9.14] V.S. Anishchenko - *Dynamical Chaos in Physical Systems: Experimental Investigation of Self-Oscillating Circuits*, Teubner-Texte zur Physik, **Bd.22**, Leipzig (1989).

[9.15] V.S. Anishchenko, A.B. Neiman - "Increase in the duration of correlations during a chaos-chaos intermittency", Sov.Tech.Phys.Lett. **13**, No.17, pp.1063-1066 (1987).

[9.16] R. Ecke, H. Haucke - "Noise-induced intermittency in the quasiperiodic regime of Rayleigh-Benard convection", J.Stat.Phys. **54**, pp.1153-1172 (1989).

[9.17] P. Manneville - "Intermittency, self-similarity and 1/f - spectrum in dissipative dynamical systems", J.Phys. (Paris) **41**, pp.1235-1241 (1980).

[9.18] W. Li - "Generating nontrivial long-range correlations and 1/f - spectra by eplication and mutation", Int.J.Bifurcation and Chaos **2**, No.1, pp.137-154 (1991).

[9.19] I. Procaccia, H.G. Schuster - "Functional renormalization group theory of universal 1/f - noise in dynamical systems", Phys.Rev. **28**, pp.1210-1214 (1983).

[9.20] H. Fujisaka - "Theory of diffusion and intermittency in chaotic systems", Prog.Theor.Phys. **71**, pp.513-523 (1984).

[9.21] V.S. Anishchenko, A.B. Neiman - "Statistical properties of intermittency in quasihyperbolic systems", Sov.Phys.Tech.Phys. **17**, pp.3-14 (1990).

[9.22] L.O. Chua, I. Tichonicky - "1-D map for the double scroll family", IEEE Trans.Circuits and Systems **38**, pp.233-243 (1991).

[9.23] Yu.L. Klimontovich, *Statistical Physics*, Hardwood Academic Publishers (1986).

[9.24] V.S. Anishchenko, A.B. Neiman, L.O. Chua - "Chaos-chaos intermittency and 1/f - noise in Chua's circuit", Int.J.Bifurcation and Chaos **4**, No.1, pp.99-107 (1994).

[9.25] R. Benzi, A. Sutera, A. Vulpiani - "The mechanism of stochastic resonance", J.Phys. **A14**, L453-L457 (1981).

[9.26] S. Fauve, F. Heslot - "Stochastic resonance in a bistable systems", Phys.Lett. **A97**, pp.5-8 (1983).

[9.27] B. McNamara, K. Wiesenfeld - "Theory of stochastic resonance", Phys.Rev. **A39**, pp.4854-4869 (1989).

[9.28] P. Jung, P. Hänggi - "Resonantly driven Brownian motions: basic conceptions and exact results", Phys.Rev. **A14**, pp.2977-2988 (1990).

[9.29] F. Moss - "Stochastic Resonance: From the Ice Ages to the Monkey's Ear", In: *Same problems in Statistical Physics*, ed. G.H. Weiss, SIAM, Philadelphia (1992).

[9.30] V.S. Anishchenko, A.B. Neiman, M.A. Safonova - "Stochastic resonance in chaotic systems", J.Stat.Phys. **70**, No.1/2, pp.183-196, (1993).

[9.31] V.S. Anishchenko, A.B. Neiman - "Structure and properties of chaos in the presence of noise", In: *Nonlinear Dynamics of Structures*, eds. Sagdeev R.Z. et al., pp.21-48, World Scientific, Singapore (1991).

[9.32] W. Horsthemke, R. Lefever - *Noise-Induced Transitions. Theory and Applications in Physics, Chemistry and Biology*. Springer-Verlag (1984).

[9.33] V.S. Anishchenko, M.A. Safonova, L.O. Chua - "Stochastic resonance in Chua's circuit", Int.J.Bifurcation and Chaos **2**, No.2, pp.397-401 (1992).

[9.34] D. Ruelle, F. Takens - "On the nature of turbulence", Comm.Math.Phys. **20**, pp.167-192 (1971); **23**, pp.343-344 (1971).

[9.35] S. Newhouse, D. Ruelle, F. Takens - "Occurrance of strange axiom A attractors near quasi-periodic flows on T^m, m = 3", Comm.Math.Phys. **64**, pp. 35-40 (1978).

[9.36] L. Landau - "On the problem of turbulence", Dokl.Akad.Nauk USSR **44**, pp.339-342 (1944).

[9.37] V.S. Afraimovich, L.P. Shilnikov - "Invariant two-dimensional tori, their destruction and stochasticity", In: *Methods of Qualitative Theory of Differential Equations*, pp.3-26, Gorky Univ., Gorky, Russia (1983).

[9.38] T. Matsumoto, L.O. Chua, R. Tokunada - "Chaos via torus breakdown", IEEE trans. Circuits Syst. **CAS-34**, No.3 (1987).

[9.39] V.S. Anishchenko - "Destruction of quasi-periodic oscillations and chaos in dissipative systems", Sov.Phys.Tech.Phys. **31(2)**, pp.137-144 (1986).

[9.40] V.S. Anishchenko - *Complicated Oscillations in Simple Systems*, Moskow, Nauka Publ. (1990).

[9.41] V.S. Anishchenko, M.A. Safonova - "Mechanisms of distruction on invariant curve in a model map at the plane", Radiotechnika i Elektronika **32**, pp.1207-1216 (1987).

[9.42] V.S. Anishchenko, M.A. Safonova - "Noise-induced exponential dispersal of phase trajectories in the neighborhood of regular attractor", Sov.Tech.Phys.Lett. **12(6)**, pp.305-306 (June 1986).

[9.43] V.S. Anishchenko, H.-P. Herzel - "Noise-indused chaos in a system with homoclinic points", ZAMM **68**, pp.317-318 (1988).

CHAPTER 10

[10.1] C. Meunier, A.D. Verga - "Noise and bifurcations", J.Stat.Phys. **50**, pp.345-375, (1988).

[10.2] M. Mackey, A. Longtin, A. Lasota - "Noise-induced global asymptotic stability", J.Stat.Phys. **60**, pp.735-751, (1990).

[10.3] C.W. Gardiner - *Handbook of Stochastic Methods for Physics, Chemistry and Natural Sciences*, Springer, Berlin, (1984).

[10.4] Yu.L. Klimontovich - "Ito, Stratonovich and kinetic forms of stochastic equations", Physica **163A**, pp.515-532, (1990).

[10.5] H. Haken - *Advanced Synergetics: Instability Hierarchies of Self-Organizing Systems and Devices*, Springer-Verlag, Berlin, (1983).

[10.6] W. Horsthemke, R.Lefever - *Noise-Induced Transitions: Theory and Applications in Physics, Chemistry, and Biology*, Springer-Verlag, (1984).

[10.7] A.I. Khibnik - *Periodic Solutions of the Differential Equations System*, Algorithms and Programs in FORTRAN 5, Center of Biolog.Res., Pushchino (1979).

[10.8] H. Risken - *The Fokker-Planck Equation: Methods of Solution and Applications*, Springer-Verlag, Berlin, (1984).

[10.9] J.M. Sancho and M. San Miguel - "Langevin equations with colored noise", In: *Noise in Nonlinear Dynamical Systems*, eds. F. Moss and P.V.E. McClintock, **1**, pp.72-109, Cambridge University Press, (1989).

[10.10] P. Hanggi - "Colored Noise in continuous dynamical systems: a functional calculus approach", In: *Noise in Nonlinear Dynamical Systems*, eds. F. Moss and P.V.E. McClintock, **1**, pp.307-328, Cambridge University Press, (1989).

[10.11] J. Casademunt, R. Manella, P.V.E. McClintock, F. Moss, J.M. Sancho - "Relaxation times of non-Markovian processes", Phys.Rev. **35A**, p.3000, (1987).

[10.12] R. Fox - "Numerical simulation of stochastic differential equations", J.Stat.Phys. **47**, pp.1353-1366, (1989).

[10.13] R. Manella, V. Paleschi - "Fast and precise algorithm for computer simulation of stochastic differential equations", Phys.Rev. **40A**, pp.3381-3386, (1989).

[10.14] N.N. Nikitin, V.D. Razevig - "Methods for numerical simulation of stochastic differential equations and estimation of their accuracy", J.Comp.Math. **18**, pp.106-117, (1978).

[10.15] R. Fox, I. Gutland, R. Roy, G. Vemuri - "Fast accurate algorithm for numerical simulation of exponentially correlated colored noise", Phys.Rev. **38A**, pp.5938-5940, (1988).

[10.16] W. Ebeling, L. Schimansky-Geier - "Transition phenomena in multidimensional systems - models of evolution", In: *Noise in Nonlinear Dynamical Systems*, eds. F. Moss and P.V.E. McClintock, **1**, pp.279-306, Cambridge University Press, (1989).

[10.17] V.S. Anishchenko, A.B. Neiman - "Poincare return period in dynamical chaos regime", Sov.Phys.Tech.Phys. **59**, pp.117-118, (1989).

[10.18] O. Martin - "Lyapunov exponent of stochastic dynamical systems", J.Stat.Phys. **41**, pp.249-261, (1985).

[10.19] L. Arnold - "Stochastic systems: qualitative theory and Lyapunov exponents". In: *Fluctuations and Sensitivity in Nonequilibrium Systems*, eds. W. Horsthemke, D. Kondepudi, pp.11-18, Springer, Berlin, (1984).

[10.20] I. Shimada, T. Nagasima - "A numerical approach to ergodic problem of dissipative dynamical systems", Prog.Theor.Phys. **61**, pp.1605, (1979).

[10.21] M.A. Safonova - *Noise influence on the attractor bifurcations of auto-oscillation systems*. Ph.D. dissertation, Saratov, (1987), Russian.

[10.22] R.L. Stratonovich - *Nonlinear Nonequilibrium Thermo-dynamics*, Moscow, Nauka Publisher, (1985).

[10.23] R.L. Stratonovich - "Some Markov methods in the theory of stochastic process in nonlinear dynamical systems". In *Noise in Nonlinear Dynamical Systems*, eds. F. Moss and P.V.E. McClintock, **1**, pp.16-71, Cambridge University Press, (1989).

[10.24] A.N. Malakhov - *Cumulant analysis of stochastic non-gaussian processes and their transformations*, Sov.Radio, Moscow, (1978).

[10.25] W. Just, H. Sauermann - "Ordinary differential equations for nonlinear stochastic oscillators", Phys.Lett. **131A**, pp.234-238, (1988).

[10.26] S. Kai, H. Fukunaga, H. Brand - "Structure changes induced by external multiplicative noise in the electrohydrodynamic instability of nematic liquid crystals", J.Stat.Phys. **54**, pp.1133-1152, (1989).

[10.27] G. Ahlers, Gh. Meyer, D. Cannel - "Deterministic and stochastic effects near the convective onset", J.Stat.Phys. **54**, pp.1121-1130, (1989).

[10.28] R. Fox - "Stochastic calculus in physics", J.Stat.Phys. **46**, pp.1145-1157, (1987).

[10.29] N.G. Van Kampen - "Langevin-like equation with colored noise", J.Stat.Phys. **54**, pp.1289-1308, (1989).

[10.30] G. Debnath, F. Moss, Th. Leiber, H. Risken, F. Marchesoni - "Holes in the two-dimensional probability density of bistable systems driven by strongly colored noise", Phys.Rev. **42A**, pp.703-710, (1990).

[10.31] Th. Leiber, F. Marchesoni, H. Risken - "Colored noise and bistable Fokker-Planck equations", Phys.Rev.Lett. **59**, pp.1381-1384, (1987).

[10.32] Th. Leiber, F. Marchesoni, H. Risken - "Numerical analysis of stochastic relaxation in bistable systems driven by colored noise", Phys.Rev. **38A**, p.983, (1988).

[10.33] V.S. Anishchenko, A.B. Neiman - "Bifurcational analysis of bistable system excited by colored noise", Int.Journal of Bifurcation and Chaos **2**, pp.979-982, (1992).

[10.34] P. Hanggi, P. Jung, F. Marchesoni - "Escape driven strongly correlated noise", J.Stat.Phys. **54**, pp.1367-1380, (1989).

[10.35] C.T. Sparrow - *The Lorenz Equations: Bifurcations, Chaos and Strange Attractors*, Berlin, Springer, (1982).

[10.36] E.V. Astashkina, A.S. Mikhailov, A.V. Tolsopyatenko - "Noise-induced instability in Lorenz model", Izv.Vuz. Radiofizika **24**, pp.1035-1037, (1981).

[10.37] J. Crutchfield, B.A. Huberman - "Fluctuations and the onset of chaos", Phys.Lett. **77A**, pp.407-410, (1980).

[10.38] J. Crutchfield, J.D. Farmer - "Fluctuations and simple chaotic dynamics", Phys.Rep. **92**, pp.45-82, (1982).

[10.39] M. Napiorkowski, U. Zaus - "Average trajectories and fluctuations from noisy, nonlinear maps", J.Stat.Phys. **43**, pp.349-368, (1983).

[10.40] J. Crutchfield, N.H. Packard - "Symbolic dynamics of noisy chaos", Physica **3D**, pp.201-223, (1983).

[10.41] H. Svensmark, M.R. Samuelsen - "Influence of perturbations on period-doubling bifurcation", Phys.Rev. **36A**, pp.2413-2417, (1987).

[10.42] B. Shraiman, C.E. Wayne, P.C. Martin - "Scaling theory for ncisy period-doubling transition to chaos", Phys.Rev.Lett. **46**, pp.935-939, (1984).

[10.43] F.B. Vul, Ya.G. Sinai, K.M. Khanin - "Feigenbaum universality and thermodynamics formalism", Russian Math.Surveys **39**, pp.3-37, (1984).

[10.44] K. Wiesenfeld - "Noisy precursors of nonlinear instabilities", J.Stat.Phys. **38**, pp.1071-1097, (1985).

[10.45] V.S. Anishchenko, A.B. Neiman - "Statistical analysis of period doubling bifurcation in presence of noise", Radiotekhnika i Electronica **35**, pp.1666-1673, (1990); A.B. Neiman, V.S. Anishchenko, J. Kurth - "Period-doubling bifurcations in the presence of colored noise", Phys.Rev.E **49**, No.5, pp.3801-3806, (1994).

CHAPTER 11

[11.1] Yu. Kifer - "On small random perturbations of some smooth dynamical system", Izv.AN SSSR Matematica **38**, pp.1091-1115, (1974).

[11.2] Yu. Kifer - "Attractors via random perturbations", Commun.Math.Phys. **121**, pp.445-455, (1989).

[11.3] Ya. Sinai - "Stochasticity of dynamical systems", In: *Nonlinear Waves*, Eds. A.V.Gaponov-Grekhov, pp.192-211, Nauka Publisher, Moscow (1979).

[11.4] F.B. Vul, Ya.G. Sinai, K.M. Khanin - "Feigenbaum universality and thermodynamics formalism", Russian Math.Surveys **39**, pp.3-37, (1984).

[11.5] V.S. Afraimovich, L.P. Shilnikov - "On strange attractors and quasiattractors", In: *Nonlinear Dynamics and Turbulence*, Eds. G.I. Barenblatt, pp.1-34, Pitman, Boston-London-Melbourn (1983).

[11.6] L.P. Shilnikov - "Bifurcation theory and turbulence", In: *Nonlinear and Turbulence Processes*, **2**, pp.1627-1635, Gordon and Breach, Harwood Academic Publishers(1984).

[11.7] Yu.L. Klimontovich - *Turbulent Motion. Structure of Chaos. The New Approach to the Statistical Theory of Open System*, Kluwer Academic Publishers, Dordrecht, (1991).

[11.8] D.V. Turaev, L.P. Shilnikov, S.V. Gonchenko - "On models with non-rough Poincare' homoclinic curves", Physica **D**, (1992).

[11.9] W. Horsthemke, R. Lefever - *Noise-Induced Transitions: Theory and Applications in Physics, Chemistry, and Biology*, Springer-Verlag, (1984).

[11.10] V.S. Anishchenko, M.A. Safonova - "Noise-induced exponential divergence of phase trajectories in the neighbourhood of regular attractors", Sov.Tech.Phys.Lett. **12(6)**, pp.305-306, (1986).

[11.11] V.S. Anishchenko, M.A. Safonova, V.V. Tuchin - "Bifurcations and stochasticity induced by an external noise in a laser with a nonlinear absorption", Sov. J. Quantum Electron. **18(9)**, pp.1178-1183, (1988).

[11.12] D.M. Vavriv, G.A. Gromov, V.V. Ryabov - "Interaction of chaotic and noisy oscillations in quasi-linear systems", Sov.Phys.Tech.Phys. **66**, pp.1-10, (1990).

[11.13] V.S. Anishchenko, M.A. Safonova - "Bifurcations of attractors in the presence of fluctuations", Sov.Phys.Tech.Phys. **33(4)**, pp.391-397, (1988).

[11.14] V.V. Bykov, A.L. Shilnikov - "On the boundaries of Lorenz attractor existence", In: *Methods of qualitative theory and bifurcation theory*, pp.151-159, Gorky (1989).

[11.15] A.R. Osborne, A. Provenzale - "Finite correlation dimension for stochastic systems with power-law spectra", Physica **35D**, pp.357-381, (1989).

[11.16] G.M. Zaslavskiy - *Stochasticity of Dynamical System*, Nauka Publishers, Moskow, (1984).

[11.17] A.N. Kolmogorov - "On the entropy per time unit as a metric invariant of automorphisms", Dokl.Acad.Nauk USSR **124**, p.768, (1959).

[11.18] V.S. Anishchenko, M.A. Safonova - "Correlation time and entropy of chaos during inverse period doubling bifurcations", Sov.Tech.Phys.Lett. **14(8)**, pp.639-641, (1989).

[11.19] A.J. Lichtenberg, M.A. Liberman - *Regular and Stochastic Motion*, Springer-Verlag, New York Heidelberg Berlin, (1983).

[11.20] V.S. Anishchenko, H. Herzel - "Noise-induced chaos in a system with homoclinic points", ZAMM **68**, p.317, (1988).

[11.21] E.V. Astashkina, A.S. Mikhailov, A.V. Tolsopyatenko - "Noise-induced instability in Lorenz model", Izv.Vuz. Radiofizika **24**, pp.1035-1037, (1981).

[11.22] V.S. Anishchenko, A.B. Neiman - "Noise-induced transition in Lorenz model", Sov. Tech.Phys.Lett. **17**, pp.43-45, (1991).

[11.23] P. Mannevile, Y. Pomeau - "Different ways to turbulence in dissipative dynamical systems", Physica **1D**, pp.219-226, (1980).

[11.24] V.S. Anishchenko, A.B. Neiman - "Structure and properties of chaos in the presence of noise". In: *Nonlinear Dynamics of Structures*, Eds. R.Z.Sagdeev et al., pp.21-48, World Scientific, Singapore (1991).

[11.25] H.G. Shuster - *Deterministic Chaos*, VCH Verlagsgesellshaft, Weinheim, (1989).

[11.26] P. Berge, Y. Pomeau, C. Vidal - *L'Ordre Dans le Chaos: Vers une approche deterministe de la turbulence*, Hermann, Editeurs des sciences et des arts, (1988).

[11.27] P.S. Landa, R.L. Stratonovich - "On the theory of intermittency", Izv.Vuzov. Radiofizika **30**, pp.65-69, (1987).

[11.28] J.P. Eckmann, L. Thomas, P. Witter - "Intermittency in the presence of noise", J.Phys. **14A**, pp.3153-3158, (1981).

[11.29] G. Mayer-Kress, H. Haken - "Intermittent behavior of the logistic systems", Phys.Lett. **82A**, pp.151-155, (1981).

[11.30] J.E. Hirsh, B.A. Huberman, D.J. Scalapino - "Theory of intermittency", Phys.Rev. **25A**, pp.519-532, (1982).

[11.31] B. Hu, J. Rudnick - "Exact solutions to the Feigenbaum renormalization group equation for intermittency", Phys.Rev.Lett. **48**, pp.1645-1648, (1982).

[11.32] C.W. Gardiner - *Handbook of Stochastic Methods for Physics, Chemistry and Natural Sciences*, Springer, Berlin, (1984).

[11.33] V.S. Anishchenko, A.B. Neiman - "Statistical properties of intermittency in the systems with quasiattractors", Sov.Phys.Tech.Phys. **60**, pp.3-14, (1990).

[11.34] L. Pontryagin, A. Arnold, A. Vitt - "On the statistical treatment of dynamical systems", In: *Noise in Nonlinear Dynamical Systems*, eds. F. Moss and P.V.E. McClintock, **1**, pp.329-347, Cambridge University Press, (1989).

[11.35] V.S. Anishchenko - "Interaction of strange attractors and chaos-chaos intermittency", Sov.Tech.Phys.Lett. **10**, pp.266-270, (1984).

[11.36] F.T. Arecci, R. Badii, A. Politi - "Low frequency phenomena in dynamical systems with many attractors", Phys.Rev. **29A**, pp.1006-1009, (1984).

[11.37] E.G. Gwinn, R.M. Westervelt - "Intermittent chaos and low-frequency noise in the driven dumped pendulum", Phys.Rev.Lett. **54**, pp.1613-1616, (1985).

[11.38] V.S. Anishchenko, A.B. Neiman - "Increase in the duration of correlations during a chaos-chaos intermittency", Sov.Tech.Phys.Lett. **13**, No.17, pp.1063-1066, (1987).

CHAPTER 12

[12.1] P. Grassberger, I. Procaccia -"Characterization of strange attractors", Phys.Rev.Lett. **50**, p.345 (1983).

[12.2] P. Grassberger, I. Procaccia - "Measuring of strangness of strange attractors", Physica **9D**, p.189 (1983).

[12.3] M. Casdagli -"Nonlinear prediction of chaotic time series", Physica **35D**, p.335 (1989).

[12.4] J.D. Farmer, J.J. Sidorovich - "Predicting of chaotic time series", Phys.Rev. Lett. **59**, p.845 (1987).

[12.5] J.P. Crutchfield, B.S. McNamara - "Equation of motion from a data series", Complex system **1**, p.417 (1987).

[12.6] M. Kuchler, W. Eberl, A. Hubler, E. Lusher - "The description of complex systems by simple maps", Helv. Phys. Acta **61**, p.232 (1988).

[12.7] H.D.I. Arbanel, R. Brown, J.B. Kadtke - "Prediction in chaotic nonlinear systems: Methods for time series with broadband Fourier spectra", Phys.Rev. **A41**, p.1782 (1990).

[12.8] J. Cremers, A. Hubler - "Construction of differential equations from experimental data", Z. Naturforsh. **42A**, p.797 (1987).

[12.9] Ju.A. Kravtsov - "Randomness, determinisity, predictability", Uspekhi Phys. Nauk **158**, p.93 (1989).

[12.10] T. Kapitaniak - *Chaos in systems with noise,* World scientific, Singapore (1988).

[12.11] V.S. Anishchenko, A.B. Neiman - "Structure and properties of chaos in presence of noise", In: *Nonlinear Dynamics of Structures,* ed. R.Z. Sagdeev et. al., p.21, World Scientific, Singapore (1991).

[12.12] V.S. Anishchenko - *Dynamical Chaos in Physical Systems. Experimental Investigation of Self-Oscillating Circuits,* Teubner-Texte zur Physik, **Bd.22**, Leipzig (1989).

[12.13] V.S. Anishchenko, M.A. Safonova - "Homoclinics in the reconstruction of dynamic systems from experimental data", Appl.Mech.Rev. **46**, No.7, p.361 (1993).

[12.14] V.I. Lukjanov, L.P. Shilnikov - "On some bifurcations of dynamical systems with homoclinic structures", Soviet Math. Dokl. **19**, p.1314 (1978).

[12.15] S.V. Gonchenko, L.P. Shilnikov - "On dynamical systems with structurally unstable homoclinic curves", Soviet Math. Dokl. **33**, p.234 (1986).

[12.16] D.V. Turaev, L.P. Shilnikov, S.V. Gonchenko - "On models with non-rough Poincare homoclinic curves", Physika **D**, (in press).

[12.17] S.E. Newhouse - "The abundance of wild hyperbolic sets and non-smooth sets for diffeomorphisms", Publ. Math. IHES **50**, p.101 (1979).

[12.18] V.S. Anishchenko, M.A. Safonova - "Bifurcations of attractors in presence of fluctuations", Sov.Phys.Tech.Phys. **33(4)**, p:391 (1988).

[12.19] V.S. Afraimovich, L.P. Shilnikov - "Invariant tori, their destruction and stochastisity", In: *Methods of qualitative theory of differential equations*, p.3, Gorky State University (1983).

[12.20] V.S. Anishchenko, M.A. Safonova - "Noise induced exponential extension of phase trajectories in the regular attractors' neighborhood", Sov.Phys.Tech.Phys. **12**, p.740 (1986).

[12.21] V.S. Afraimovich, A.M. Reiman -"Dimensions and entropies in multidimensional systems", In: *Nonlinear waves. Dynamics and evolution*, p.238, Nauka, Moscow (1989).

[12.22] *Dimension and Entropies in chaotic systems.* Ed. G. Mayer-Kress, Springer, Berlin-Heidelberg (1986).

[12.23] T.S. Parker, L.O. Chua - *Practical numerical algorithms for chaotic systems*, N.Y., Springer-Verlag, (1989).

[12.24] G.G. Malinetskij, A.B. Potapov - "On calculation of strange attractors' dimension", Journal of computed mathematics and mathematical physics **28**, p.1021 (1988).

[12.25] I.S. Aranson, A.M. Reiman, V.T. Shehov - "Measurements for correlation dimension", In: *Nonlinear waves. Dynamics and evolution*, p.238, Nauka, Moscow (1989).

[12.26] P.S. Landa, M.G. Rosenblum - "Time series analysis for systems identification and diagnostics", Physica **48D**, p.232 (1991).

[12.27] D.S. Broomhead, G.P. King - "Extracting qualitative dynamics from experimental data", Physica **20D**, p.217 (1986).

[12.28] F. Takens - "Detecting strange attractors in turbulence", Lecture Notes in Mathematics **898**, p.336 (1981).

[12.29] N.K. Gavrilov, L.P. Shilnikov - "On three dimensional dynamical systems close to systems with structurally unstable homoclinic curve", part I, Math. USSR Sb. **17**, p.467 (1972).

[12.30] N.K. Gavrilov, L.P. Shilnikov - "On three dimensional dynamical systems close to systems with structurally unstable homoclinic cureve". part II, Math. USSR Sb. **19**, p.139 (1973).

Index

A

Algorithm
- for calculating the multipliers, 44
  - – – dimensions, 64, 65
  - – – the LCE spectrum, 58, 59
  - – – the Poincaré maps, 35
  - – – the separatrixes, 61
  - – reconstruction of attractors, 340-341
  - – – – dynamical systems, 341
  - Henon –, 37
  - Newton –, 43

Attractor, 10
- double-scroll –, 229
- – of Lorenz type, 302
- quasihyperbolic (quasiattractor) –, 146, 229, 302
- robust hyperbolic –, 302
- Shilnikov –, 128
- strange –, 146

B

Bifurcation, 15
- bands merging –, 101-102, 135
- nonlocal –, 20-21, 29, 31
- – of doubling for 2D-torus, 151, 196
- – of doubling for 3D-torus, 197
- – of equilibrium states, 16-17
- – of homoclinic tangency, 134
- – of limit cycle birth (Andronov-Hopf –), 18, 80
- – of periodic solutions, 22, 79, 83
- – of two-dimensional torus birth, 25
- period-doubling – for cycle, 23, 47, 84, 91
- pitchfork –, 268
- symmetry-breaking –, 27

Bifurcation diagram, 16, 42, 149, 227
Bifurcation point, 15, 79
Bifurcational transition
- abrupt –, 19-20, 162, 180, 186
- soft –, 16-18, 158, 161, 164

Bistable systems, 248

C

Chua's circuit, 220
Codimension of bifurcation, 16, 22, 45
Correlation integral, 66
Cumulant analysis, 274

D

Diffusion matrix, 270
Dimension, 67
- correlation –, 65, 142
- embedding –, 67
- fractal (metric) –, 64, 142
- information –, 64, 142
- Lyapunov –, 98, 144

Discrete-time system, 10
Dissipative nonlinearity, 116
Distribution density
- – of phase spectra difference, 204, 205, 216

– of residence times, 237, 238, 250
Drift coefficients, 270

E

Entropy
　– of distribution, 52
　Renye –, 65-66
　Shannon –, 65
Equation
　Chapman-Kolmogorov –, 270
　Chua's circuit –, 221
　Fokker-Planck –, 269
　Langevin's –, 269
　Lorenz –, 71, 289
　Ornstein-Ulenbeck –, 272
　Pontryagin's –, 326
　Rössler –, 46, 71
　stochastic differential –, 269

F

Feigenbaum's constant, 91, 96
Fibonacci numbers, 167
Filtering of noise, 344
Function
　correlation –, 272
　– of ordinary coherence, 204
　separatrix splitting –, 20-21

G

Golden section, 166

H

Heteroclinic trajectory, 226
Homoclinic
　– structures, 147
　– trajectories, 43, 122, 133, 224
　　– – of saddle-focus separatrix loop type, 124
Hysteresis, 87, 88, 93

I

Intermittency
　modulation –, 110, 112
　noise-induced –, 251
　– of "chaos-chaos" type, 113, 231, 245, 333
　– of "cycle-chaos" type, 154, 328
Invariant closed curve, 160, 181, 185

L

LCE spectrum, 3
　– of equilibrium state, 6
　– of periodic solution, 7
　signature of –, 3
　signature of – of chaotic solution, 10
　signature of – of periodic solution, 3
Limit cycle, 7, 8, 79, 82
Lorenzian, 238-239
Lyapunov Characteristic exponents, 2-3
　– for discrete systems, 12
　k-dimensional –, 56
Lyapunov quantity, 19, 81

M

Map
　logistic (Feigenbaum's) –, 181, 328
　Henon –, 181, 338
Matrix
　fundamental –, 2
　monodromy –, 6-7
Multiplier of periodic solution, 7

N

Noise
　$1/f$ –, 231
　additive –, 270
　colored –, 272, 297
　multiplicative –, 270, 281

white –, 272, 281

O

Oscillations
　chaotic –, 9
　quasiperiodic –, 8
　periodic –, 6
Oscillator with inertial nonlinearity, 75
　modified –, 76, 139
　nonautonomous equations for –, 148

P

Poincaré
　– map, 11, 234
　– secant, 11
　– section, 11, 230
Probability density distribution, 51, 268

R

Realization, 337
Resonance
　– on torus, 149-150
　stochastic –, 248
Robustness, 15

S

Saddle quantity, 31
Silver section, 166
Smoothness loss
　– by invariant curve, 161
　– by torus, 158
Stability
　– according to Lyapunov, 2, 5, 9
　– according to Poisson, 9

– of chaotic solutions, 9
– of discrete system solution, 13
– of equilibrium states, 5-6
– of periodic solutions, 6-7
– of quasiperiodic solutions, 8-9
structural –, 15
Synchronization, 159
　mutual –, 199, 211
　forced –, 199, 205
　– via frequency locking, 159, 207
　– via frequency supression, 199, 208
　– of chaos, 199, 206, 210, 241-244

T

Theorem
　Takens –, 67, 340-341
　– on torus breakdown, 256
Time
　correlation –, 137
　mean first passage –, 270, 325, 326
Transition to chaos
　noise-induced –, 110, 112, 319
　– via two-dimensional torus, 159
　– via three-dimensional torus, 189
　– via a series of period-doubling bifurcations, 91

V

Variational equations
　– for differential system, 2
　– for discrete-time system, 12

W

Winding (rotation) number, 148, 259